Computer Telephony Integration

Second Edition

For a complete listing of the *Artech House Telecommunications Library*,
turn to the back of this book.

Computer Telephony Integration

Second Edition

Rob Walters

Artech House
Boston • London

Library of Congress Cataloging-in-Publication Data
Walters, Rob.
Computer telephony integration / Rob Walters. — 2nd ed.
p. cm.
Includes bibliographical references and index.
ISBN 0-89006-969-7 (alk. paper)
1. Digital telephone systems—Automation. 2. Computer networks.
I. Title.
TK6397.W35 1998
384.6—dc21 98-41587
CIP

British Library Cataloguing in Publication Data
Walters, Rob
Computer telephony integration. – 2nd ed.
1. Digital telephone systems 2. Digital communications
I. Title
621.3'85

ISBN 0-89006-969-7

Cover design by Elaine K. Donnelly

685 Canton Street
Norwood, MA 02062

International Standard Book Number: 0-89006-969-7
Library of Congress Catalog Card Number: 98-41587

10 9 8 7 6 5 4 3 2 1

To my mother and Margaret,
the mother of my children

Contents

	Preface	**xv**
	Acknowledgments	**xvii**
1	**Integrating the Two Worlds**	**1**
1.1	Renting a Car	1
1.2	Collecting a Debt	2
1.2.1	Paying the Bill	4
1.3	What is Computer Telephony Integration?	4
1.3.1	The Two Worlds	7
1.3.2	The Linkage Required for Functional Integration	18
1.3.3	Application Basics and the CTI Bridge	21
1.3.4	Workgroup Integration	24
1.3.5	Approaches to CTI: The Computer System View	25
1.4	The History of CTI	29
1.4.1	Prehistoric CTI	29
1.4.2	The Foundation Decade	38
1.5	The Delivery Decade	47
	References	50
2	**The Computer and Telephony Environment**	**51**
2.1	It's All Bits Anyway	52

2.2 The Telephony Environment 52
2.2.1 Telephony Basics 53
2.2.2 Telephone Switches 74
2.2.3 Telephone Traffic 77
2.2.4 Telephony Features 80
2.2.5 Call Processing and Call Modeling 83
2.2.6 Evolving Telephone Networks 85
2.3 The Computing Environment 86
2.3.1 Computer Basics 87
2.3.2 Types of Computers 89
2.3.3 Computer Terminals 91
2.3.4 Computer Software 92
2.3.5 Computer Networks 95
2.4 Trends in Telephony and Computing 100
2.5 Convergent Networks 102
2.5.1 The ISDN 102
2.5.2 Asynchronous Transfer Mode Networks 103
2.5.3 Mobile Networks 104
2.6 Multimedia 105
References 105

3 Integration Technology 107

3.1 Don't Mention Technology 107
3.2 Introducing Small CTI Technology 108
3.3 Integration Architecture 109
3.4 First-Party Architecture 109
3.4.1 The Application Program and API 110
3.4.2 The Switch 110
3.4.3 The Computer 110
3.4.4 The Telephone 111
3.4.5 The Terminal 111
3.4.6 The Attachment Interface 111
3.4.7 The Signaling Intercept 113

3.4.8 Line Signaling 120
3.5 Networked Computing and First Party 122
3.6 Third-Party CTI Architecture 126
3.6.1 The Application Programs and API 127
3.6.2 The Switch 127
3.6.3 The Computer 128
3.6.4 The Telephone 128
3.6.5 The Terminal 128
3.6.6 The CTI Link 129
3.7 CTI Protocols Under the Microscope 130
3.7.1 What is a Protocol? 131
3.7.2 Models 132
3.7.3 General Requirements of a CTI Protocol 134
3.7.4 Proprietary CTI Protocols 138
3.7.5 Protocol Converters 139
3.8 Primary, Dependent, and Merged Solutions 141
3.9 Networked Switching and Third Party 144
3.9.1 Single Server Connected to the LAN 145
3.9.2 Single Server Connected to All Switches 145
3.9.3 Server per Switch 146
References 146

4 Media Processing Technology 147

4.1 What are Voice Processing Systems? 147
4.1.1 Voice Messaging 148
4.1.2 Voice Response Units 151
4.1.3 Interactive Voice Response Systems 152
4.2 Voice Processing Technology 160
4.3 Fax and Image 162
4.3.1 Fax Processing 163
4.3.2 Document Image Processing 163
4.4 Interworking 164
4.4.1 The Integration Interfaces 164

4.4.2 Unified Messaging 166
4.5 Multimedia 168
4.6 Video Processing 168
References 171

5 Application Elements and Application Creation 173

5.1 Generic Functions of CTI 173
5.1.1 Screen-Based Telephony 174
5.1.2 Call-Based Data Selection 174
5.1.3 Application-Controlled Routing 175
5.1.4 Voice and Data Call Association 177
5.1.5 Data Transport 177
5.1.6 Coordinated Call Monitoring 178
5.1.7 Integrated Voice Response 178
5.1.8 Messaging Exchange 179
5.2 Application Generation 179
5.3 How Do You Create Integrated Applications? 181
5.3.1 CTI in the Real World 182
5.3.2 Who Creates the Integrated Application? 183
5.3.3 Overview of Approaches 186
5.4 Application Programming Interfaces 187
5.4.1 Functions 189
5.4.2 Examples of Application Programming Interfaces 190
5.4.3 Media and Message Processing APIs 222
5.5 Application Creation in CTI 223
5.5.1 Application Enabler Layers 226
5.5.2 CTI Middleware 228
References 235

6 Applications and Case Studies 237

6.1 Finding the Right Application 237
6.2 Application Integration Potential 239

6.3	Application Categories	240
6.4	General Applications	242
6.4.1	Desktop Communications	242
6.4.2	Inbound Call Handling	242
6.4.3	Outbound Call Handling	243
6.4.4	Automated Inbound Call Handling: Interactive Voice Response	243
6.4.5	Automated Outbound Call Placement: Power Dialers	244
6.4.6	Soft Automatic Call Distributors and Call Routing	246
6.4.7	Multimedia Messaging	247
6.5	CTI Application Examples	247
6.5.1	Call from Teleint	248
6.5.2	Phone from Microsoft	249
6.5.3	Phonetastic from CallWare	251
6.5.4	FastCall from Aurora	253
6.5.5	Group PhoneWare from Q.Sys	256
6.5.6	Unified Messenger from Octel/Lucent	258
6.6	Sample Case Studies	260
6.6.1	Automating Banking Services	261
6.6.2	Campus Emergency	264
6.6.3	Car Company Collections	265
6.6.4	Efficient Charity	266
6.6.5	Quicker Help Desk for Fast Cars	268
6.6.6	Selling Books	269
6.6.7	Automated Holiday Bookings	270
6.6.8	Campus Emergency and Integrated Message Center	271
6.6.9	Operator Assistance in the Public Telephone Network	272
6.6.10	The Tour Operator	274
6.6.11	Telemarketing Bureau	275
6.6.12	Happy Travel Agents	277

6.6.13 Customer Service from a Utility 278
6.6.14 Investment Management 280
6.6.15 Personal Service in Collections 281
6.6.16 Preventing Car Theft 283
6.6.17 Voice Response in Banking 284
6.6.18 Fully Integrated Finances 285
References 286

7 The Market for CTI 287

7.1 How Many Fish in the Sea? 287
7.2 Market Definition 287
7.3 Market Drivers 290
7.3.1 Expectations 291
7.3.2 Customer Service Demands 292
7.3.3 Operational Efficiency 294
7.3.4 Competitive Edge 295
7.3.5 Business Diversification 296
7.3.6 Desktop Standardization 296
7.3.7 The Internet 297
7.3.8 Application and Solution Availability 298
7.3.9 Compatibility and Strategic Partnerships 299
7.3.10 PC Card Technology 299
7.4 Market Inhibitors 300
7.4.1 Standards 300
7.4.2 Regulation 301
7.4.3 Ignorance 301
7.4.4 Culture 303
7.4.5 Integration Expertise 303
7.4.6 Cost 303
7.5 CTI Benefit Analysis 304
7.5.1 CTI Benefits 304
7.5.2 Quantification of CTI Benefits 309
7.5.3 A User's View 314

7.6 CTI Market Quantification 315
7.6.1 Who Buys CTI? 316
7.6.2 What Does CTI Cost? 318
7.6.3 Who Sells CTI? 320
References 321

8 Computer Telephony Integration System Engineering 323

8.1 What is System Engineering? 323
8.2 Implementation 324
8.2.1 Formal Implementation Methodology 325
8.2.2 Risk and its Containment 331
8.2.3 User Concerns 333
8.2.4 Networking Options for Third-Party CTI 335
8.3 Performance Constraints 336
8.3.1 Delay 336
8.3.2 Bus Capacity 339
8.3.3 Throughput 340
8.3.4 Reliability Implications 342
8.4 Management 343
8.4.1 Link Management 344
8.4.2 Database Synchronism 344
8.4.3 Coordinated Call Monitoring 345
8.4.4 The CTI Manager 346
8.5 The Regulatory Environment 346
8.5.1 North American Regulation 347
8.5.2 European Regulation 347
8.5.3 European Approval Issues 348
8.6 Standards 349
8.6.1 CTI Linkage 352
8.6.2 Voice Over Packet 375
8.6.3 Messaging Standards 376
8.6.4 Standards for Collaborative Working 377
References 378

9 Merging on All Fronts 379

9.1 Review of CTI 379
9.2 History and Evolution 380
9.3 The Basics of CTI 385
9.4 Convergence Problems 387
9.5 Technology and Terminology 390
9.6 Openness and the Production of CTI Applications 392
9.7 The Market and the Benefits 394
9.7.1 The User's View of CTI 395
9.8 Delivering CTI 396
9.9 The Patent Situation 397
9.10 Merging Desktops: The PC as a Telephone 401
9.10.1 PC Phone Options 403
9.11 Merging Servers: The PC-Based CTI System 404
9.11.1 Server-Independent Architectures 405
9.11.2 Server-Dependent Architecture 407
9.11.3 Boxed Solutions 408
9.12 Merging Networks: IP Telephony and All That 409
9.12.1 Voice Over Internet Protocol 409
9.12.2 IP Phones and Gateways 412
9.12.3 Standards and Gatekeepers 414
9.13 The Future of CTI 415
9.13.1 The Last Word 417
References 417

CTI Glossary 419

About the Author 449

Index 451

Preface

I actually like integration! That is why I felt the need to write this book. It all began in my days as a switch designer. My career commenced at the time when computers were only just beginning to be used in telephone exchanges. Many people in those days thought that the computer would never be capable of controlling the switching of telephone calls. Surely the computer would not have enough power, surely it would break down too often. They were quite right! The early "stored-program-controlled" private automatic branch exchanges were not too good. However, technology came to our rescue. Computers got faster and became more reliable. Good software practice reduced the bugs to a reasonable level.

It was then that we began to dream about the possibilities of running business applications within the PBX! Foolish thoughts.

In the 1980s I became more and more involved in voice processing and integrated systems. It was an uphill battle, and I still have the scars to prove it. However, we learn more from failure than success, and I entered the 1990s with both my liking for integration undiminished and an overpowering need to share my experiences with others. This launched me into the higher orbits of "gurudom" in the CTI world. It is said that in the valley of the blind, the one-eyed man is king. Fortunately, the 1990s saw CTI eyes becoming widely distributed. As a result, "gurus" are now unnecessary.

This book is intended to be an introductory text and a manual. It is difficult to achieve these two extremes without producing the ultimate cocktail—the drink that contains something that everyone likes but has a taste nobody likes. I hope to have achieved something that is much nearer to a lucky dip—a lucky dip in which you can line all the presents up and open them in sequence or just put your hand in and see what comes up.

I am not going to define computer telephony integration in this preface; that is done in Chapter 1. However, I would like to say a few words about the second chapter, in which the worlds of computing and telephony are introduced. I would not expect anyone with wide experience in telephony or computing to spend much time on Chapter 2. On the other hand, when writing about a topic that bridges two worlds, it would be a dereliction not to spend some time introducing those two worlds—especially when, and I strongly suspect this will be the case, the people who really need to know about computer telephony integration are not computer or telecommunications experts.

I believe that anyone with an interest in this topic needs to know something of the technology used to connect the two worlds, and this technology is explained in the third chapter.

Chapter 4 introduces media processing systems. The first book that I wrote concentrated on voice processing systems and wrote about computer telephone integration as an adjunct; here the situation is reversed. Furthermore, CTI has now embraced media in addition to voice—image and video, for example.

Computer telephony integration is all about making business applications work better by incorporating telephone functions. In Chapter 5, I explain how integrated applications are created, and in Chapter 6, I have collected a number of case studies to show how and where computer telephony integration is used.

The seventh chapter looks at the commercial side of integration. Who will buy it and how much will they buy? What is the market likely to be in future and what are the benefits? Realizing those benefits is the difficult part. Chapter 8 explores the problems of implementing integrated systems and provides some pointers toward best practice and to CTI standards.

The last chapter contains a song—the "Computer Telephone Integration Song." Only those of you who successfully read and absorbed all the other chapters should sing this song. It is a revisionary piece—not very musical, I'm afraid. Chapter 9 also contains coverage of the extension of CTI into the Internet world—and suggests an end to CTI as we know it.

So, good luck with this cocktail. I hope that you enjoy it and that you receive a little of my liking for the subject from it. And remember that however much of this particular mixture you imbibe, you should not get a hangover.

Acknowledgments

You cannot create a book like this alone. Lots of people and lots of companies help—some of them unwittingly. One of the most difficult and the most rewarding tasks in producing the first edition was researching the history of computer telephony integration. Apparently, there is nothing new under the Sun, and the following people helped me to prove it. I would like to thank them very warmly for their support.

Bill Deas, of IBM, U.K., for historic information about early IBM installations; Keith Bellamy, for information on the Delphi system and its development; Peter Newman, for details on the architecture of that splendid attempt to dam the river of time—the Plessey 2150.

This was my second book and there is one person who helped and encouraged me in producing it. Thank you very much, Hugh Daglish, for reading through the emerging manuscript and adding to the quality of it in so many ways, not the least of which was in spotting the many grammatical errors of which the author is so fond. Hugh died suddenly in 1995. I am sure that this second edition would have been improved by Hugh's literary eagle eye.

In the early months of 1992, I was commissioned by Jo Piggot of Schema to write part of the Schema study on computer-supported telephony. I think it was at that time that the idea of writing the original edition of this book really emerged. Thank you, Jo, for the idea and for your permission to use material from the Schema study. Schema has continued to research the CTI market—and has been joined by many others. Thanks to all of them for material referenced in this second edition.

Finally, two ladies in my life must be thanked—Annabella, for her Spanish translation, and last but by no means least, my wife Margaret. She groaned when I announced that this book had been accepted by the publisher—but has helped in so many ways to make it happen.

1

Integrating the Two Worlds

1.1 Renting a Car

"Good morning. This is Autorent and I'm Liz. How can I help you?"

"I want to rent a car for next weekend, please."

"Yes, that's the 24th and 25th. Autorent provides a special offer for that weekend. Could I have your name and address, please?"

"My name's Gary King and I live at number 32, Enderby Close, Tavistock. Can you tell me how much the rent will be?"

"Well, I need a few more details first of all. Will you pick the car up from our Tavistock branch office? That's on Brook Street."

"Yes."

"And are you over 21 with no accidents or convictions for the last five years?"

"Yes."

"And what type of car do you wish to rent? Do you want a compact car or a larger model?"

"Well, I actually want an estate car. You see I've got these long . . ."

"Ah, I'm sorry, Mr. King. I only deal with the normal cars. I'll have to transfer you to specials. Could you hold, please? I'm just transferring you now."

(Pause)

"Hello. This is Autorent Specials Department. I'm Kate. How can I help you?"

"I want to rent an estate car for next weekend."

"We should be able to help you there. Where do you live?"

"I've just told the other lady that!"

"Oh, sorry about that. Could I have your name and address, please?"

Has this ever happened to you? Has your call ever been passed from department to department, and at each new department have you been asked to relate why you are calling all over again? Frustrating, isn't it?

It this really necessary? Is there some law of nature that allows the free movement of telephone calls yet prevents the data associated with them from following? If Liz and Kate were sitting next to each other, Gary's problems would not have occurred. Kate could have had instant access to Liz's notes, whether they were on paper or on a computer screen.

Of course, there is no law of nature that prevents the association and movement of data with telephone calls. But there *are* barriers—barriers that are rooted in history, in incompatible products, and in social attitudes. Computer telephony integration (CTI) is all about removing or surmounting those barriers. Assuming that Liz had entered Gary's details into the office computer, and nowadays this is much more likely than not, then what prevents that information from arriving at Kate's terminal when the call is transferred? The barriers are not physical. Liz and Kate will be using the same computer system for rental bookings—even if the connections to that computer are different. The telephone system may not be physically connected to the computer system, but even if they aren't connected there are many ways in which they could be. The barrier is not a physical one. Exactly what it is and how CTI overcomes it is the subject of this book.

Before plunging into the detail of CTI, it is worthwhile to consider an example of the use of the telephone and computer that illustrates a more congenial coexistence.

1.2 Collecting a Debt

Andrew fitted his headset and made himself comfortable in front of the screen. A few keystrokes and he had logged into the system. The call arrival icon immediately began to flash and the screen filled with details about the debtor to whom Andrew was about to speak, a Mr. Ostaccini of Lovett Drive, Desoto.

"This is Andrew of the Spring Valley Collection Service. Can I speak to Mr. Ostaccini, please?"

"No, this is not Mr. Ostaccini."

"Can I speak to him, please?"

"Who are you?"

"My name is Andrew, and I'm from the Spring Valley Collection Service."

(Pause)

"Buenos dias. Señor Ostaccini."

"Ah, Mr. Ostaccini. My name is Andrew, and I'm phoning about the bill for $37.50 that we sent you two months ago. We are still awaiting payment. Is there some problem with the bill?"

"No comprendo. No English."

"Ah, no English. Please wait, Mr. Ostaccini. I will connect you to our Spanish representative."

Andrew presses a key and the call leaves his screen—to be replaced by an entirely new one. Meanwhile, what has happened to Mr. Ostaccini? After a short pause, during which he hears music from the telephone, he is connected to an agent called Jaime who happens to be working in a different building.

"Hola Señor Ostaccini. Me llamo Jaime. Quisiera hablarle sobre la cuenta por el valor de $37.50 la cual le enviamos hace dos meses y que todavia aguarda su pago. ¿Es que hay algun problema con esta cuenta?"

Those of you who do not read Spanish may rest assured that Jaime said precisely the same thing to Mr. Ostaccini that Andrew had previously said in English.

Somewhere in this story there is a little touch of magic. It occurs at the point that Mr. Ostaccini's call is picked up by Jaime. How does Jaime know:

- Mr. Ostaccini's name?
- How much money Mr. Ostaccini owes?
- How long the debt has been outstanding?

Andrew didn't tell him. When he left the call, Andrew simply began to deal with the next—there was no contact between the two agents.

One of the sad things about magic is that it is all illusion. One of the great things about it is that, if the illusion is good enough, the effect really is magical. "How did he do that?" asks the crowd. In this book, the magic of communicating the data that Andrew had on Mr. Ostaccini to Jaime—in synchronism with the hand-over of the telephone call—will be explained. And of course, as in all magic, there are a number of ways of achieving the same effect. Here is a little more magic.

1.2.1 Paying the Bill

Hillary amazed herself. She disliked shopping almost as much as she hated computers. She hated the first because it brought her into contact with too many people. She hated computers because she considered them impersonal. Yet here she was sitting in front of the screen—shopping through the Web!

The whole thing had become easy to do. In fact, it seemed that her hatred of computers stemmed not from the fact that they were impersonal, but because they were difficult to use. They intimidated her. This shopping thing was a piece of cake, however. She had to do so little to get started.

She stopped at the wine "counter." The supermarket was offering a special deal on cased wine selections. Hillary did not know a great deal about wines—but she liked trying different ones. She clicked on the "French collection" and read about some of the wines that were included. They seemed nice. She decided to order and began to fill in the order form. Then she came to the bit about payment—and stopped.

Hillary was not keen on sending her credit card details across the Internet. Friends told her that it was okay, but she was not convinced. Then she spotted a button that said, "Credit Card Callback." She clicked the button and was asked to enter her telephone number. She was also asked to indicate when she would like to be called. Hillary typed "anytime," closed the computer, and began to get ready to go out.

A few moments later the telephone rang. When Hillary answered, a friendly young person from the supermarket repeated her order and then asked for the credit card details. Two days later, the wine arrived. It was very nice.

1.3 What is Computer Telephony Integration?

In previous writings on related topics, I have quoted Donald A. Treffert, who, in his book on extraordinary people, writes, "The beginning of wisdom is to call things by their right name [1]." Inverting that very wise statement, one could say, "If there is no right name for a thing, then there can be no wisdom." Unfortunately, there is no right name for CTI—as yet. On the other hand, someone else once said, "Show me where the language is changing and I'll show you where things are really happening."

CTI is where things are happening at present. It represents a synthesis of the lessons learned in the 1980s, when so many attempts were made to converge telephony and computing. Consequently, it is a subject area that is littered with terminology, some of it deliberately misleading. So, the "right names" are still evolving. Currently, there are so many to choose from that confusion abounds and wisdom seems a long way off. I hope that this book will begin to redress that imbalance.

The all-embracing term used throughout this book is computer telephony integration (CTI). One of the best overall definitions of CTI that I can find was produced by Lois B. Levick of the Digital Equipment Company: "A technology platform that merges voice and data services at the functional level to add tangible benefits to business applications [2]."

That definition was coined in 1990; a lot of water has flowed under the CTI bridge since then. Whole industries that had a clearly separate existence have congregated under the CTI umbrella. CTI is now regarded as much more than a functional integration: It embraces automation, and it is used to describe physical integration. These are very different things, but the trend is clearly toward a combination of all three—automation, functional integration, and physical integration. Although this book covers all aspects of CTI, it emphasizes integration, which is sometimes called "small CTI." *Computer-mediated communications* [3] is an all-embracing term that brings together all aspects of "big CTI" and extends the field to messaging. On the other hand, the term *computer telephony* is used to cover many aspects of CTI. This term has its origins in the PC-based voice messaging world and is mostly concerned with voice processing and physical integration in the PC. I sometimes—partly with tongue in cheek—describe CTI as complete and total integration rather than computer telephony integration. Perhaps my tongue should not be in my cheek.

Small CTI should be thought of as an enabler. In some way, both voice and data services are linked to create a platform upon which integrated business applications can be produced—integrated, that is, in the sense that the applications can use functions from both the telephone world and the computing world quite seamlessly.

One of the problems of defining small CTI is that it is not a system. You cannot point to a box and say "That is CTI supplied by IBM." What you can say is "That is a CTI application." And why is it a CTI application? Because it depends on telephony functions and on computing functions to do its job. Thus, CTI provides the platform on which these integrated applications rest. In this sense, it is comparable to a computer operating system. An operating system provides an environment for application creation that gives access to the functions of the computer. Small CTI provides an environment for application creation that gives access to the functions of both the telephone system and the computer system.

Autocalling, or autodialing, enables telephone calls to be placed via the computer screen. You select a number, or even a name, and use the keyboard to indicate that a telephone call should be made to that party. In some way the computer causes the call to be placed and indicates that you should lift the handset. The progress of the call is not monitored by the computer. Is simple autodialing a CTI application? Many PCs provide this capability through the addition of a modem or a simple telephone card. It does represent a functional merging of voice and data services, albeit at a basic level. Autocalling can be regarded as the baseline for a definition of CTI—everything above this level is CTI. It is the applications above this level and their realization that will be of interest here, especially where the central business application is being extended into the telephone world.

Autocalling is to CTI what the telephone answering machine is to a voice mail system. The answering machine offers many functions that are similar if not identical to those of its bigger brother. It is, however, less sophisticated and limited to single-line applications, and it displays a very definite cut-off in functionality.

The comparisons between answering machines, voice mail systems, autocalling, and CTI are useful in other ways. When absorbing a new concept, we try desperately to fit the new idea to some existing model. When voice mail is described, people say, "Ah, you mean a telephone answering machine," or "You mean a cassette recorder." Of course you do not, so you have to convince your audience that what you are talking about is something different. This is not always easy and is best done by relating the new concept to existing models. For example, one new facility that a voice mail system does have is the ability to broadcast voice messages—and this can be likened to multiple calls made to lots of different answering machines. So too with CTI. The autocalling function is a simple concept and therefore a good starting point for appreciating the more extensive and sophisticated fuctionality of CTI. How can you achieve, from the autocalling function, the magic utilized by Andrew in dealing with Mr. Ostaccini? It is too early in this book to ascertain whether, or to describe how, this could be achieved. Nonetheless, the example does serve to illustrate the difference between simple autocalling and a more sophisticated CTI function.

In summary then, small CTI is a platform on which business applications can be created—business applications that embrace the worlds of telephony and computing. Integration must not be confused with the elements, connections, and signaling schemes that make the platform possible. Since the application software is most likely to exist within the computer environment—because that is where the major business applications exist—many computer vendors have invented their own terms for CTI. And they were not the only group to lend their inventiveness to this task. Table 1.1 provides a list of the terms that have been used, together with the source of each term. There are more. Inventiveness in this sphere appears to have no limits. Many of these terms will be explored in further detail later, when the approaches of various suppliers are described. Some have already become part of the history of CTI.

Table 1.1
The Varied Terminology of CTI

Term	Acronym	Source
Applied computerized telephony	ACT	HP
Call applications manager	CAM	Tandem
CallPath services architecture	CSA	IBM
Computer-aided communications	CAC	Davox
Computer-aided telecommunications	CAT	Dutch PTT
Computer-assisted telephony	CAT	GPT
Computer-integrated telephony	CIT	DEC
Computer telephony	CT	ECTF
JavaTel	—	Sun
Methods of interfacing switches and computers	MISC	ISO
PABX computer teaming	PACT	Siemens
PABX host interfacing	PHI	Probe Research
Switch-to-computer interfacing	SCI	Dataquest
Netware telephony services	NTS	Novell
Windows telephony	—	Microsoft

1.3.1 The Two Worlds

Computing and telephony represent two very different worlds. They began life in quite distinct eras and so were originally based on very different technologies.

1.3.1.1 The World of the Telephone

Early automatic telephone systems, often called telephone exchanges or simply switches, were almost entirely analog, and the control algorithms that determined the way in which calls were routed were hardwired. They were born out of necessity. One of the earliest and by far the most successful designs is credited to Almond B. Strowger of Kansas City in the United States. Patented in 1889, the concept is quite simple and is still in use in some parts of the world to this day. Strowger was an undertaker, and the story goes that a rival company was collecting nearly all the available funeral business. Naturally, the telephone system in those far-off days was manually operated. Fortunately for Strowger's rival, this telephone system was run by his wife. Hence, if some bereaved person called and wished to be connected to a funeral director, it wasn't Strowger that got the call!

The system Strowger designed was entirely electromechanical. The building blocks were electromagnets and electromagnetic relays. Figure 1.1 gives an insight into the overall appearance of the equipment. Calls were routed under direct control of the user's dial. The system has no intelligence except that which is implied by the way individual Strowger switches are wired together. Strowger switches are called two-motion selectors. The contacts that carry the telephone call can be moved up vertically or swept around horizontally. It is this movement that "steers" a call through the system. There is no central control. Each dialed digit operates a different switch, and each switch can deal with only one call at a time. Control moves across the entire exchange as the call is set up. It is a distributed system, and Strowger exchanges are often called step-by-step exchanges. This is a good description of its operation. If a caller dials a "2," the Strowger switch steps up two levels, then searches for the next free switch—by sweeping over all of the possible outlets. The contacts stop on the outlet connected to the first free switch and that switch then receives the next dialed digit. In this way the call routes itself through the whole exchange.

Crossbar switches were invented by Swedish engineers in the 1930s as an improvement over Strowger systems, although they are based on a 1912 patent by Gotthief Betulander. Electromechanical switches are used, which is something that Strowger and the crossbar system have in common. However, in the crossbar system, each switch can carry a number of calls and all the dialed digits are collected in a routing register of some kind before the switch is set up. The crossbar exchange therefore has some form of central control, even though this too is electromechanical.

Modern designs evolved from the crossbar system. First, the switches themselves have been miniaturized—initially through the use of reed relays and later through the use of solid state devices. Second, the control area has been progressively converted to solid state such that the modern controller is very similar to, and often is, a digital computer. This combination represents the modern electronic telephone exchange. Its evolution would seem to owe much to the availability of technology. In fact, the reverse is true in many areas. Telecommunications has driven technology rather than being driven by it. Remember that the transistor, the founding element of the modern electronics industry, was discovered at Bell Telephone Laboratories.

The next step in this evolution is the conversion of the switch itself to digital technology. This represents the arrival of the modern digital exchange. This is a key element in CTI and will be examined in more detail in Chapter 2. It is followed by an absorption of switching functions into the computing infrastructure and into communications networks.

That is a very brief overview of the evolution of telephone *switching*—which is all about routing a call between two people. The other aspect of

Figure 1.1 An early Strowger telephone exchange.

telephony is how you ensure that what is said at one end is heard at the other. This aspect is known as transmission and it is complex. All telephone systems degrade speech to some extent. Certain degradation can be measured, for example, progressive loss of power or frequency band limitation, but the only way of assessing the precise effect on quality is by subjective experimentation. There are at least five major impairments that degrade speech quality. All of these can interact and, once again, the combined effect can only be ascertained by subjective experiments, using a large number of subjects to get reliable results. Only people can determine quality levels in speech—engineers can then relate these levels to measurable things.

Modern systems use digital transmission. Speech is converted from analog to digital and transmitted in this form. Consequently, speech quality can be much improved. However, there is also a trend towards economizing on transmission cost by squeezing multiple conversations into one channel. This technique is called digital speech compression. Compression always degrades speech quality to some extent. Once again, only subjective experimentation can ascertain the quality of transmission over compressed channels.

This, then, is a very different world to the bits and bytes of a computer program. The telephony world is, in relation to the information it is designed to carry, a subjective world. This fact has a major, if subliminal, influence on telephony, because, in the final analysis, there are no absolutes here, performance is subjective.

Telephones and telephone exchanges are connected by the telephone network. Almost all of the very large telephone networks are public—in the sense that they provide a service to the public as a whole rather than to one company or group of companies. Public networks are characterized by a number of points:

- A very high degree of connectivity with other public networks;
- Adherence to international standards;
- A very high reliability requirement;
- Heavy regulatory control;
- Many generations of equipment.

All of these points are interrelated. They combine to produce the most open network in existence. It is a network that allows telephone calls to be made between remote corners of the world without punitive charges and with a fair chance of a speech channel of good quality. However, the same points conspire to make it technically difficult to connect new equipment to the public network and to make the network slow to respond to new advances in technology. Meanwhile, an invention of the computer world is beginning to challenge most of the accepted tenets of telephony.

1.3.1.2 The World of the Computer

Computer technology is almost entirely digital and is essentially flexible. A computer can do nothing without its program; on the other hand, it can do many different things with diverse programs. The earliest computing engines were mechanical. Charles Babbage's analytical engine lay unfinished at the time of his death in 1871. The architecture of this machine was quite remarkable [4]. It was a programmable machine. It used punched-card input for instructions and data. Results were delivered on punched card, graph plotter, or printer. It was capable of looping under programmed control and of conditional branching. It was to have worked to an accuracy of 40 decimal places and was microprogrammable. That description is so similar to a modern computer it is almost unbelievable—but here the similarities end. Babbage's design was entirely mechanical; it was powered by the turning of a handle. Figure 1.2 shows a portion

Figure 1.2 A partial reconstruction of Babbage's analytical engine. (*Photo courtesy of:* Institution of Electrical Engineers)

of the analytical engine. Complete, it would have stood some 4.5 meters high and up to 6 meters long!

Historically important though Babbage's machines are, they are not the forerunners of the modern computer in the way that Strowger's exchanges were the forerunners of the modern digital exchange. For many reasons the analytical engine was not completed. There is no continuous thread that links Babbage's designs to the modern digital computer. Digital computing is a relatively modern activity. It had to await the birth of the electronic age. It is, to some extent at least, the child of the telecommunications industry.

The history of modern computing begins with Claude Shannon who, perhaps coincidentally, worked for Bell Telephone Laboratories. Shannon joined Bell Telephone Laboratories after receiving his Ph.D. at Massachusetts Institute of Technology (MIT) in 1941. Most famous for his work in information theory, Shannon also showed how electrical circuits could be employed to add, subtract, multiply, and so on. He also drew attention to the fact that the circuits were much simplified if the binary system was used. It is this combination of electronics and binary calculation that marks the beginning of the development of the modern digital computer. At about the same time that Shannon was publishing his work, another Bell Labs employee was developing the first binary adder—an essential element in any digital computer. This man's name was George Stibitz, and in 1940 he actually demonstrated an elementary computer working across the telephone line! Like the Strowger telephone system, Stibitz's circuits were based on electromagnetic relays.

A relay-based computer called the Harvard Mark 1 was designed by Howard Aiken, funded by IBM, and first demonstrated in 1943. It was a monster—55 feet long and eight feet high. It was also very slow. Perhaps the first electronic computer was the British Colossus designed by a very learned team that included one of computing's most famous figures—Alan Turing. Colossus was designed for a specific purpose: to crack the German ciphers in use during the second world war. Constructed secretly at the British Post Office research station, it proved to be very successful. Based on electronic valves or tubes and using high-speed paper tape input, it was said to be over a thousand times faster than the electromechanical Harvard Mark 1.

There followed a number of famous machines, beginning with ENIAC, which, demonstrated in 1946, was possibly the first programmable electronic computer. Famous as the capabilities of these machines were, they were also known for their unreliability, size, and power consumption. Computers needed the transistor and its descendant, the integrated circuit, to provide reliable economical computing power for the world in general. The transistor was supplied by Bell Telephone Laboratories in 1948. It was the transistor that led to the first general mainframes and then to the minicomputers of the 1960s and 70s.

In the 1970s, the first microprocessors became available. These were seized upon as quickly by PBX designers as they were by the creators of *personal computers* (PCs). For the first time, both worlds were using identical technology in the very heart of their systems. Since that time, the microprocessor has become the core element of a larger and larger slice of the information technology market.

A related trend that also began to gain sway in the 1970s was that of computer networking. Users needed computers to communicate so they could access shared facilities: databases, printers, gateways, and so forth. Initially, data communications used the telephone network—by converting the data into voicelike signals via a modem. This represents the earliest form of *physical* voice and data integration. Later, public and private data networks were created, offering higher capacity and more reliable transmission. Within buildings, *local area networks* (LANs), based on special wiring, became common. Often the LAN had a gateway onto the public or private data network. More recently, the public telephone network operators began to offer access to integrated digital circuit and packet networks. This will be covered in Chapter 2.

In parallel with the evolution of computer network infrastructure has been the growth of the Internet. From a quiet beginning at the end of the 1960s, it has stealthily spread its tentacles around the world, erupting onto the international scene in the 1990s. More powerful than a simple communications network, it consistently challenges the computer/telephony divide.

Perhaps the greatest revolution in the computing world has been the increase in performance of the underlying technology and the way that software has evolved to take advantage of it. The level at which application programmers work is forever moving upward, away from the basics of the computer and its hardware and toward user-perceived requirement.

1.3.1.3 Similarities and Differences

Three trends in telephony have brought the two worlds much closer together:

- Telephone exchange designs have increasingly adopted stored-program control to provide flexible call routing and enhanced customer facilities.
- The means by which the telephone exchanges communicate have increasingly converged upon the use of data link technology.
- The telephone systems of the world are rapidly converting to digital transmission—to improve voice quality and to provide enhanced data transmission capabilities.

Thus, with the exception of the periphery of the system (that which interfaces with the analog lines), the basic technology used for computing and telephony is identical. But much of this comparison and its conclusion may be made with, for example, an automatic washing machine! Here, all of the control elements are now microprocessor-based. It's merely those difficult bits on the periphery that are still electromechanical-based, like the drum and so forth. Shared technology is not a reason in itself for linking things together. There has to be some sharing of function to really begin to create benefits.

Having casually introduced the topic of automatic washing machines, it may be worthwhile to look at their evolution in a little more detail. Early attempts to automate the clothes washing task were similar to Mr. Babbage's engines. The washing "machine" was simply an eccentrically mounted barrel into which was placed the water, soap, and dirty clothing. Turning the handle caused the whole lot to slosh around inside the barrel so that the dirt gradually left the clothing and entered the water. Rinsing could be dealt with in a similar way—by refilling the barrel with fresh water. The first stage of drying was also machine-aided. Clothing was squeezed between the rollers of a mangle, powered, as always, by the turning of a handle.

The next stage of automation was to replicate these functions, but using an electric motor to power them. Hence, the early washing machine was invented, followed later by the spin dryer. Wet clothing was transferred between one and the other, ensuring that the floor also got wet! The first stage of *integration* was to clip the spin dryer to the washing machine. This of course led to the twin-tub, a very good example of physical integration. In the twin tub there is no functional integration. The machine performs no differently from the separate units. The units simply share the same housing.

The sharing of transmission capacity for both data and voice is a physical integration. It implies no sharing of function. Its purpose is to save money and to take advantage of an existing network. In a sense, it also prevents the floor from getting wet!

Of course the next logical step in the evolution of the washing machine is *functional* integration of washing, rinsing, and spinning. The fact that this step requires a further physical integration of the units that performed the functions of washing and spinning—a single tub—is unfortunate. All analogies have their breaking point, and the washing machine analogy breaks down here because it omits a step. The fully automatic washing machine is analogous, in the CTI world, to the creation of a single telephone switch and computer—a fully integrated machine. This is a step that the world may not be ready for. The step that is missing in the evolution of the washing machine is the straightforward linking of functions—the linking of the washing, rinsing, and spinning controllers. However, washing, spinning, and rinsing are serial activities. The rinse controller enters the stage only when washing is complete. There is nothing it needs to know about the washing process while the two take place in separate tubs. Telephony and computing are not normally serial tasks. They are parallel and linked in many applications, and the linkage is supplied by a person. It is this linkage and this step in the evolutionary chain of integration that concerns small CTI, and it is a very important step.

Within the worlds of telephony and computing there are major differences and significant similarities. The overlap of function between the washing machine and the spin dryer is almost complete or can be made so. Figure 1.3 illustrates the degree of overlap in telephony and computing functionality.

Though overlap does occur, there are large areas that have no commonality of function. Hence, the single tub solution does not necessarily apply here. Those that tried the fully integrated approach in the 1980s failed. Some companies tried the twin-tub approach; this too was unsuccessful. Placing a computer and a telephone switch in the same cabinet placed constraints on each, constraints that provided no palpable user benefit. Such systems provided a solution of the "Look at the clever things we can do" type. The failures of the 1980s indicated that a computer is more suited to personal, departmental, and corporate data processing tasks, and a telephone system is best at dealing with telephone calls. Physically combined systems make compromises in both worlds and were therefore not as good at any one task as the separate systems are. However, what was wrong for the 1980s may be right for the 1990s. Costs, technology, and markets change. The integrated solution will be considered later.

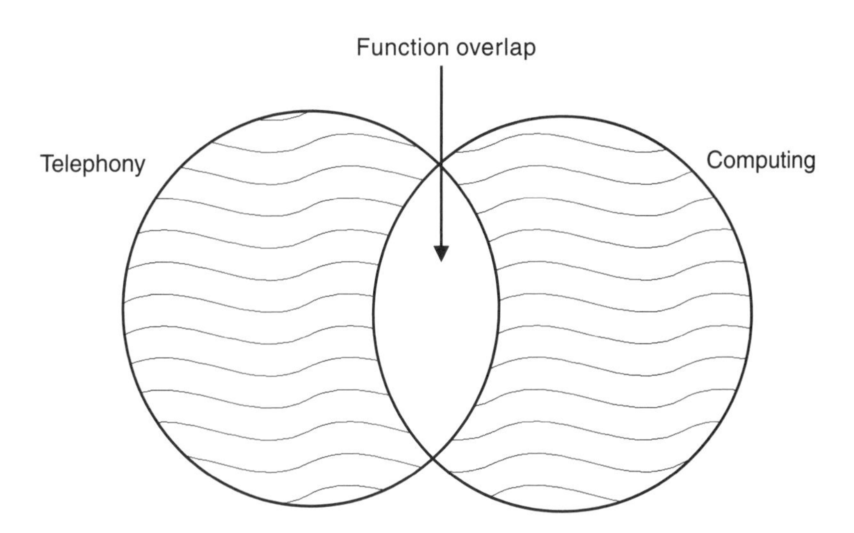

Figure 1.3 The functional overlap between the two worlds.

For now, what is needed is a closer look at the two worlds and their functions to understand where the overlap of Figure 1.3 occurs and what might be in it. Table 1.2 is a basis for such an examination. It is not an exhaustive description of the two worlds. An exhaustive table would be very large indeed. Table 1.2 lists the main points but does not, for example, extend to messaging.

Obviously, the physical overlap is not complete. The area of overlap lies in control. Though identical technology may be used here, the performance aspect noted in Table 1.2 places quite different demands on the telephone controller over those on the computer.

It is often stated that the telephone switch is a *real-time* system. Though this is not entirely accurate, the performance required of the controller within the switch is more demanding than that required of most computers. Also, the workload of the switch controller needs to be more predictable than that of most computers. Telephone switches are dimensioned according to anticipated traffic loads. Though these are not predictable in detail, there is a vast collection of statistical data that allows the telecommunications engineer to make reasonable assumptions about the daily ebb and flow of traffic and hence, the performance required of the controller. This process of dimensioning is made easier

Table 1.2
Some Comparisons of the Two Worlds

Characteristics	Telephony	Computing
Function	Person-to-person communication	Processing data
Business use	Routing calls	Creating documents
	Carrying voice information	Storing data
	Carrying data	Accessing data
	Establishing and maintaining relationships	Manipulating data
Input means	Microphone, dial, keypad	Keyboard, mouse
Output means	Earpiece, headset, loudspeaker, telephone displays	Screen, printer, plotter
Peripheral technology	Analog transducers, mechanical, liquid crystal display (LCD)	Mechanical and electro-mechanical, CRT
Central technology	Digital logic	Digital logic
Performance	Response less than one second	Response greater than one second
Reliability expectation	No failures	Some failures

because the control program is itself fairly stable. Though facilities may be added and malfunctions corrected, basic call set-up time remains roughly the same, and it is call set-up that is the central task in a telephone exchange.

Contrast this with the computing environment. Though the task may be relatively stable, the application software is likely to be undergoing regular changes as improvements are included. Also, the variation of workload is affected by many factors besides the time of day. Complexity of application varies considerably, and the computer may be simultaneously running a number of different applications. Predicting an application computer's response-time behavior is therefore a far more difficult and imprecise task.

Perhaps computer applications should be resident in the switch controller. After all, it is a "near-real-time" environment, and it is often a more reliable one than that provided by conventional computers. I clearly remember a serious discussion that took place some time in the late 1970s. We were busily designing a digital PBX at that time and the first models were commencing field trials. Blissfully unaware of the support issues that were soon to hit us, we were gazing into the future. "The central processor of our PBX is only busy for one or two peak hours of the day. For the rest of the time it could be running applications; it could be providing payroll services for the office or shared word processing, for example," we said. It is a sobering thought to recall that the processor we were talking about was an early 8-bit device—a processor that had the speed of a bicycle compared to today's supersonic devices. But our misunderstanding was not really based on the issue of computer power. It reflected an ignorance of the environment required to run business applications. We worked in an era in which, if you wanted an operating system you wrote one! And we blithely assumed that we could suddenly become experts in payroll systems or even word processors!

So, although the telephony controllers and computing processors controllers use the same technology, the tasks they face are rather different—as is the performance required of them. Hence, though there is an overlap here, it does not occur in the realm of functionality.

The overlap of Figure 1.3 is therefore less about a single point of control and more about shared data and shared control. Table 1.2 indicates that the telephone network carries computer data—but this is a physical rather than functional overlap. The Mitel definition of CTI, which reads as follows, helps to convey this concept.

> "CTI links computer technology with telecommunications to deliver services that can combine the communications signaling of the telephone network with the database resources and programming capabilities of computers."

Functional integration is not about data transport; it is about the linking of control data. Functional overlap actually occurs when one or more of the following conditions apply:

1. Data in the computer is of value in routing incoming or outgoing calls;
2. Call-related data available at the switch can be used to select appropriate items from the computer's database for display on the computer screen;
3. The input peripherals of the computer provide a preferable method of controlling call set-up;
4. The output peripherals of the computer provide a superior method of monitoring the progress of a call;
5. The input peripherals of the telephone system provide a useful alternative means of data collection for the computer;
6. The output peripherals of the telephone system provide a useful alternative method of conveying output data to people;
7. The data of the computer and the routing of a call can be bound together in some useful way (Andrew's magic?);
8. The data processing capability of the computer can be used to process data from the telephone system;
9. The management of the telephone system and the computer system is best performed at one point.

These are somewhat abstract areas of overlap. Their practical use in providing CTI applications will be examined later. The point to be made here is that there *is* significant overlap. The extent of that overlap is not demonstrated in any absolute way by Figure 1.3. Obviously, as the degree of overlap grows, the argument for a single tub solution—a combined telephony/computer system, also grows. Clearly, the similarities are sufficient in number to suggest that a gain in functionality will result from some form of linkage. The question is, what form or forms should that linkage take?

1.3.2 The Linkage Required for Functional Integration

If the list of overlapping conditions given above is complete, and I doubt that, then it should be possible to use the list to tease out the nature of the linkage required. It is unlikely that the list is complete for a very good reason. In examining form and function, it has been observed that the inventor of a new form, a new platform in this context, cannot envisage all the ways that platform will be used—unless, that is, the platform is capable of only a few functions. A good platform will be used for many things that the original designer never imagined. The computer is such a platform, particularly the microprocessor. Who could have dreamt of all the uses to which a PC has been put? Ingenious applications for this very flexible platform are still being created and will continue to evolve. CTI provides an even more flexible platform. Combining the form of the PC and of a telephone switch suggests that CTI applications could be more numerous than those provided for the PC itself.

The question then is, is the list of overlapping functions of any value, given that it is incomplete? The answer is yes, because the list is very general and because it is reasonable to assume that functions that are not listed are going to benefit from a similar linkage. To use more scientific jargon, the list is not complete, but it is sufficient.

The next step is to examine the list to determine who does what and to whom, in order to create a CTI platform. In this context the *who* and the *whom* are the computer and the switch, and the *what* is the information flow between them. That information flow will consist of requests or commands and responses to these, plus spontaneous data that needs to be conveyed to indicate happenings at either end.

Table 1.3 begins the process of mapping the overlapping functions into the sort of information flow that will be needed between the computer and switch and vice versa.

The mapping of Table 1.3 does not define the exact nature of the required linkage. However, it does indicate the nature of the requests, responses, and data flow that need to be coped with to achieve CTI platform functionality. It would be possible to go further at this stage, but such a level of detail would be inappropriate for four reasons:

1. It would be premature in the light of other aspects of CTI yet to be examined.
2. The derivation of the overlap list has already made one implicit assumption—that the nature and performance of applications that might use the CTI platform are already known.
3. The comparison is biased toward call control and excludes other media.
4. Messaging has not been fully considered.

Table 1.3
Mapping the Overlap Functions to Linkage Requirements

			Direction	
Function	**Overlapping Function Description**	**Information Flow**	**Computer**	**Switch**
1.	Data in the computer is of value in routing incoming or outgoing calls.	Route request trigger point Request routing data Return routing data	⇒ ⇐ ⇒	
2.	Call-related data available at the switch can be used to select appropriate data for display on the computer screen.	Call data, for example, the origin of the call	⇐	
3.	The input peripherals of the computer provide a preferable method of controlling call set-up.	Call request Call routing	⇒ ⇒	
4.	The output peripherals of the computer provide a superior method of monitoring the progress of a call.	Call status	⇐	
5.	The input peripherals of the telephone system provide a useful alternative means of data collection for the computer.	Collected data, for example, stock figures for a particular department	⇐	
6.	The output peripherals of the telephone system provide a useful alternative method of conveying output data to people.	Data output, for example, current sales conversion rates	⇒	
7.	The data of the computer and the routing of a call can be bound together in some useful way.	Screen and call data association	⇔	
8.	The data processing capability of the computer can be used to process data from the telephone system.	Call information statistics	⇐	
9.	The management of the telephone system and the computer system is best performed at one point.	Configuration requests Management information	⇒ ⇐	

Though the tales recounted earlier in this chapter have sketched a few applications, the second reason does include a big assumption. In later chapters, applications will be examined in detail, and the assumption can then be tested. However, it is probably apparent that, in some cases, the linkage provided in practical implementations is heavily influenced by the applications they support. In other cases, a more general approach to the provision of linkage needs to be taken—to ensure that a wide range of applications can be supported.

Table 1.3 takes the more general approach. Within it, a definition of the nature of the information itself and of the direction of information flow has been indicated. Without going into too much detail, a number of points can be extracted from this analysis:

1. A both-way request and status linkage is required.
2. A number of the information flows identified are needed for purposes other than functional integration (an obvious example being *call information statistics,* shown under function 8).
3. The data that is to be shared implies that the computer "understands" the world of telephone call processing.

Point 2 raises the question, do the computer and switch already possess the interfaces necessary to provide the linkage required or is something special needed? This is a question that is far easier to ask than to answer. With ingenuity and imagination it is possible to achieve an awful lot of capability from existing interfaces. There are a number of different approaches to linking computer and telephone systems—some specialized, others not. It is necessary to define these approaches as part of this introduction because through these definitions a further step will be taken in answering the question that heads this section of the book: What is computer telephony integration? First, however, it is desirable to rise above the issue of linkage and view CTI in the wider context—from the all-important applications viewpoint.

1.3.3 Application Basics and the CTI Bridge

The most important deliverable of CTI is the collection of applications that it makes possible. The linkage between the computer system and telephone system is key to the functionality of a CTI implementation; however, though it tends to interest the technically minded, it is generally transparent at the application level. As physical integration proceeds, it becomes irrelevant. An application is a task—something that needs to be carried out within a business. Collecting a debt is an application. In the tale about debt collection, the application involves Andrew the collection agent, a computer, and a telephone exchange. The computer supports *application software.* Application software is an implementation of those parts of the application that run on the computer—it is not the application itself. Indeed, the application Andrew works on is not confined to the computer alone—it is also supported by the software within the telephone system.

Bearing this in mind, Figure 1.4 attempts to provide a clear picture of what an integrated application is. It straddles the telephone and the computer system, bringing the two together functionally. This concept is useful.

Figure 1.4 clearly suggests that a single CTI application is created, not two applications communicating across a CTI link, one in the computer, the other in the switch. When discrete systems are integrated, there are of course—physically and actually—two application *programs* and each has to be produced separately. However, from the concept level—and from the application specification level—there is only one application. It just happens that the application requires two separate machines to implement it, in much the same way that washing clothes requires a washing machine and a spin dryer.

Figure 1.4 is also useful in suggesting a bridge. The bridge allows information to flow from the telephone system and vice versa. For example, the telephone system often knows the directory number of an incoming caller; bridged to the computer, that raw information can be translated into the caller's name, relevant data on previous dealings with the caller, the caller's geographical location, and so forth. The computer can then display such information on the

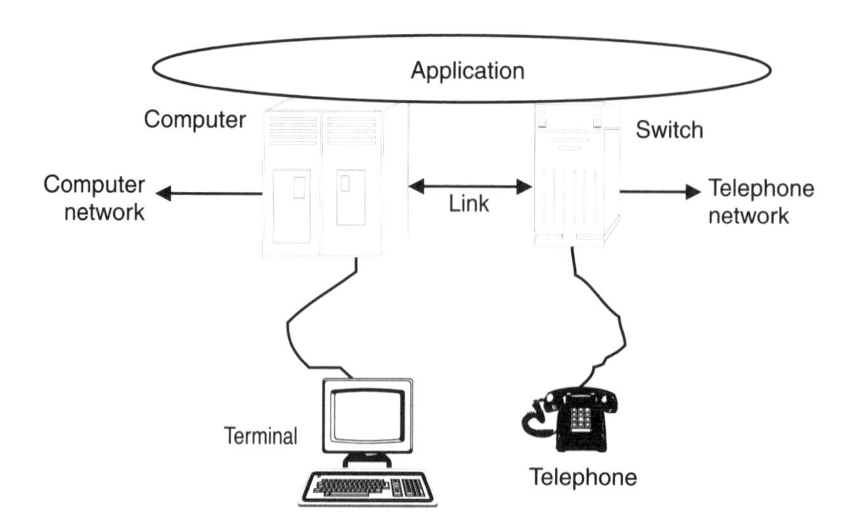

Figure 1.4 The applications bridge.

user's terminal prior to answering the call. It is this bridge that allows system designers to overcome the barriers that bedevil so many integrated systems.

Bridges and integration are all very well—but what does this functional integration really do for business applications? The generic functions that can be identified are:

- Screen-based telephony;
- Call-based data selection;
- Application-controlled routing for incoming calls;
- Application-controlled routing for outgoing calls;
- Voice and data call association;
- Data transfer;
- Coordinated call monitoring.

Some of these have already been discussed. All telephony-integrated applications use at least one of them; many use more than one. The detailed operation of each of these generic application functions will be explored in Chapter 5. For the moment, note that they do suggest a bidirectional bridging of the switch and computer.

1.3.3.1 CTI Application Areas

There are three major areas of applications for CTI: the call center, the workgroup, and the desktop. These are not mutually exclusive—they can overlap and merge into one another

1.3.3.2 The Call Center

One of the key application areas for CTI is the call center. Not surprisingly then, this is a term that is used many times in this book, and it is therefore appropriate to define the call center in this, the introductory chapter. A call center is based on a group of people, usually called agents, whose sole or main task is to do business via the telephone. The business that is transacted is almost always automated via a computer system. That is to say, the computer contains the business application software and the business database. Each telephone call, whether it be incoming or outgoing, will require access to data from the computer system and usually will require that the data be modified. Agents are normally equipped with a headset and computer terminal. The whole process is usually heavily monitored. With that definition it should already be clear that call centers form an important area of CTI application—but they do not form the only area. There is no such thing as a typical call center, but Figure 1.5 provides one example.

1.3.4 Workgroup Integration

1.3.4.1 The Workgroup

A workgroup is a group of people with a clear work relationship that entails communication. The most obvious example of a workgroup is a group that is working on a specific project. The group members are more likely to share information and communicate with each other than with personnel outside of the group. A workgroup may be a department within a company—the human resources department, for example.

1.3.4.2 Desktop Integration

The desktop provides another key application area for CTI. Here, the CTI applications are more concerned with personal productivity than the group productivity gains realized by CTI within the call center. The provision of screen-based telephony from a personal computer provides a simple example of desktop CTI; there are many others. Some observers describe the combination of telephone and terminal provided by CTI as a *desktop-area network* (DAN).

Figure 1.5 A call center in action. (*Photo courtesy of:* Pathway PR.)

1.3.5 Approaches to CTI: The Computer System View

There are two distinct approaches to the practical achievement of CTI. The two approaches are called *first-party CTI* and *third-party CTI.* In physical terms the first approach can be likened to plugging the computer into a normal telephone line; in the second approach, the computer is plugged into a special socket on the switch. In functional terms, the approaches can be related to the normal telephone user and the attendant/switchband operator, respectively.

In first-party CTI, the computer can do exactly what a telephone can do—and no more. Similarly, it can only know what a telephone can know. In third-party CTI, the computer's view of the telephone system is very similar to that of an attendant. The attendant sets up calls for other people, usually dropping out of the call once it is satisfactorily established. It is in this sense that the attendant is regarded as a third party. Often the attendant is privy to a lot more information about users and the system itself. Feedback is usually achieved through a screen and includes details of the progress of call set-up, configuration details of called and calling parties, and so forth. The attendant position is often multiline, or at least it appears to be so. This means that an attendant can deal with two or more calls completely and independently. All of this applies to the third-party computer.

So, in first-party CTI, the computer's view of the telephony world is limited to the telephone's view. Generally, calls can be made from the line associated with the first party and received from it. Calls cannot be made on the behalf of other parties except via the conventional call, inquiry, transfer, and release sequences available at each phone.

In third-party CTI, the computer has a wider view of the telephone world. It has more information and can make calls for other parties. Indeed, that is its purpose, since it normally takes no part in the speech phase of the calls that it deals with.

It might seem that first-party CTI maps directly onto desktop integration, as described above, and third-party CTI maps directly onto the call center. Though there may be good architectural reasons for providing the integration in this way, it would be misleading to suggest that the relationship is exclusive. First-party CTI can be used in call centers and desktop integration can utilize third-party CTI. Either can be used for the workgroup.

Having briefly defined first- and third-party CTI, the next step is to examine the practicalities of implementing the two approaches. Figure 1.6 illustrates both approaches and will provide the basis of the comparisons that follow.

The physical connectivity of the links shown in Figure 1.6 is not of particular interest at this stage. However, the point in the overall architecture of the telephone system that the computer connects to is important. It is this point

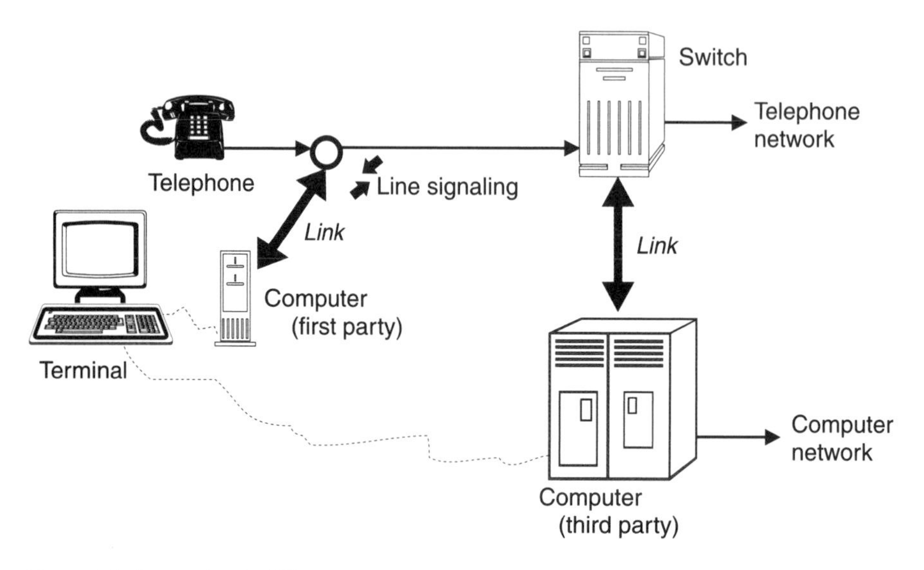

Figure 1.6 Approaches to CTI.

that, to a large extent, differentiates first- and third-party CTI. Note that this discussion and Figure 1.6, upon which it is based, are a simplification of the real world. The computer system might be distributed; the switch can be one of many different types and may even be a part of the public network. Voice processing and image processing units are not included, though they may well perform a key function within some applications. All of this is deliberately excluded at this point—in order that the wood can be seen without the trees. Here, the focus is on simple call control and functional integration.

1.3.5.1 First-Party CTI

Figure 1.6 shows a first-party computer linked to a telephone line. The linkage allows access to the line signaling that normally passes between the telephone and the switch. Signaling is the means by which the telephone communicates something to the switch or receives some indication from it. For a simple telephone the signals are basic, for example:

- Handset lifted;
- Dial tone;
- Digit keyed or dialed;
- Ringing current.

Note that the signals flow in both directions. Lifting the handset causes a signal to be sent from the telephone to the switch. If the switch is ready to deal with a new call, it sends a dial tone back to the telephone. Thus, assuming a suitable interface from the first-party computer to the line, all the signals that a telephone produces can be generated by the computer. Similarly, all the responses and signals from the switch can be detected and interpreted by the computer. Using this information, telephone calls can be controlled from the keyboard and monitored via the screen associated with the first-party computer.

There are two categories of telephone, basic and enhanced. Basic telephones utilize the conventional signaling conditions used by all telephones that are *public switched telephone network* (PSTN) compatible. There is no separately identifiable signaling channel or protocol, and functionality is limited—particularly in the switch to telephone direction. Enhanced telephones are sometimes called feature phones or feature sets. They use a more sophisticated signaling system. Unfortunately, the signaling systems used are almost entirely proprietary. They vary from switch to switch and specifications are not always released by the switch supplier. Key system telephones form a special group here.

Since there are two categories of telephone and therefore two types of signaling, then first-party CTI must break down into two subclasses. These are termed *first-party basic* and *first-party enhanced.* The capabilities of each subclass will be explored later.

1.3.5.2 Third-Party CTI

Figure 1.6 indicates that the third-party computer connects to the switch through an interface that is not related to the telephone line. This link is best thought of as a processor-to-processor link. Particularly in relation to call control, it allows the computer to request that calls are connected, cleared, conferenced, and so on. It also supports the return of status messages so that the computer can keep track of the calls it is interested in. It can also allow the switch to make requests and receive data from the computer. Some example messages might be the following.

- Make a call;
- Status of call;
- Clear call.

One notable difference between third-party CTI and first-party CTI relates to the association between computer terminal and telephone. For first-party CTI, this relationship is implicit; both connect to the same line. For third-party CTI, there is no relationship unless the application software establishes one. An implication of this is that third-party CTI allows for applications that utilize any telephone on the switch and any terminal on the computer. Also, a third-party CTI application can, in principle, associate any telephone with any terminal, if required.

Because switches vary in CTI capability, subclasses of third-party CTI must be identified. First of all, most switches have specialized linkages for CTI purposes. The computer and switch communicate on equal terms. The linkages are mainly proprietary and vary considerably in their capabilities. Nevertheless, they are all grouped into the subclass called *third-party compeer.* The second subclass contains those switches that are *only* capable of processing calls when connected to an external computer. These switches are sometimes called dumb switches, which seems unnecessarily pejorative; they are also called matrix switches and sometimes specialty switches. In any event, they require a different linkage to the switches that are innately capable of call processing. This subclass is termed *third-party dependent.* The third subclass contains those implementations that do not use a specialized linkage, merely taking advantage of existing system interfaces provided by the switch; this is the *third-party primary* subclass. Here, the computer may emulate another switch or a group of telephones.

1.3.5.3 Overall Taxonomy of CTI Approaches

Summarizing the classifications of CTI, there are two possible points of connection within a switch environment: at line level and at system level. These have a definite effect upon the view that the computer has of the telephone world. First-party and third-party CTI are derived from these two connection points. Figure 1.7 displays the categories of CTI that will be used in this book, whereas Figure 1.6 introduced physical integration into the taxonomy. When each element of a CTI system is separate, the term discrete is used. When the elements are placed in the same box, the term merged is used. A third-party merged system is equivalent to the single tub washing machine. Both switch and computer are in the same box, and there is only one controller for telephony and computing. Hence, there is no concept of a link. Accordingly, the approach must be third-party-dependent.

Where the telephone becomes part of the terminal or PC, basic or enhanced first-party still applies since this is dependent on the signaling system provided by the switch.

Possibly the two most important areas, and therefore the ones that take up much of this book, are first-party enhanced and third-party compeer. It is the latter that has so far received the most attention from standards organizations and implementers in the 1990s. As physical integration proceeds, third-party dependent will assume greater importance.

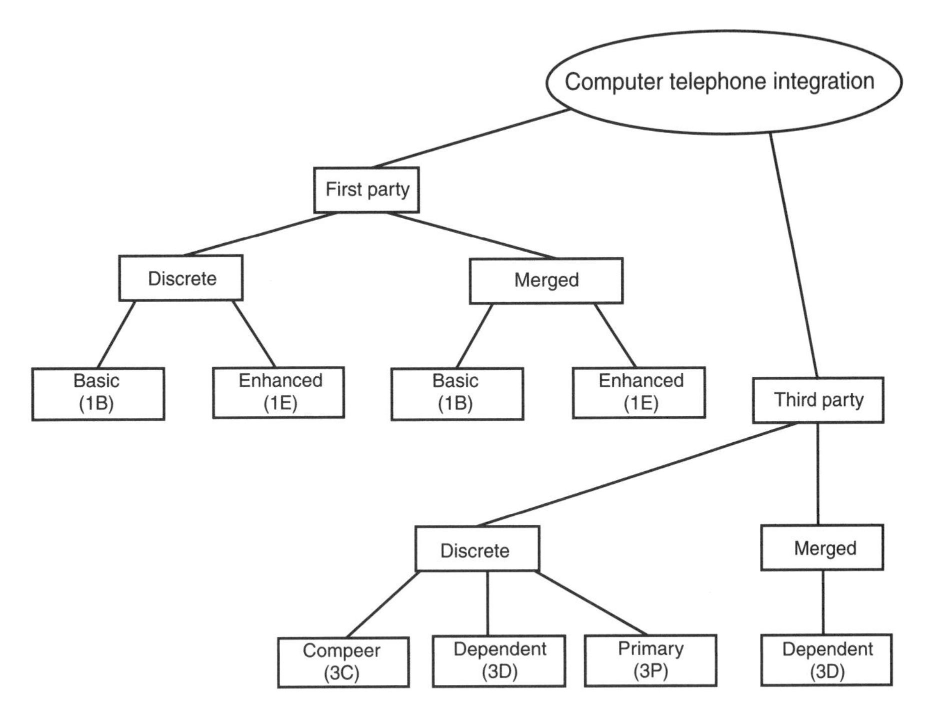

Figure 1.7 The taxonomy of computer telephone integration.

This book is primarily directed towards CTI applications in which the computer half of the platform is customer-premises-based. Intelligent networks, in which the computer is embedded within the telephone network, are covered in outline and referred to regularly, but here they are not regarded as part of CTI.

1.4 The History of CTI

It is a surprise to some people that CTI has a history—it seems to be a very recent development. However, CTI is certainly not new, and examining the history of the technology provides a useful introduction to its current status.

1.4.1 Prehistoric CTI

As far back as the 1960s and early 1970s, IBM was installing computer/ telephone solutions based on a linkage of the company's two key products: the 360 mainframe computer and the 2750 PBX. Many of these applications were in Germany. At first sight, this may seem strange since Germany, in those days at least, presented a highly regulated telecommunications environment. In the words of someone who was there at the time, "Things were either compulsory or forbidden!" Yet this seems to be the birthplace of CTI. This puzzle is not so difficult to resolve as it seems. The Bundesposte clearly had no wish to encourage IBM as a supplier of switches. The method chosen to regulate this was to restrict IBM to selling switches only in circumstances in which there was a related data application. Necessity was then, as it is now, the mother of invention, and the next story should demonstrate IBM's inventiveness at the time.

1.4.1.1 Keeping the Shelves Full

After lunch, Horst always insisted on a brisk walk. The route was always the same. Leaving Kramerstrasse, he would pass the Market Church, pausing only to check that his watch and the tower clock corresponded. Weaving his way through the traffic in Friedrichswall, he entered the welcoming silence of the Maschpark. Walking the entire perimeter of the lake, he then left the park by the Town Hall exit, reentered Friedrichswall, and so returned to the KNO bookstore in Kramerstrasse.

As usual, business was reasonably good. While Horst chose to spend his lunch time striding through the park, many of the office workers and tourists preferred to spend theirs browsing through the vast collection of books. The bookstore was large, the largest in the city. It was part of a chain of similar stores throughout Germany. Each store in the chain exhibited an individuality, partly derived from the different buildings that housed them—all old and large—and partly from the varied reading tastes of their customers.

Horst went immediately to the office, placed his umbrella in its usual position, then removed a piece of equipment from its housing beside the telephone. The equipment was about the size of a large book and in appearance resembled a calculator. It was a simple hand-held terminal. Horst checked the date and time on its screen—2:00 p.m. on Thursday the 13th of February 1970. Horst was precisely on schedule. He left the office and ascended the stairway to the top floor, the music and art department. He entered the correct designation for this department via the keyboard, then made his way to the beginning of the arts shelves. There were a number of customers milling around in this department, and Horst was forced to deal with three queries as he attempted to traverse the floor.

Finally arriving at the shelf that held the "A's" of the arts section, he began to examine each stack of books. If they were depleted, he entered the ISBN code for the book and the number of books required to restock the shelves into his terminal. At 3:20 p.m. the sales assistant came to report that two popular novels had sold out. Horst asked her for the ISBN numbers. She didn't have them, so Horst respectfully asked her to get them. Upon her return, Horst entered the numbers and indicated that an order of 12 each should be generated. He then returned to the shelves and examination of the art books.

At the end of the day, Horst returned his hand-held terminal to the receptacle on his desk and began a tour of the building. By 5:45 p.m. everyone had left; Horst switched off all the lights and locked the doors of the store for the night.

Two hours later in the darkened office the telephone rings. Of course there is no one there to answer it. But it is answered. The equipment that Horst has been using begins to glow slightly as an indicator lamp turns on. A subdued intermittent whistling is heard for some minutes, then the lamp goes out, and silence returns to the office.

The next day Horst spends the morning rearranging the book display in the travel department. He lunches at his usual time, then returns from his lakeside excursion to find a delivery lorry blocking the entrance to Kramerstrasse. Horst approaches the driver and indicates that the lorry should approach the store via the Burgstrasse entrance. He then returns to his office, almost three minutes late due to the incident with the driver. He informs the porter of the imminent arrival of the lorry and asks to be informed when the unloading is finished.

As Horst sorts through a pile of pamphlets describing new books, he is interrupted by the telephone. The porter informs him that unloading is complete and he hurries down to the ground floor. He checks through the driver's delivery list; all is complete. The driver departs and Horst and the porter move the delivered books to the top floor. There they begin to fill in the gaps in the arts section starting with the "A's." This takes some time. Finally, there are 24 books left. Horst and the porter take them to the foyer where the current best sellers are displayed.

Is Horst another magician? Of course not. However, he was one of the earliest users of CTI. The head office staff of the bookstore chain had purchased an IBM 2750 PBX. They had worked with IBM to create an application that would allow their existing 360 mainframe to collect orders from each of the company's many bookstores. The application would run at night, when capacity was available on the mainframe and the telephone lines were not in use. It required that the computer and the PBX be linked in some way.

1.4.1.2 Teleprocessing Line Handling

The linkage between the IBM 2750 PBX and the 360 mainframe was known as *teleprocessing line handling* (TPLH). TPLH required special software within the 3750 and the 360. The messages that made the linkage possible were carried across *a binary synchronous communication* (BSC) link using a slightly modified remote job entry protocol (2780/3780).

TPLH was used to control the PBX from the computer. At a specified time at night, the 360 computer selected the telephone number of a bookstore, then instructed the 3750 PBX to make a call to that number. The line was associated with the receptacle holding Horst's terminal. The receptacle performed a modem function. It answered the call from the 3750 PBX and then transferred the data that Horst had entered into the terminal to the line. This it did by sending the data in the form of touch-tone digits. The PBX's normal tone-digit receivers could then be used to capture the data and, in turn, transfer this data across the link to the 360 computer. The computer could then select the number of the next store and extract that store's book orders for the next day.

Thus, by the opening of business on the following day, the computer could print out a complete list of the orders for all stores, together with delivery addresses. In this way, Horst was filling the shelves with books he had ordered only the day before. Clever stuff, especially recalling that this system was in use over a quarter of a century ago! Nor was the bookstore the only application of remote automatic order entry. A similar arrangement was implemented for pharmaceutical shops, once again in Germany.

The problems that beleaguered IBM in those days and probably prevented the widespread adoption of such solutions were that

- It was difficult to provide financial justification for a wide variety of applications at that time, given the prevailing levels of hardware costs inherent in the computer and the PBX.
- Though the applications described worked well enough, there was a lack of robustness in both hardware and software components that demanded a reasonable level of human back-up.
- Software development tools did not posses the level of sophistication they now have. This added to implementation costs and reduced overall reliability.

The cost element was significant. When the 2750/3750 PBXs were introduced into the United Kingdom, they were up to six times more expensive than ordinary PBXs. Hence, though the IBM PBXs undoubtedly represented a great leap forward in functionality, their adoption did require that the customer have deep pockets and a sense of adventure.

The second bulleted point mentions human back-up. In fact, these early implementations did cater to this lack of robustness. If Horst's terminal had not responded for some reason, the 360 computer would have printed an exception report for an operator's attention the next day. Horst could then be contacted directly and his order taken manually.

1.4.1.3 Data Collection by Telephone

The 2750 PBX was also used in data collection applications that were initiated by human operators. SKF, the Swedish company best known for its range of mechanical bearings, installed such a system in October 1970 at their German plant in Bad Cannstaat. To quote from a promotional leaflet of the time:

> In 1969, SKF studied their job and bonus reporting procedures, which were then based on an IBM System/360 with batch entry of data. The company wanted to speed up the collection of data so that production could be more tightly controlled with up-to-date information. Individual factory departments were found to be using over 300 official and unofficial forms with 100 different reports. Data was collected at an estimated DM 1.5m ($480,000) by 79 timekeepers, production schedulers, clerks, and key punch operators over three shifts. The old procedures have been summarized as having had several problems:
>
> - *Time:* Production reports were produced too late to be of great value.
> - *Volume:* There were too many reports to read, and too many cards to punch and process.
> - *Cost:* Existing cost of manual data collection was high and increasing [5].

The 2750 PBX was linked to the company's System/360 computer and the production data was then entered via push-button telephones. Production reports from the 360 were then fully up to date. Also, exception reports could be printed as soon as they were identified. Significant manpower savings in data collection were also reported.

The 2750 PBX did some simple error checking on entered data. If the entry was successful, a continuous tone was returned. If not, an interrupted tone was returned to indicate an error. If the link between the 2750 and the 360 failed, the PBX could punch the data onto paper tape. The notable points about this very early example are the following:

- It used the link for data transfer—one of the generic functions of CTI.
- It is a simple application requiring very little functionality in the linkage between the computer and the switch.

- Status messages from the computer were transformed into telephonic signals (tones) by the switch.

1.4.1.4 Electromechanical CTI

When IBM launched the 3750 PBX in the United Kingdom in the early 1970s, the indigenous manufacturers of telephone switches became concerned. They were still producing electromechanical PBXs of either Strowger or Crossbar design. Their products could not offer the advanced facilities and features of the 3750 because, unlike that switch, their offerings were not programmable.

The Plessey company reacted strongly to the potential threat. Instead of designing a new, computer-controlled switch, the company decided that there was still some mileage in the existing crossbar switches. So the Plessey engineers designed an interface that allowed the connection of a computer to their crossbar PBX [6]. The computer was an Interdata 7/16, a minicomputer of that era that was operated in dual-processor mode for reliability. The interfacing was extensive. Every line of the switch was wired to the computer so that the status of each extension could be monitored. Furthermore, each register in the control area was intercepted (the registers are the components that collect dialed digits for the switch). The computer examined the digits that were dialed, then translated them into a command that would cause the crossbar switch to perform whatever routing the Interdata's program determined. In this way the computer took control but still allowed the normal controller to make the connections.

The resulting product was called the System 2150. Two of them were installed in the mid 1970s: one in British Petroleum's offices in London and the other in British Steel's facility at Corby. However, the development was, to a large extent, overtaken by events. Plessey decided in the latter part of the decade to adapt the Rolm PBX for the British market. This became the PDX. It was a computer-controlled switch, so the ancillary processor of the 2150 no longer had a role to play.

The 2150 probably represents the nearest thing to an early example of a third-party dependent CTI implementation. It had a lot of capabilities that were unknown to conventional electromechanical PBXs. On its 1.5-MB disks (which were some 18 inches in diameter), it stored the complete telephone directory for the PBX. People in the directory could be called by name, and the 2150 provided sophisticated call information logging facilities. The system also offered data options, including data entry and retrieval from a low-cost terminal.

1.4.1.5 Early Inbound Installations: Telephone Answering Bureaus

The role of a telephone answering bureau is to answer a company's calls, often out of normal working hours, responding to the call in the company's name and taking messages. In 1975, four men at Bell Labs took a close look at the way existing telephone answering bureaus worked. These were paper-based systems. The bureau held a card containing the details of each client. They were slow because it took the operators a significant amount of time to find the information card for the company they were answering for, and they were not interactive. The four men decided that the technology was available to automate much of this process and set up the Delphi Corporation to design and implement a suitable system.

The first Delphi system was called Delta 1 and proved the founders to be correct. By 1978/79, Delta 1 was providing telephone answering services in the San Francisco Bay area. Calls were answered by the fourth ring and in that time a database was consulted, the correct company details selected, an appropriate agent selected, and the agent's screen filled with the relevant answering script. After greeting the caller, the agent took a message, read a specific script, or did both. Subscribers to the service rang a separate line to collect messages, change the greeting, and so forth. The system proved significantly cheaper to operate than the manual bureau: its agents were around 30% more efficient.

This first installation was a great success. It was estimated that Delphi had taken 50% of the telephone answering market in the San Francisco area within two years. But the technology that was used in Delta 1 was very much out of date. The system was a prototype. Based on discrete logic mounted on meter square boards, it needed reengineering to become a production system—and that caused an increase of staff from the original 10 to 250! And that is where Delphi's troubles began.

The problem that Delphi was trying to solve all those years ago should already be familiar—how do you functionally link the processing of a call with a computer database? The link to the subscriber was established in a very direct way. A physical connection was made to the subscriber's line and extended to the bureau! There are more sophisticated ways of achieving this association nowadays. However, the point is this: the system knew which company was being called because individual lines were physically tied to that company. The system then had to allocate an agent—an automatic call distribution task; interrogate the database; write to the agent's screen—a computer task; and, finally, switch the call through—a PBX task. The Delphi team designed the hardware and wrote the software to do all of these things and more. Delta 1 was therefore a fully integrated system, so the linkage between the computer application software and the call-processing software was implicit.

Why did Delphi do it that way? Why wasn't a separate computer used for the database application? After all, there were plenty of minicomputers around in the mid 1970s. The reason seems to have been performance. A conventional computer running a timeshared database application could not be relied upon to bring up the greeting data on the agent's screen within the required time. In addition to this, the integrated system had a single call manager that simplified the entire business of synchronizing call and data events.

Around 1979, the design of Delta 2 began, funded by a significant infusion of capital from the oil company Exxon. A little later in that year a company called Nexos, funded by the British government, became a preferred distributor for the planned system in Europe and began to take an active part in its design and particularly its adaptation for the British regulatory environment. At one point, 20 engineers from Nexos were working at Delphi's Los Angeles office. Things did not progress well. By the end of 1981 there was still no product, so Nexos withdrew, taking its engineers with it. A year later, with still no product and therefore no customers, Exxon also withdrew, and the saga drew to a close. Though the market was there, manufacturable technology was not.

1.4.1.6 Early Outbound Installations

Though the history of CTI is dominated by inbound applications, it is interesting to observe that the earliest of the applications described was outbound—in that the computer originated the calls to the bookstores. Power dialing forms one of the major outbound application areas. Power dialing and the related approach, predictive dialing, is described in much more detail later. For the moment, a glimpse of its origins will be useful.

Early Power Dialing

Place yourself in the late 1970s in Dallas, Texas, for a moment, and picture a secretary with eight tape recorders, four telephones, and a list of potential customers in front of her. The telephone operator picks up a phone, rings someone on her list, and says, "Please hold on, I have a message from Zsa Zsa Gabor for you." The called party then hears the voice of Zsa Zsa Gabor explaining how splendid a particular perfume is and asking for their name and address.

To play the message from Zsa Zsa, the secretary started a cassette player, which, when it got to the end of the recording, automatically started a tape recorder that then recorded the person's name and address. As soon as the secretary started the sales call, she could leave it and start a similar one on another line. This was a very efficient method of selling and was applied to a number of sales campaigns. However, it did have a flaw—what happens if part of the way through the interesting message from Zsa Zsa Gabor a disaster occurs at the home of the called party? A coal falls from the fire and onto the carpet. The carpet immediately bursts into flame. The caller shouts at Zsa Zsa that she must get off the line because they have to ring the fire brigade. Zsa Zsa merely continues to deliver the message and amplify the benefits of the particular perfume that she is selling. The called party slams the receiver down, then immediately lifts it up again expecting to hear dial tone. However, all they hear is Miss Gabor continuing to describe the virtues of her perfume. By this time the room is engulfed in flames!

To overcome this problem, a young man named Skip Cave was brought onto the scene. His job was to detect the click that occurs when the called party hangs up, so that Zsa Zsa Gabor could be silenced and the line cleared. It so happens that detecting the hang-up click is very similar to detecting the answer click, so Skip designed (and patented) a means of doing both.

Having provided the technology to do this, the next thought was, why not automate the entire process by removing the secretary? This step brought the computer onto the scene, to hold the list of prospective customers, and required development of early "grunt mode" detection, the ability to detect silence or the presence of speech. In fact, Skip Cave developed a method of detecting both long-duration speech and short-duration speech so, for example, if someone spoke for too long they could be interrupted or if they didn't speak at all they could be prompted to do so.

This level of sophistication enabled the company for which Skip worked, then called *Telephone Broadcasting Systems* (TBS), to construct a system in 1983 to help a radio ministry preacher raise funds. Calls were placed automatically, and when an answer was detected, a recording of the preacher's voice asked for a contribution to some needy cause, after which the called party was switched to a live agent who completed the transaction.

All of this technology finally led to the production of an early predictive dialer, which was installed in Kenton, Ohio, in 1984 for the General Electric Credit Corporation. This company collected outstanding payments for a wide range of domestic items, such as refrigerators and cookers. Collections were carried out over the telephone and required 200 agents. The installation of the predictive dialer reduced this requirement to 30 agents!

In 1987, TBS was purchased by Davox, which had previously supplied it with dualport terminals for predictive dialer installations. Davox was still supplying predictive dialers at the end of the century.

1.4.2 The Foundation Decade

In 1980, VMX (later part of Octel and then Lucent) installed the first voice messaging system on a customer site. The customer was 3M, and the system probably signaled the true beginnings of the voice processing industry. The first interactive voice response installation appears to have been installed by Periphonics in 1982. The first predictive dialer was installed in 1983. Thus, the 1980s laid the foundation for what has become known as CTI.

In the early 1980s, a supplier-driven surge of interest in the convergence of voice and data began. There is little evidence that users wanted the products that were envisaged and plenty of evidence that they did not purchase the products that were offered. Clearly, there were insufficient purchases to provide a viable convergent marketplace at that time.

The drive seemed to be stimulated by the need to protect an established market against predators, coupled with a misinterpretation of who the predator might be. The computer supplier thought that the telephone system supplier would take away a significant proportion of its traditional market. The telephone system supplier thought that the computer supplier was busy designing switches and would soon be making inroads into its traditional markets. The extent to which these beliefs were fueled by IBM's adventure into the PBX world in the 1970s remains a matter of speculation.

Interestingly, the threat to traditional computer suppliers came from within their own midst—in the guise of the PC and distributed computing. Meanwhile, the threat to telephone system suppliers has been the slowing of market growth as the conversion from the big old electromechanical systems to the modern compact digital systems showed signs of saturation.

Nevertheless, suppliers from both worlds did actively pursue convergence in the 1980s. "After all," everyone said, "the technologies are now the same." As if that were all there was to it!

1.4.2.1 Integrated Voice and Data Terminals

Suppliers from the computer and telephone worlds, but predominantly from the latter, developed physically integrated telephones and terminals. In some instances these were simple box-sharing developments, and often the telephone handset was integrated with the keyboard. There was no sharing of function. These were similar to the twin-tub washing machines spoken of earlier. Their justification lay in ergonomic rather than functional fields. Arguably, these physically integrated machines consumed less desk space than two separate devices. Also arguably, it was easier to transfer data from screen to mouth or from ear to keyboard when the two machines were physically linked. For example, it may be easier to read data from the screen to a caller or to obtain a telephone number from the screen and transfer it to the keypad.

However, many *integrated voice data terminals* (IVDTs), as they were generally called, did provide some form of functional integration. Not surprisingly, that integration tended to emanate from the telephone directory. The computer part of the IVDT maintained a personal database; the telephone part transmitted a selected number.

Table 1.4 lists several IVDTs from the 1980s. These devices were a mixed bag as far as architecture is concerned. In general, they consisted of analog telephone components from an existing product (no one designs new telephone circuitry from scratch if it can be avoided) plus a microprocessor-based PC. IVDTs certainly looked unusual. It's quite amazing how many different places a telephone handset can be attached to a terminal. Probably every variant that can be thought of was tried by at least one IVDT designer! The STC Executel was one of the most interesting IVDTs in appearance. Its heredity was obviously the telephone rather than the terminal. It was rather like an overgrown telephone. Similar in dimensions to a PC keyboard, it sported a small (perhaps 4-inch) screen thrusting out of the surface, mounted on a little pole. Many people liked it, but not enough to save it from the ignominy suffered by most IVDTs—bargain basement sell-offs at well below manufacturing cost!

Claimed functionality in IVDTs varied from telephone support and enhancement, to a personal organizer, to full-blown office automation. The apex was the ability to annotate text with voice. Targeting depended as much on the salesperson as on the product. Applications were generally nonstandard, simply because the hardware platforms were.

Table 1.4
Some Integrated Voice and Data Terminals

IVDT	Supplier	Observations
Displayphone	Northern Telecom (Nortel)	One of the more successful IVDTs
Executel	STC, now part of Northern Telecom	Offered manager/secretary working
IMP Workstation	ITL	Developed from an office automation terminal that offered voice annotation
One per Desk	ICL and BT	Based on Electron home computer
Davox Workstation	Davox	IBM 3270 terminal emulator and telephone combined

Although some IVDTs did achieve a degree of market success, these were the exception rather than the rule. IVDTs fell into that most common of integration traps: the telephone part was not as good as the best of telephones and the computer part was not as good as the best of computers. Worse still, the computer was not a part of the office computing environment. It provided a new environment and an alien one.

Most of the IVDTs used simple signaling to the telephone system so they could be used on any line. They can thus be categorized as merged first-party basic CTI implementations.

1.4.2.2 Fully Integrated Systems

Probably the most integrated and most advanced system to be put into use in customer's offices in the 1980s was the system originally designed by Sydis of San Jose, California. The Sydis system centered on a UNIX-based information manager that provided an application environment for centralized office packages, such as the following:

- Word processing programs;
- Spread sheet programs;
- Diary managers;
- Database managers.

The information manager also supported text and voice mail and was able to switch telephone calls. Architecturally, it front-ended a PBX on the extension side. Each extension line was intercepted and converted from conventional analog transmission to high-speed digital. An integrated voice and data terminal, unusual in its triangular shape [7], replaced the extension telephone and connected to the digital link. The terminals could be used with or without a keyboard. Advanced facilities included the ability to annotate text documents with voice and the provision of a multimedia in-tray.

Sydis, the original designer of the system, sank into liquidation in 1986. The technology was purchased by British Telecom and the system was in production under the new name of Mezza in England by the summer of 1987. The system was sold into a series of high-profile locations, including:

- The Canadian prime minister's office;
- The office of the Italian president;
- The Pentagon in Washington.

Although considerable sums were sunk into the system's development, it never did sell in sufficient numbers to assure its future and was finally dropped by British Telecom in 1989. One of the greatest difficulties encountered in attempting to sell the system stemmed from the fact that prospective purchasers did not know what it was—computer, switch, voice-messaging system, e-mail system, or whatever. That, coupled with the problems of identifying the target purchaser within a company, the difficulties of identifying sales people capable of operating in the integrated market, and the high cost of low-volume systems conspired to truncate the development of this interesting product.

Meanwhile, Northern Telecom (now Nortel) was actively developing similar systems over much the same time scales, though its products were more tightly coupled with its own switch range. Two systems are worthy of historical note. The first was originally called SL1C and later the Packet Transport Equipment; later still it received a number within the Meridian range of products. Similar in many respects to the Mezza system, it differed in the point of connection to the PBX. Northern Telecom's product connected digitally to the switch and required an intelligent signaling link (the command and status link) to the switch's processor. This link was a proprietary item, so the equipment could only work with the Northern Telecom SL1 switches. It is interesting to note that this represents one of the first third-party compeer links.

The terminal was a large, integrated voice and data terminal labeled the M4000. The system was able to offer very similar facilities to Mezza. Every attempt was made to mirror telephone facilities in the computing environment. Coordinated transfer of voice and data and data conferencing were implemented in this environment. Though Northern Telecom invested vast sums in its development, no details are available of commercial application of this product, but field trials and demonstration systems did exist.

The other Northern Telecom product did enter production and the field, but it was not a commercial success. This was the DV1. The DV1 shared technology with its larger brother. The products used the same M4000 terminal and both were UNIX-based. However, the DV1 had lower capacity and was a standalone product. It did not require connection to a PBX; it actually operated as a PBX itself. Also, the level of integration provided by the DV1 was lower, particularly at the application level.

1.4.2.3 Voice and Data PBXs

In the very early 1980s, telephone switch suppliers thought the market was crying out for data-switching capabilities via the PBX. Many offerings were developed, from simple asynchronous circuit-switching options to gateway provision and even embedded packet switching. Generally, customers continued to keep their data and voice networks separate and were often critical of the data capability of the PBX, particularly in dealing with a wide variety of data interfaces. Nevertheless, AT&T, Northern Telecom, and Plessey provided multiplexed interfaces that could be connected to a computer. These allowed the connection of multiple terminals to the computer via the PBX. Signaling information was passed for routing purposes and so forth. The AT&T interface used the term digital multiplexed interface (DMI), the Northern Telecom offering used computer PBX interface (CPI), and Plessey used an ECMA standard for this method of data interfacing. These interfaces are not to be confused with CTI links. They are not functional links; they simply provide the means of connecting one data channel to another data channel.

Meanwhile, other telecommunications suppliers were making claims that their PBXs were not only capable of carrying data, they could also run applications. CXC of Los Angeles, California, was one of these. Formed in 1981, the company had trials of its innovative Rose PBX running in September 1984. The Rose had a distributed architecture that incorporated application processors, although these were not available in the early releases of the switch. CXC designers regarded the supply of applications for these processors as more appropriate to distributors and system integrators than themselves. Although the switch had some successes, the only integrated application that was known to be developed for the application processors dealt with statistics processing for automatic call distribution applications.

Ztel, another United States-based start-up in the PBX business, was located in Boston. Ztel's PNX had a distributed architecture in common with the CXC Rose, though quite different in implementation. This switch also supported embedded application processors, though once again they were not much used before the company's demise in the mid 1980s. The applications that were supplied were communications-specific: an online telephone directory and simple e-mail.

Mitel offered a similar capability by embedding a PC environment in the switch, though the company's attention became more focused on third-party compeer than embedded applications as the decade progressed. It is worthy to note that Mitel is the only survivor of those that offered embedded application processing.

1.4.2.4 Telemarketing Bureaus

In the early 1980s, British Telecom in the United Kingdom began to think about the potential business available from establishing a telemarketing bureau. Interest centered on inbound calls. A number of operators would answer calls, provide information, take orders, and so forth. The service was to be a general one, offering what was in effect an advanced answering service for hire. Operators needed to deal with a number of different campaigns, often for different companies. The scripts that indicated what the operator should say were held on a DEC VAX computer, the telephone calls were dealt with by a Mitel Regent PBX.

The designers were faced with a classic CTI problem. The PBX had information the computer needed. The minimum the computer needed to know in order to provide the correct scripting information for the agent was the last few digits dialed by the incoming caller.

Some of the design team had been to the United States to visit the Delphi Corporation and, in particular, to see the Delta system. The approach used in Delta had much in common with the system British Telecom needed.

The service was up and running by 1983. It was named Telecom Tan and was very successful. The PBX was linked to the computer via a Macintosh computer. The Macintosh did not, of course, run the telemarketing application; it simply provided protocol conversion functions for the VAX computer on which the application actually ran. It also buffered the real-time world of telephony from the VAX.

The PBX had no special interface for this application. The Telecom Tan team used a line display output from the Regent console; this provided the status of each line, and the information from this source was processed by the Macintosh to provide the necessary data to the VAX. This then is an early example of third-party primary CTI.

1.4.2.5 Personal Paging

During 1982 a company called Aircall needed to provide a sophisticated paging system. The system was required to provide personal answering to incoming callers who wished to page someone. Aircall approached a company called *Special Telephone Systems* (STS), which was an early producer of automatic call distributors in the United Kingdom. Aircall's requirements included transfer of the received digits from the switch to the computer and the ability to simultaneously transfer telephone calls and data screens. The former provided the operator with an immediate display of the identity of the paged person and the latter provided the ability to transfer difficult calls to a more experienced operator.

STS was able to implement a suitable interface and, with Aircall's help, gained regulatory approval for connection of the automatic call distribution switch to the network (no mean feat in the early 1980s). Roughly twelve of these systems were sold. Early sales were to the Helsinki telephone company in Finland, closely followed by the Swedish PTT; later systems were delivered to Switzerland and Australia.

This implementation provides an early example of the use of two important generic functions of CTI: call-based data selection and voice and data call association.

1.4.2.6 Convergent Industries

The early 1980s saw the great flirtations and marriages between key industry players in both worlds. The really significant relationships involved the then dominant computer suppliers. Not the least active was IBM. The company's involvement in private switching systems was a European phenomenon. The developments took place at La Gaude in France and were only marketed in Europe, though there were a number of internal IBM installations in the United States. Development was discontinued in 1983 at about the time that IBM completed its search for a suitable telecommunications-based partner.

One company that caught IBM's eye at a very early stage in that search was the fast-growing Canadian company, Mitel. Mitel had begun life in the small-end analog PBX market and had done extremely well. Flushed with this success, the company began designing and heavily promoting an advanced digital switch, the SX2000. It was this switch that caught IBM's attention. IBM moved in force to Mitel's Kanata development facility. Huddled in a hastily constructed portacabin village, the IBM people began to pore over the SX2000 documentation. Presumably, they found something not to their liking because after a while and amid much speculation and rumor, the IBM contingent left Kanata.

A little later, IBM and Rolm announced their partnership. That partnership began as a marriage but ended as a complete takeover of Rolm by IBM. One issue of that marriage was AS/400 Telephony Applications Services, announced in September 1989 [8]. Figure 1.8 presents the general architecture. Based around the Redwood 16–100-line telephone system, this basic CTI implementation offered the following features:

- Computer-assisted dialing—the AS/400 could initiate a call for any associated Rolmphone;
- Collection of call detail records for processing by the AS/400;
- Alarm processing by the AS/400 for the Redwood;
- Voice response via the associated voice unit.

Probe Research called this offering "a rather creative use of existing products." It is in fact another good example of third-party primary CTI. It does allow calls to be placed from the AS/400 terminal. This uses the telephone connection server, as shown in Figure 1.8. The device interfaces to the same line unit as the Rolmphone and therefore behaves in just the same way. Calls made from the AS/400 are placed by the server, then transferred to the associated telephone. The Rolmphone is forced into answer mode by activation of its loud-

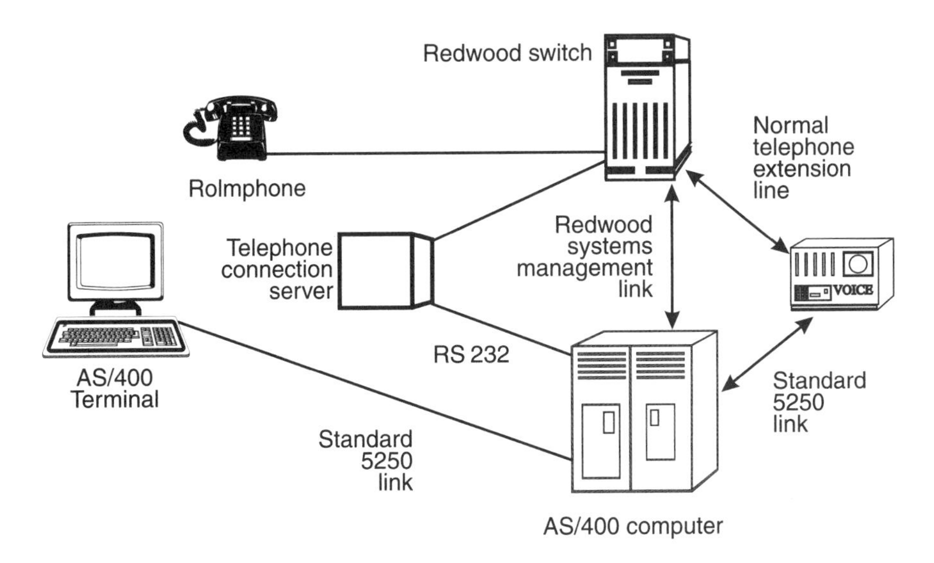

Figure 1.8 AS/400 Redwood telephony application services.

speaker, and it is then up to the user to monitor the call. The AS/400 and server are then free to deal with other calls. The AS/400 does not monitor the progress of the calls.

Another offspring of this relationship was the 8750/9750 PBX range, which was one of the first to offer an IBM-compatible third-party compeer interface. Interestingly, there is a strange reversal of fortunes here. Whereas IBM's own PBX range (the 3750 and others) was never released in the United States, the 8750 was never released in Europe, though some field trials were held in the United Kingdom in the late 1980s.

During most of the decade, the company that came second to IBM in computer market share was DEC, though this position was lost when Boroughs and Sperry joined forces to form Unisys. However, during the foundation decade of CTI, DEC was probably as active as IBM if not more so. Although there were many rumors of closer relationships with various telecommunications companies, DEC played the field. The company worked with AT&T, Northern Telecom, Mitel, and others to develop links that would allow interworking and, at the same time, it actively encouraged the creation of interworking standards.

One of the more historically interesting activities was the joint venture between DEC, Mitel, and British Telecom. This resulted in the demonstration, at Telecom '87 in Geneva, of a system that was code-named Stanza by British Telecom. Figure 1.9 provides a picture of the architecture and components. Any telephone or terminal could be used. The telephone system was a

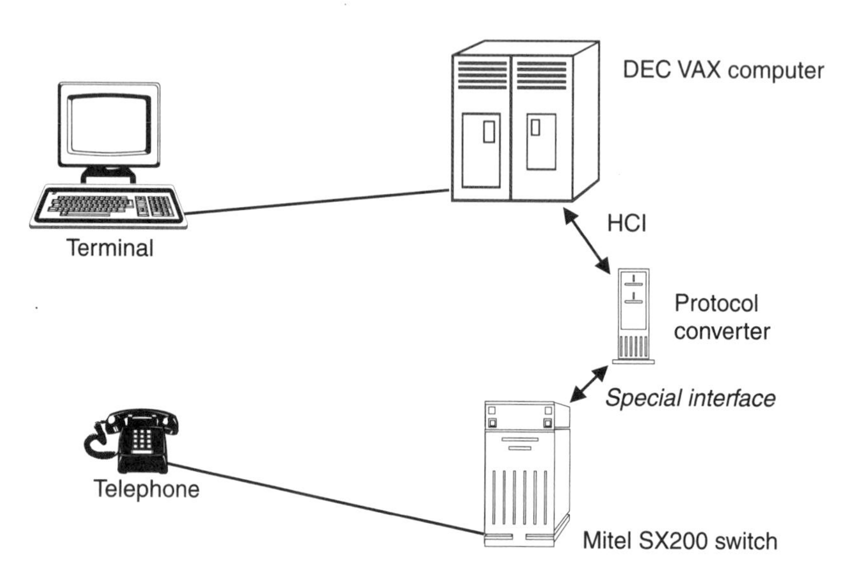

Figure 1.9 The Stanza system.

medium-sized PBX from Mitel marketed by British Telecom under the name Regent. Mitel called this switch the SX200; it was an analog switch and one of the company's most popular in the 1980s. Stanza can be regarded as a more general approach to providing the Telecom Tan solution described earlier; it was expected that production systems would be based on Mitel's digital switch, the SX2000.

As Figure 1.9 shows, the computer and switch were not directly connected; a protocol converter was needed. This device was based on a PC, and its task was to translate the messages flowing to and from the switch into a Mitel standard protocol called the *Host Command Interface* (HCI). The HCI protocol was not implemented on the SX200, hence the need for the protocol converter. The interface from the SX200 was a modified attendant console interface; the modifications were performed within the software of the switch. The HC1 protocol was, however, later implemented directly on Mitel's SX2000 switch, which was regarded as the mainstream switch for CTI applications.

The applications demonstrated at Geneva were advanced and impressed a number of the exhibition attendees. One of the applications was an integrated telemarketing package. It displayed most of the key CTI functions, including the ability to simultaneously transfer telephone calls and data sessions.

Though some field trials were installed in the United Kingdom, Stanza was never formally launched. British Telecom, whose role in the venture was always a little unclear, switched its funding from Stanza to the Mezza system. Mitel continued to develop the SX2000 as its strategic product in this area, and commercial applications were later created that did link this PBX and DEC computers.

There were many other companies actively engaged in the CTI area in the 1980s. Wang, for example, produced a number of products to allow its minicomputers to interwork with the telephone. Wang was also financially linked to Intecom, the PBX supplier. Olivetti, the Italian computer supplier, forged strong links with the part of AT&T that is now called Lucent, and AT&T formed an industry-based forum to specify and implement computer/switch interworking protocols. And so the list goes on. Unfortunately for the companies and the people involved, the list of successes in the foundation decade is very short.

1.5 The Delivery Decade

Although there were a number of examples of integration in the 1980s, the idea did not reach the ignition point on a wider scale. This came about in the 1990s—the decade of delivery.

In the early part of this decade, a number of events conspired to popularize and promulgate CTI. Not the least of these was the adoption of the term *CTI.* In the 1980s, CTI was the thing with no name. Many people used *computer-integrated telephony* (CIT) as a generic term, but this was DEC's name for its own particular approach to integration. In Europe, the term *computer-supported telephony* (CST) flourished for some time. Fortunately, this ridiculous state of affairs ended with the gradual adoption of the term CTI all over the world. At first, the 'T' stood for telephone; later telephony became the common expansion.

Who invented the term CTI? ACTAS (see later) was the first industry group to use it in the United States, and ACTIUS certainly popularized it in Europe. Some say that Jim Burton, a United States-based consultant and commentator on all things CTI-related, first coined the term. He certainly believes so. One thing is for certain; CTI now covers a lot more territory than it did when it was first conceived! This is why I often refer to it today as complete and total integration.

In 1991, the standards group ECMA published the first real standard for a CTI link protocol. Some people say that this was too little and too late for CTI standardization, but it helped. It was followed by yet another standard for the same thing by ANSI—on the basis that two standards are better than one!

Early in the 1990s, the software world had changed so much that people were openly saying, "Microsoft owns the desktop and Novell the network." True or false, the dominance of these two companies in the world of client server computing was so great that any moves by them into the CTI world were momentous.

In 1992, Novell announced its move. At first a little vague, the initiative was quickly given form as NetWare Telephony Services and its application programming interface TSAPI—more on this later. Novell worked with AT&T on this initiative, and the companies' use of the ECMA standard as a basis for the API helped to give that protocol the credibility that it needed. Hot on the heels of Novell came Intel and Microsoft with their joint announcement of the Windows Telephony and TAPI also in 1992.

The effect of these announcements, and the subsequent rollout, has been threefold. First, CTI has been popularized and accepted in many areas of the market that were not previously exposed to it. Second, expectations, and to a certain extent, realization of CTI cost levels have fallen. Third, many software companies have formed simply to produce CTI applications.

Although the activities of Novell and Microsoft have seized the limelight, traditional suppliers, such as IBM, have been busily building an extensive base of installations. The number of system integrators has increased, and the range of packaged CTI software solutions for the call center and the desktop has expanded significantly. Meanwhile Sun, the inventor of Java, has joined the CTI scene with the 1996 announcement of JavaTel and its own application programming interface JTAPI.

During the 1980s, voice processing systems began to migrate toward PC platforms. In the 1990s, that trend rapidly became the norm. There were many repercussions. First, the actual cost of systems did fall, but this merely continued an ongoing trend. More importantly, the trend has introduced "Leggoland" to voice processing. Voice mail, voice response, and audiotex systems can be rapidly developed by almost anyone. These systems are now mostly created from off-the-shelf hardware and software components. Each succeeding generation requires less added value than the last to produce systems that can be delivered to the end user. Board densities have increased rapidly so that really large systems can be created in a single PC chassis—and the chassis can be linked together to create yet larger systems. Meanwhile, the switching of calls between boards within the PC has become an accepted function—leading to merged CTI systems at the server level.

Application creation has moved away from the code level and onto graphical generators that simply require the developer to pluck system functions from a palette and join them together—at least according to the promotional material.

All of this gave birth to a new industry that has grown very quickly indeed. That industry needed a name, as it also wanted to be perceived as part of the CTI supply industry. Hence, the term *computer telephony* (CT) emerged. That this is confusing is beyond doubt—so is the fact that the term has taken root, fertilized and watered by a very large exposition and accompanying magazine in the United States. Often, CTI and CT are used interchangeably, although some analysts do attempt to define them. It is in the nature of a fast-growing industry that terminology keeps pace with the changes; the dust will settle sometime. Key players in the creation of the PC telephony industry are the PC card suppliers, and the market leader throughout the growth of this industry has been Dialogic. This company has been innovative not only in the design and manufacture of PC voice cards but also in the promulgation of methods of connecting those cards together and in the creation of the software environment needed to control them. The CT industry is certainly not limiting itself to voice processing. It is active in integration and in extending the Leggo approach to other platforms and media—particularly fax.

If the creation of CTI systems on the PC platform has been a great mover and shaker in the 1990s, it has not been alone. Enter the Internet and everything begins to turn on its head. "What is the relationship between CTI and the Internet?" someone asked me some years ago. "Not a lot," I said hesitantly. "The Internet is an information network, not a computer." Then someone said that the network is the computer, and that got me thinking. I took each of the generic functions of CTI and reviewed them with the Internet and Web in mind. Some of it made a lot of sense. It is possible to position the Internet in so many different roles. One of those is as an alternative to a voice processing system for automation, and just as you want to transfer from the machine to a person in the one, so you may want to in the other—hence the idea of Web access to the call center, something that has spread like wildfire since the first announcement by Rockwell at the end of 1995.

Of course, there are many other ways in which the Internet and CTI do relate, and this book will describe some of the most relevant later. However, there have been other developments on the Internet that have had a significant influence on CTI—not the least is Internet telephony. Considered by some as the incomprehensible chasing the incorrigible, it certainly excited some response from the established telephone companies. Coupled with a tendency to bundle in collaboration software tools, this development causes some head scratching in the CTI world. Thoughts turn toward the network computer and the incorporation of Internet telephony into it. Is this third-party or first-party CTI? Those definitions will need to be jettisoned if the NC does take hold, but that is not history—it is future gazing.

References

[1] Treffert, Donald A., *Extraordinary People: An Explanation of the Savant Syndrome,* Bantam Press, 1989.

[2] Levick, Lois B., "PBX/Host Interfaces: What's Real, What's Next?" *Probe Research Conference Digest,* 1990.

[3] Walters, Rob, *Computer-Mediated Communications: Multimedia Applications,* Norwood, MA: Artech House, 1995.

[4] Swade, Doron, "Charles Babbage and the First Computer," *IEE Review,* June 1991.

[5] *IBM 2750 Voice and Data Switching System,* IBM marketing brochure, 1970.

[6] Shoobridge, E. H., "System 2150 Add-On Facilities for Private Telephone Exchanges," *Systems Technology,* No. 22,1975.

[7] Walters, Robert E., *Voice Information Systems,* NCC Blackwell, 1991.

[8] Gladstone, Steven J., *PBX/Host Interfaces: The Products, Applications, Markets, Strategies,* Probe Research, Inc., 1989–1993.

2

The Computer and Telephony Environment

The computer and telephony environments are different, though they have a lot in common. In this chapter the emphasis is on the differences. The material presented here is introductory; readers already familiar with the two worlds may wish to skip directly to the technology of CTI, which is the subject of Chapter 3. However, the remainder of this book assumes that the background and terminology introduced here are known, so you are encouraged to read on. Readers with some grounding in computers or telecommunications might find that a few misconceptions will be exposed as the two worlds are examined.

The bias of this chapter is towards the telephone world. There are many reasons for this, not the least of which is the belief that it is more difficult for the computer expert to comprehend telephony than it is for the telephony expert to understand the computer world. Computing is a general topic; using computers requires at least a superficial understanding of their operation. *Using* the telephone is more widespread—however, it requires no understanding of the operation of the device or the network that backs it up. Interestingly, this too is changing. The introduction of the browser has created an interface than can be used with no training. Also, for basic information access, the browser requires no knowledge of the computer or the network to which it attaches. Is this more evidence of convergence?

With that introduction, it is worthwhile to consider the commonality of the two worlds before taking a detailed look at each of them.

2.1 It's All Bits Anyway

"Of course, it used to be tough in the telephone world. Everything you did interfered with the blessed calls. I even heard of some old lady that had trouble with ants!" said Brian to his ardent apprentice Walter.

"Ants? What, in her telephone?"

"No, no. In that little plastic box they used to screw to the wall. The one that the telephone cord went to before they had these plugs and sockets."

"So what did the ants do? Eat the cable or something?"

"Oh no. They didn't do any damage like that. It was more a case of them being damaged. They got killed, you see. When the telephone rang. That's a high voltage that they use for ringing the phone, you know. So it killed them."

"And how did this affect the old lady? Was she upset because the ants were killed?"

"Well, actually, she didn't know about the ants. It was the noisy telephone line that troubled her."

"Could she hear the ants dying then?"

"Not exactly, but all these bodies sort of bridged the line. And, of course, live ants kept coming to the box to carry the dead ones away."

"So what happened in the end?"

"Simple. As soon as the engineer found the problem, he vacuumed out the connection box and sealed it up with goo. No more trouble. 'Course it wouldn't happen today."

"Why's that? There are still ants about."

"Ah yes. But the telephone system's gone digital. It's only bits now. Just like a computer. Bits aren't affected by ants, you see."

2.2 The Telephony Environment

Is Brian correct? Is it all bits now? In examining the telephony environment, that question will be answered. After all, it is a question that is rather pertinent to the integration of telephones and computers. Before entering that discussion, however, this chapter is going to introduce some of the basics of telephony. In doing so it will cover a number of other fundamental questions:

- What does the telephone actually do?
- What are the measures used in telephony and what do they mean?
- Why are the calls switched?
- Why is traffic planning and performance so important?

- How does the computing world see the telephony world?
- What are the major trends in telephony?

Telecommunication is a large topic. What follows is but a toe in the water of an almost limitless ocean of knowledge. However, as in most topics, it doesn't take long to become an expert—relative, that is, to those who know nothing! The next stage is to become sufficiently expert to recognize how little you know about this vast topic. To reach this stage you will need other books, and there are many excellent texts around. Here, the intention is simply to provide sufficient background information for those who are relatively new to the telephone world and sufficient knowledge to safely address the problems, opportunities, and practicalities of computer telephony integration.

2.2.1 Telephony Basics

Telephony basics introduce the telephone itself, the way in which telephone networks carry calls, and the way that those telephone calls get to the right person. This includes the three physical elements of telephony: terminals, transmission channels, and switches.

2.2.1.1 The Telephone

Everyone knows the telephone, so what is there to say? Familiarity breeds contempt, and who can fail to feel some contempt for what is undoubtedly the most prevalent electronic device in the modern world. Surely there is nothing very interesting going on inside it. Or is there? Just think of one simple thing. A microphone has two wires coming out of it; a loudspeaker or earphone also has two. A telephone has both a microphone and an earphone, yet only two wires, not four, connect it to the telephone exchange. How do those four wires get changed into two, and how can signals flow in both directions at the same time on the two-wire telephone line yet get delivered to the right device, microphone, or earphone at the other end? Such are the basics of the telephone. This question and a few others are answered in the following introduction to the analog telephone. After covering the basics, this section examines the feature phone and the variants of digital phones that are becoming available.

The Analog Telephone

Everyone is well aware of the general form of the common analog telephone, splendidly varied though it may be. But what of its basic functions? A telephone performs the following tasks.

- Alerts the exchange that a new call is being made;
- Converts keyed or dialed numbers into a suitable form for transmission to the exchange;
- Converts sound pressure at the microphone into an analogous electrical signal;
- Converts incoming electrical signals to sound pressure at the earphone;
- Allows the transmit and receive signals to use the same path to and from the telephone exchange;
- Converts the incoming ringing signal to a suitable sound;
- Detects that the user has answered and informs the exchange;
- Alerts the exchange, should the user wish to signal something to it during a call;
- Indicates that the user has cleared.

All of these tasks are performed under extremely variable conditions. The telephone is connected to the exchange by the local loop. It might be right next to the exchange or it might be a long way from it. The simple analog telephone gets the power it needs to operate from the telephone exchange. In electrical terms, the exchange supplies the telephone from a large battery. Usually the battery is rated at 50 volts. Though a telephone that is very near to the exchange gets most of this voltage, a distant telephone will see very much less—because the line connecting it has resistance. In fact, the telephone is expected to work satisfactorily with as little as 5 volts available to it! Compare this to a conventional logic circuit within a computer (or a PBX). Mostly these circuits run off 5 volts and the supply must be within 5% of this voltage!

Not only are conditions variable, they are also arduous. Ringing voltage that is used to indicate an incoming call is nominally transmitted at 75 volts; it can peak at over 100 volts. Don't try connecting that to an RS-232 computer interface! Dial pulses are generated by shorting out the line. This can generate pulses that swing between 0 and 50 volts—ten times a second. Within all this, the signal of real interest, that which carries the telephone conversation, is minuscule by comparison—just fractions of a volt. And, after traversing the telephone network, it may be reduced to one a thousandth of its original amplitude. On the other hand, logic signals within a computer are generally represented by just two amplitudes (0 volts or 5 volts).

Of course, early telephones were designed some time before the task was understood to be impossible! Unaware of the insurmountable difficulties, the early engineers just did it. Then they fiddled around with the design until it worked. However, the serious point is that the humble telephone is not so humble after all. It does a difficult job in very difficult circumstances. How it does that job might be clarified by an examination of Table 2.1.

Functions 1 and 2 and 7, 8, and 9 result in signals that are sent to the telephone exchange. Function 6 results from a signal coming from the exchange. Together these make up the basic signaling system between the telephone and the exchange. Of course, the whole point of making a telephone call is to converse, and this is represented by functions 3, 4, and 5. However, these functions do not just contain conversation. Because the telephone is controlled by a person who can learn to understand the meaning of different sounds, call progress indications can be sent in the form of tones. So the basic signaling system is enhanced by the provision of dial tone, ring tone, busy tone, and so on.

It is the signaling system that provides control over switches and it is therefore the signaling system that is the key to understanding the role of a computer in a CTI application. In first-party CTI, the computer is directly concerned with telephone-based signaling. In third-party CTI, the relationship is not so direct, but it is still there. On the whole, the computer is doing much the same thing to the switch that a telephone does, and on the whole, it receives much the same information, though in a very different form.

Table 2.1
Telephone Functions and Their Implementation

Function	Mechanism	Line Signal
1. New call	Lift handset	Loop line
2. Send digit	Key or dial	Tone or pulse
3. Speak	Sound transducer (microphone)	Electrical analog of speech
4. Listen	Electrical transducer (earphone)	Electrical analog of speech and call progress tones
5. Simultaneous speak and listen	Two-to-four wire converter	Electrical analog both way
6. Incoming call indication	Bell, tone, lamp	Ringing voltage
7. Answer	Lift handset	Loop line
8. Recall	Press button	Short break or connection of the line to earth
9. Clear	Replace handset	Open line

One of the interesting things about the analog telephone—so called because it produces electrical analogs of airborne speech—is the degree to which speech and signaling are intermixed. Remember that the telephone exchange is connected to the phone through the two-wire line. It has to sort out the signaling from the speech and it has to do that reliably and cheaply. It is helped by the fact that it does know something about the sequence of events involved in setting up a call (i.e., it has a call model built into it in some way). It is also helped by the fact that many of the signals received from the telephone are not at all speechlike. Speech varies continuously and rapidly. The loop condition that indicates a new call will persist for the whole duration of the call and must persist for tenths of seconds before it is recognized as genuine. Dial pulses vary at the rate of 10 pulses per second compared to speech signals that vary at 300 cycles per second at the minimum. And of course the dial pulses are several orders of magnitude larger than the speech signals anyway. The recall, or flash signal is sent in the speech phase of the call. However, whether it breaks the line for a short period or connects a line to the Earth to attract the telephone exchange's attention (both schemes are used), the signal is clearly distinct from speech.

Tone signals are quite different from the other signals mentioned so far. They are chosen to be speechlike so that they can traverse the telephone network precisely as speech does. All the other signals mentioned above terminate at the first exchange (the exchange that they are directly connected to). If the call is routed to another exchange, the two exchanges will signal between themselves using some signaling system that is probably quite different from that described for telephone-to-exchange signaling. Call progress tones may be transmitted from any exchange involved in the call. Ring tone, for example, will normally be sent by the exchange on which the call terminates. The nature of call progress tones is such that a human listener can clearly distinguish between them and speech—even if the listener does not always know what the tone means!

Tone signals that carry routing digits—those from the keypad, for example—are in a different category. Here, the listener is a machine of some sort, not a person. There is a possibility that the machine might mistake a segment of speech for a digit or, at the other extreme, ignore a genuine tone digit because it seemed like speech. Two factors make these possibilities unlikely. The first is that the machine (the tone digit receiver) is not usually listening during the speech phase of the call. That is why we need a recall signal to bring it into the call if a supplementary service is required. The second reason is that the tones are chosen such that, though they are speechlike, they are unlikely to be produced by a human voice.

We speak by passing air over our vocal cords to produce a frequency-rich sound source. We vary the frequency and loudness of this source with our articulatory organs: our mouth and nose. We do not normally produce pure tones, that is, single-frequency notes, and even if we do, by whistling, we can only produce one note at a time. The frequency of a note is a direct measure of how fast the sound is changing. It is measured in hertz (Hz); 1 Hz being one cycle of change per second. The low hum that you hear from some public announcement systems is caused by leakage from the electricity supply and has a frequency of 50–100 Hz. The high-pitched tone that you hear when a fax machine answers your telephone call has a frequency of over 1,000 Hz. Adequate speech quality across the telephone network is achievable with frequencies constrained to lie between 300–3,400 Hz. This is the normal telephone channel bandwidth and the tones used for transmitting routing information are given frequencies right in the middle of this band. To prevent confusion with speech, two tones are allocated to each digit and sent at the same time. This description explains the name given to this type of signaling: *dual-tone multifrequency* (DTMF). The frequencies used for DTMF signaling are divided into two groups: one low, the other high. One tone is selected from each group of frequencies to represent a single digit. The digit 0 is therefore transmitted by sending the tone frequencies 941 and 1,336 Hz simultaneously, as shown in Table 2.2, there are 16 possible combinations, 12 of which are normally used. The ABCD digits are not often used in telephony and not often found as keys on telephones, but they are becoming of value in some voice applications.

Tone signaling is almost ten times faster than pulse signaling and has the advantage that it can travel satisfactorily throughout the telephone network. For this reason it finds many applications beyond simple call set-up. One of the applications for tone signaling is to control special telephonic features. The additional asterisk (*) and pound (#) keys are useful here because they can be used

Table 2.2
Tone Signaling Frequencies

	High Group of Frequencies			
Low Group of Frequencies	**1209 Hz**	**1336 Hz**	**1477 Hz**	**1633 Hz**
697 Hz	1	2	3	A
770 Hz	4	5	6	B
853 Hz	7	8	9	C
941 Hz	*	0	#	D

to discriminate between simple routing digit strings and, for example, a request to divert calls to another number. They are also used to segment a digit string into feature codes and telephone numbers.

If a user needs to activate a feature during the speech phase, the recall key is used to request a dial tone and thus bring the switch controller back into the call. Subsequently, for simple transfers, pulse signaling can be used. For anything more sophisticated, tone signaling will be used to specify the required service—setting up a conference call, for example.

This introduction has hardly begun to cover the complex specification of the humble telephone. Requirements vary considerably from country to country and usually specify such characteristics as:

- Send and receive loudness ratings;
- Side-tone levels;
- Terminating impedance;
- Noise;
- Stability;
- Distortion.

These are described in Table 2.4. In addition to all of this, there are often stringent safety specifications to meet when connecting equipment to the telephone network.

One last point before leaving the simple analog telephone. In the introduction to this section the intriguing problem of four- to two-wire conversion was raised. How can one pair of wires carry signals in both directions? Surely the signals get so mixed up they cannot be separated again. Yet the signal received at the telephone must be directed to the earpiece, and the signal produced by the microphone must be sent to line—not to the earpiece. Meanwhile, both devices are connected to the line. In fact, there are many ways of achieving this; the main problem is to direct the energy from the microphone to the line and not to the earpiece. It doesn't much matter that some received energy ends up at the microphone rather than the earpiece. The trick is to provide some sort of balance circuit so that any transmitted energy reflected back from the line is canceled out before it reaches the earpiece. This requires that the balance circuit provide perfect line matching—which is almost impossible. The mismatch allows some of what is spoken into the microphone to appear at the earpiece—this is called side-tone. Too little side-tone and the circuit seems dead, too much and conversation becomes very difficult—because the caller with high side-tone speaks more and more quietly.

Two- to four-wire converters appear elsewhere in the telephone network. They give rise to echo, which becomes bothersome where delay is high.

The Feature Phone

It is believed that the IBM 2750 was the first PBX to offer advanced telephony features. Since the time of its design in the 1960s, more and more switches began to sport more and more features. The length of a PBX supplier's feature list became an all-important competitive measure in the 1980s—at least in the eyes of the suppliers. A count of the features listed in the 1987 product specification for a well-known PBX yielded a total of 168 features! Admittedly, not all of these are offered through the telephone; a number of them are attendant (operator), system, or data features. However, the number is still high.

Not all the features are really useful. I clearly remember being asked by my own marketing department, "Why don't we have this desk-to-desk dialing feature on our PBX, just as this competitor switch has?" I had to carefully explain that all PBXs have desk-to-desk dialing. That, if nothing else, is what a PBX does, yet there it was, listed as a feature in the sales brochure!

Even where features are useful, they are often not used. Why is this? Take a particular example feature: call parking. If a call comes in and you answer it and then find that the information you require is in another room, why not park the call? You can then walk to the other room, find the relevant information, and pick up the parked call from the telephone in that room.

To do this, all you have to remember is the code to park the call (Recall, *7, for example) and the code to retrieve the call from the other phone (*87* plus the number of your original phone, for example). The problem is, most people cannot or will not remember feature codes, and you need to know the code at the moment you need to park the call. So, why not supply little cards listing all the feature codes? The problem here is that people lose and mislay them. In any event, they are never available at the time you need them. Added to this, people do not want to look up feature codes when they have someone on the line. Their minds are on the call and the task it brings.

Feature phones are designed to overcome these problems and, in some cases, to provide features that are not accessible from the basic analog phone. Figure 2.1 shows three types of feature phones. The most basic is the single-line standard device. This can be connected to any standard line offering tone signaling. To the switch it seems identical to a standard analog phone. This feature phone simply stores the feature codes under individual buttons. So, continuing the example, if you were to press the Park Call button, this feature phone would generate the Recall, *7 sequence. The single-line standard feature phone can therefore do no more than an ordinary analog telephone. It is simply easier to use because of the specialized buttons and the visual display device (screen,

LCD display, or whatever), which is sometimes provided to let you know what you have just done and suggest the next step.

The single-line feature phone shown in the middle of Figure 2.1 is rather different. It has a separate means of signaling to and from the switch. This immediately adds to the capability of the phone, since messages can be sent and received while a telephone call is in progress. This separate signaling capability can be provided in various ways. Perhaps the simplest is to use an extra pair of wires. Alternatively, the signaling can be multiplexed onto the speech pair such that the two do not interfere. Some suppliers send the signals at a frequency that is much higher than speech and use a filter to separate the two. The important point to note about these implementations, particularly from a first-party CTI viewpoint, is that they are proprietary.

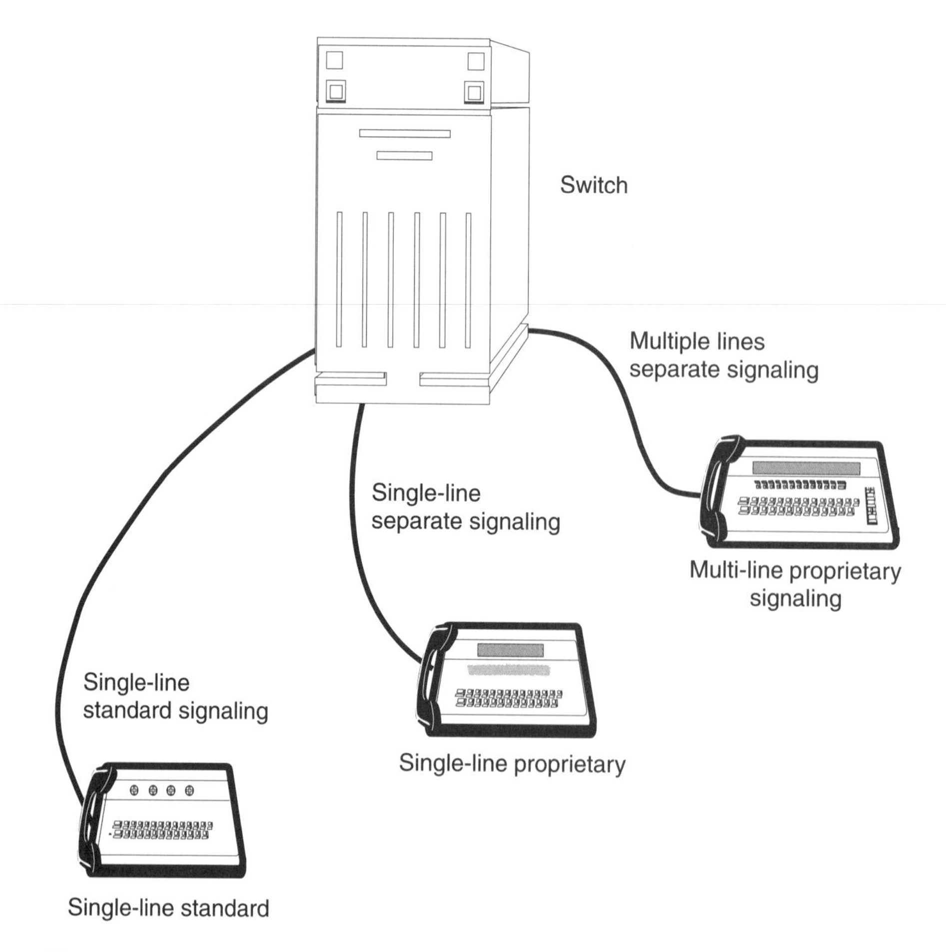

Figure 2.1 Feature phone types.

Feature phone signaling systems are not only proprietary, they are often closed. The switch supplier has little to gain by making the specification of its signaling system public. On the other hand, it has plenty to lose. Opening the feature phone interface allows other suppliers to design, manufacture, and sell feature phones into the supplier's installed base of switches. And there has been a good deal of money to be made in selling into this base. This situation is changing as open interfaces are introduced.

Figure 2.1 shows a proprietary feature phone with a screen. Most implementations have some form of display, which is an advance over the use of lamps. Because these feature phones have a more sophisticated signaling system, they do know more about the progress of calls. The status of a called party can be displayed. This can simply act as a visual alternative to progress tones—but it does go further. It is possible to indicate that a call has been diverted to another number and why. It is also possible to indicate that an incoming call has been diverted to this feature phone and where it was originally destined. And there are many other useful items of feedback that can be given at the screen—items that may also be of value in a CTI environment.

In general, the function of all the keys of the standard feature phone are prescribed. The sequence of signals produced from them might vary from switch to switch, but the function is preset. This is often not the case for the proprietary feature phone. Some keys may be programmed by the user to do specific things; a very common use for these keys is speed dialing. Regularly used telephone numbers are stored beneath speed keys and labeled accordingly. Other keys may be "soft"—their function varies according to what state the phone is in. The actual function of these soft keys at any time will be indicated by the display.

The third category of feature phone shown in Figure 2.1 often shares all the capabilities described for the proprietary feature phone but also allows access to multiple lines. There are two types of multiline phones, though the distinction becomes more and more blurred as the capability of switches broadens. Key systems are usually based on a multiline feature phone from which all trunk lines are available to all users. Key systems will be covered later. PBX-based feature phones often have a multiline capability, but the other lines are usually other extensions, not trunks, and are quite limited in number. Perhaps the most common number is two, to cover the arrangement often referred to as manager/secretary working. Here each feature phone appears to have two lines, the manager's extension and the secretary's extension. Normally, all calls are answered at the secretary's phone, but the manager can pick up the call by a single keystroke, if required. A manager or secretary can use either line for outgoing calls. One secretary might support two managers, in which case the secretary's phone would have three line appearances and each of the manager's, two.

Many other arrangements are possible to suit the requirements of particular work groups.

Of course, in most cases, the lines do not physically appear at the multiline feature phone (although there are some very small key systems that do work in this way). Generally, the switch connects the lines through, as necessary, and the signaling system allows their status to be displayed on the various phones and also allows for the connection to a line from the phone to be requested.

All the feature phones, indeed all the phones discussed so far have been analog, but it is also common for digital telephones to be used, especially in office environments. The next section examines this trend.

Digital Telephony

The first question is why? Why digital? The second is how? And the third is when? The main reason for going digital is improved voice quality—a secondary, and sometimes misleading reason is that of alignment with the world of data transmission by the telephone network. How digitizing speech improves the quality of voice connections will now be explained—by Brian.

"Well all you do, you see, is you chop what's being said into the telephone into little bits. That's not the same as the computer sort of bits, of course—they're even smaller. Let's call it chunks. You chop what's said into chunks, then you turn each chunk into a number, Then you turn each number into bits and send it down the wire. When it gets to the other end, you do the opposite. You change the bits into numbers and the numbers into chunks. Then you sort of stick the chunks together and squirt it down the telephone earpiece. Perfect. Just like a CD. No noise, nothing."

"And it's all on account of the bits, you see. They're nearly indestructible! You can send a bit across a piece of wet string and it would be all right. At the other end, you would just look at that bit and say, 'Well it's very small—but it's still a bit.' Then you could put it back to its normal size and send it on its way. Perfect—no damage done."

Brian's belief in the bit is a little strong, almost religious. But his description of analog-to-digital conversion is not far off the mark. Analog and digital forms of speech are similar to the analog or digital time displays on watches. Speech is a continuous thing. Though there are pauses between words, the signal that is produced by an analog telephone varies continually. The amplitude of the signal can be at any level between the quietest and the loudest. This is quite the opposite of the digital speech signal; here there are only a certain number of sound levels that can be carried—commonly 256.

Similarly for the watch. The hands of an analog watch sweep continually around the face, whereas the numbers of a digital watch jump at regular intervals, usually in quanta of one second. When analog speech is converted into digital form, each sample is represented by the nearest allowable level. This is called quantization. It is this approximation that distorts the signal a little—not the "chopping up into chunks," normally called sampling. If the analog speech is sampled fast enough, no distortion is introduced by sampling.

Telephone speech is normally sampled 8,000 times per second. This is more than double the highest frequency telephone speech, which is usually constrained to 3,400 Hz, and this sampling rate is sufficient to ensure perfect reproduction. Each sample is then converted to a number and encoded as an 8-bit binary number for transmission. Thus, the bit rate at which the digitally encoded signals are sent is 8,000 × 8 = 64,000 bits per second (64 Kbps).

An 8-bit number can only represent the decimal numbers 0–255, that is, 256 different levels. So, if the maximum voltage that could be converted was 2.56 volts and the step between each level is equal, the step size is 10 millivolts. The accuracy of each encoded sample is therefore ±5 millivolts. Figure 2.2 is a pictorial representation of the process. First the speech signal is sampled, the actual value is then seen to lie between two quantization steps. In this case, the

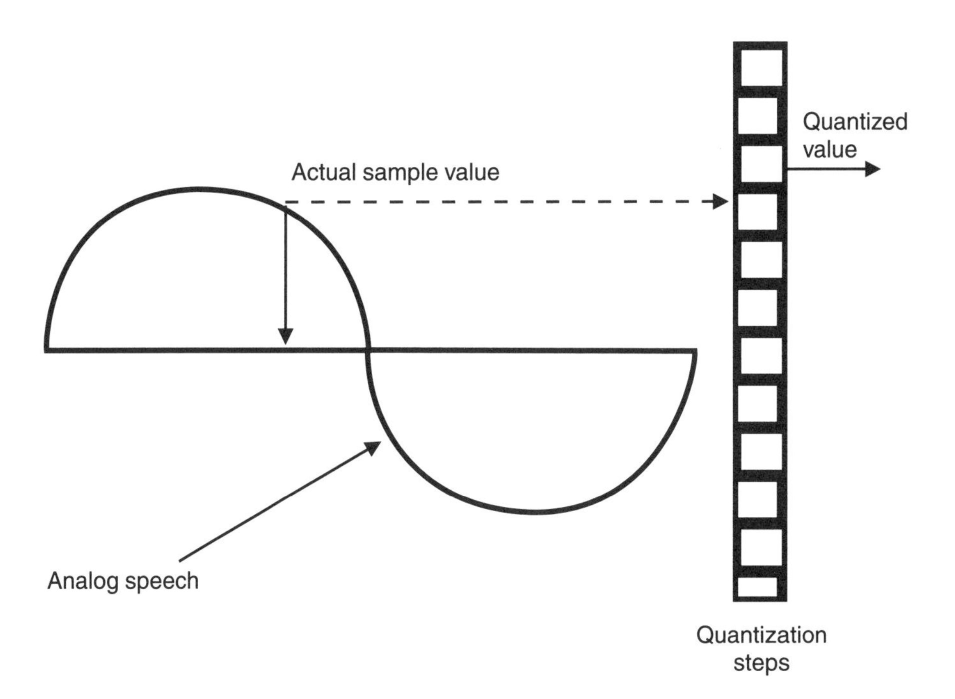

Figure 2.2 Sampling and quantization error.

quantized value is chosen as the next level above the actual value. The difference between the two represents the error. This approximation appears to the listener as noise. It is called quantization distortion.

So, Brian is not entirely correct in his claim that the quality is perfect. The process of converting analog speech to digital does introduce some distortion. Looking at Figure 2.2, it can be seen that the error is likely to affect quiet speakers far more than loud ones. As the swing in speech volume gets smaller, the quantization gets more coarse in relation to that swing. Fortunately, *pulse-code modulation* (PCM), the process usually employed for this conversion, incorporates a technique that counteracts this problem. The technique is called *companding,* a term produced by combining the words compression and expansion. In PCM, compression is produced by making the quantization steps unequal—by making them small for the quiet sounds and big for the loud. In this way the quantization distortion is spread equally between the loud and quiet speakers. There are two different companding methods used in the world. A-law is used in Europe, and μ-law is used in the United States and elsewhere.

As described above, PCM converts analog signals to digital by sampling the speech 8,000 times in each second and producing an 8-bit number from each sample. Hence the bit rate of the encoded speech is 64 Kbps. The bits are binary digits with a value of either 1 or 0. They are fairly immune to damage on the line. Provided the receiver can detect that a 1 or 0 was sent, the speech signal can be reproduced. There will be no further distortion beyond that introduced in quantizing the signal. Quantization distortion itself is almost undetectable to listeners, unless a number of analog-to-digital conversions are encountered on a call. The International Telecommunications Union (ITU) recommends a maximum of 14 conversions for an international call. Note that this figure refers to the standard method of conversion—PCM. It is possible to compress the speech signal into a digital channel with a capacity of less than 64 Kbps, but only by reducing speech quality.

For digital telephones, analog to digital conversion is incorporated into the telephone itself. The telephone transmits and receives digital speech from and to the telephone exchange over a digital link. The need to share the line between transmit and receive signals still exists—just as it does for the analog telephone. There is also a requirement for a signaling capability in each direction. The former is provided by time dividing the link between the send and receive signals or by canceling the echo of the transmitted signal. The latter is provided by a separate signaling channel that is itself digital.

The standard arrangement is for each telephone to have two 64-Kbps speech channels and a bidirectional, 16-Kbps common signaling channel. Most implementations conform to this arrangement; where they differ is in the signaling protocol. Most suppliers have their own set of messages, so a digital phone from one switch may not work on another—in precisely the same manner that proprietary analog phones are incompatible. However, it is now usual for switches to provide a standard digital interface as an option.

Because digital telephones have direct digital connections to the telephone exchange, data can be carried over the link, either via one of the 64-Kbps channels directly or via packets in the 16-Kbps signaling channel. Thus, a number of digital telephones offer data capabilities together with the enhanced signaling and improved transmission.

Digital phones range in capability from a direct replacement for the plain old telephone to an advanced multiline feature phone with one or more data channels. What they all have in common is at least one 64-Kbps digital channel that provides a digital path for the speech, and a separate signaling channel. It is this separate signaling channel that is of greatest importance in a CTI context, especially first-party enhanced CTI.

2.2.1.2 Transmission

Whereas the task of the telephone is to convert voice into transmittable signals, the task of the telephone network as a whole is to ensure that those signals arrive at the destination reasonably undamaged and at minimum transport cost. In telephony this is usually called transmission.

Transmission is a wide and potentially complex topic. It embraces detailed mathematical analyses of the behavior of electrical and light signals in different media, through to the subjective effect of damaging those signals in some way. The intention here is to introduce the topic at a very high level, reaching down just far enough to produce some familiarity with the terms and measures used.

Subjective Assessment

One thing that should be realized from the outset is that transmission quality is not simple to measure. Transmission is not simply good or bad. An engineer can measure the characteristics of a circuit quite precisely, but it is an entirely different matter to relate those measurements to the quality perceived by a human being. Though there are many different things the engineer can measure, the only thing humans are able to provide is an opinion about the quality of the circuit, and that opinion will, in some way, combine the effect of all the things the engineer can measure.

That phrase "in some way" is important. Subjective opinion is affected by any number of things—some of them having nothing to do with the circuit itself. Take, for example, speakers and listeners. Some speakers are very quiet;

some listeners are hard of hearing. If you happen to select a pair from this group, their opinion of all circuits might be poor. To perform subjective experiments correctly, you must select subjects randomly—so such a combination may occur, just as it may in the real world.

There are many ways of quantifying and measuring subjective opinion. For measuring the quality of a conversation, subjects are usually installed in two acoustically isolated chambers and given a task to perform. The task is chosen such that it stimulates conversation and is diversionary; the subjects should not be particularly aware that they are being tested. An example of a suitable task that has been used extensively requires each subject to examine an identical set of picture postcards; these might contain pictures from an art gallery, for example [1]. The subjects arrange the cards in order of personal preference, then, over the telephone, they engage in a discussion that leads to a compromise preference list. At the end of this task, each subject expresses his or her opinion of the overall quality of transmission. They do this by selecting the most suitable description from an opinion scale. Two examples of opinion scales are given in Table 2.3 [2]. The first is most relevant to the conversational test described. The second is for listening-only tests and gives a more complete definition of each level.

By carrying out tests with a selection of subject pairs and constructing the overall experiment to remove unwanted dependencies, a mean opinion score for a circuit can be ascertained. The effort required to perform reliable subjective testing is not to be underestimated. Even relatively simple experiments will require a significant number of subjects and tests to achieve an acceptable level of confidence in the result.

Subjective experimentation uses statistics to achieve results and so does the overall specification of transmission performance. It is difficult to design a telephone network economically if all calls must provide the same level of transmis-

Table 2.3
Subjective Opinion Scales

Quality Scale	Score	Listening Effort Scale
Excellent	5	Complete relaxation possible, no effort required
Good	4	Attention necessary, no appreciable effort required
Fair	3	Moderate effort required
Poor	2	Considerable effort required
Bad	1	No meaning understood with any feasible effort

sion quality—unless that level is very low. Transmission quality is therefore specified statistically. An example target might be that 95% of calls will be good or better, that is, 95% of calls will have a mean opinion score of four or greater.

Objective Measurement

The reduction in energy that occurs as speech travels across a telephone network is called *loss.* Loss is normally measured using a tone with a single frequency. Loss is a simple thing to measure. It is usually expressed as the ratio of output power to input power. In telephony, this ratio is usually converted to a logarithm and expressed in decibels (dBs). Unfortunately, speech is not a single frequency; it is made up of lots of frequencies spread across the speech band. Speech is therefore sensitive to the variation of loss across the frequency band; what is more, our ears are more sensitive to some frequencies than others. Thus, two paths that have the same single-frequency loss may have very different subjective performance.

Fortunately, a measure has been found that is objective yet still gives good subjective results. The measure is the *loudness rating.* It is expressed in decibels and is a weighted sum of a number of single-frequency losses. The weightings derive from subjective experiments and relate to the variable sensitivity of our ears with frequency. These weightings have been standardized internationally. Loudness ratings can be added together and the overall loudness rating of the sending telephone plus the transmission path and the receiving telephone will correlate well with the subjective opinion of the connection.

Absolute-loss and loss-versus-frequency characteristics are two impairments that can damage speech signal as it traverses a network. Table 2.4 lists a number of others.

Most of the items listed in Table 2.4 are critical for an analog network but of less concern for a purely digital network. Digital calls transmitted across the network suffer little or no impairment, so transmission quality is not a problem. However, to save money, voice compression techniques are employed in certain applications. Voice compression inevitably lowers quality to a greater or lesser extent and increases delay; it can therefore offset some of the advantages of digital transmission.

Multiplexing

Transmission media used for telephony include

- Twisted-pair cables;
- Coaxial cable;
- Optical fiber;

Table 2.4
Transmission Degradations: Characteristics That Damage Speech

Impairment	Description	Measure
Delay	The time it takes for speech to travel from one telephone to another. Excessive delay may produce echo problems; it can be caused by the use of satellite links or by heavy speech compression.	Milliseconds
Stability	Unstable circuits whistle—just as public announcement systems sometimes do. Circuits approaching instability "ring."	Stability margin (dB)
Quantization distortion	This is an inevitable effect of analog-to-digital conversion. It appears as noise—but only while speech is present.	Quantization distortion units (QDU)
Noise	There are many sources and types of noise, from random, to single frequency, to impulsive. Subjective effects vary but single frequency noise is the most annoying.	Signal-to-noise ratio (dB)
Errors	Digital errors can result in clicks and bangs in the speech channel—a form of impulsive noise.	Bit error rate
Side tone	The amount of your own voice that is audible in the earphone.	Side-tone masking ratio (dB)
Cross talk	The amount of other people's conversations that can be heard on your line.	Loss (dB)
Attenuation distortion	Different levels of speech experience different losses across the network.	Variation (dB)
Frequency distortion	Speech at different frequencies experience different losses across the network.	Loudness rating (dB)

- Radio links;
- Satellite links.

For a connection of any length, individual links become prohibitively expensive for carrying single calls, added to which the capacity of individual circuits is vastly underutilized in transporting a single call. Multiplexing is the term that describes the packaging of a number of telephone calls through one link.

In the constant battle to reduce transmission costs, many different approaches to multiplexing voice channels have been invented. In the analog world the prime method that was used is called *frequency division multiplexing* (FDM). This works in much the same way as domestic radio. There is only one path between the various transmitters owned by the radio stations and the radio in your home. All the stations transmit through the air, but they each group their signals around different "carrier" frequencies. To select a particular station, you tune to the station's carrier frequency by turning the tuner on the radio. Multiplexing telephone conversations onto a cable uses this principle. Each voice channel is raised to a different frequency at the input to the cable, then recovered at the distant end by tuning to the correct frequency for a particular channel. More than 1,000 voice connections can share the same circuit in this way. FDM has been used across ordinary cable pairs, coaxial cables, and radio and satellite channels.

In the digital world the prime method of multiplexing is called *time division multiplexing* (TDM). Here each voice channel has access to the circuit for only a very short time, just sufficient to send one sample. The circuit is then used by the next channel in the multiplex and so on, until all channels have been serviced and the sequence begins again. These regular periods of ownership are called time slots. At the receiving end, the samples are sent to the correct destination because their position in the time sequence of the multiplex—the time slot—is known. TDM can be used over any media capable of digital transmission and of sufficient bandwidth. The capacity of optical fibers used in this way already exceeds many thousands of simultaneous calls and will be almost unlimited in the future. Many switches can connect directly to TDM multiplexes.

There are international standards available for multiplexing and most systems comply with them. Unfortunately, in the TDM world, there are two approaches that were used in North American and Japan which were adopted by Europe and the rest of the world. The basic TDM multiplex in North America is called T1. It contains 24 channels of 64 Kbps, all of which can be used for voice. The basic multiplex in Europe is formally called the G.704 multiplex, after the ITU standard that defines it. It contains 32 channels, 30 of which are normally available for speech, and it is often known as El to complement T1.

Higher-order multiplexes are defined so that many thousands of channels can be combined in a standard way. To ease the compatibility problems between T1- and E1-based systems, and to simplify demultiplexing, a common method for packaging the various multiplexes for transport has been defined. This is the synchronous digital hierarchy that is defined purely for optical fiber-based transmission systems.

2.2.1.3 Switching and Signaling

Besides delivering the call in an undamaged condition, the telephone network also has to ensure that the calls get to the right place. The call is switched to the right place by a telephone exchange of some sort. There are many types of exchanges or switches—all of which can be of interest in a CTI context. These are examined in more detail in Section 2.2.2. Here, it is worth asking the question, why switch? Why not connect every telephone to every other telephone? The simple reason is that, though this makes sense for small numbers of telephones, the number of links required rises rapidly as the number of telephones increases.

Figure 2.3 demonstrates this graphically. It shows a network of telephones or switches connected fully or linked in a star topology. A remarkable savings in links is demonstrated by the graph. However, this is not the full story. The link that connects the two "switching nodes" in the diagram is shared and must therefore provide more than one circuit, otherwise calls will be blocked.

For all that, the savings are significant, and all telephone systems use switching. In the PSTN there will be a switching hierarchy—with only the highest-level trunk exchanges fully interconnected and with six or more switching stages involved in long distance calls.

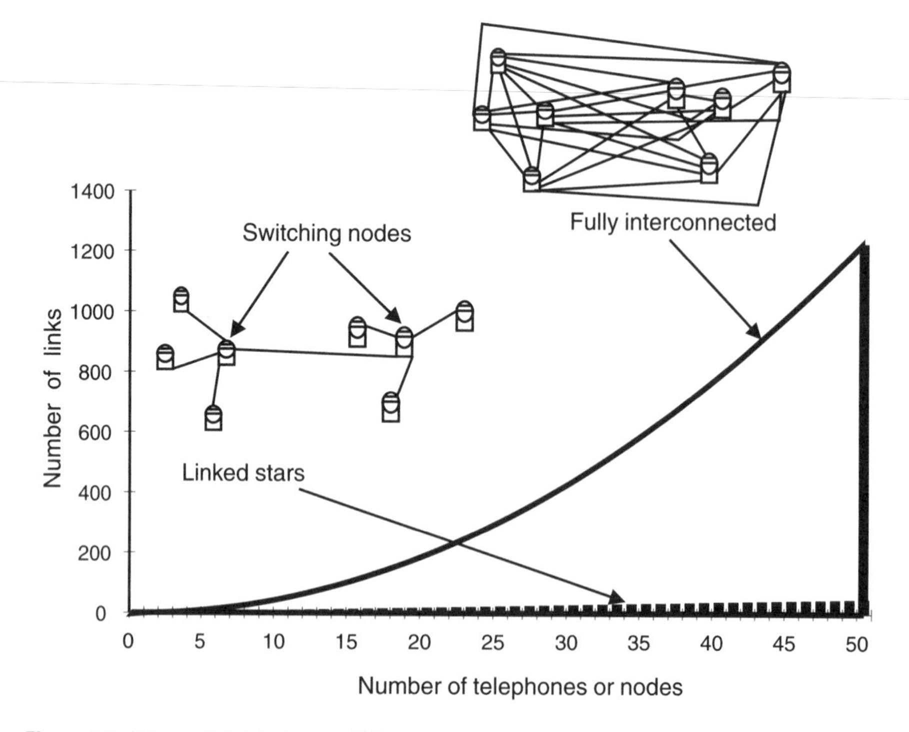

Figure 2.3 Why switch telephone calls?

Interestingly enough, optical fiber technology—offering almost unlimited capacity—could, in principle, provide full interconnection of all telephones in a country, thus obviating the need for switching in the traditional sense. Accessing the necessary bandwidth and managing such an arrangement is, at present, a research topic.

Figure 2.4 shows a telephone call in the process of set-up across a simple network. Bill has removed the handset and keyed Andrea's number. How this signaling information passes between Bill's phone and his local exchange has already been described. However, Andrea's phone is not connected to Bill's exchange, therefore the exchanges have to communicate to determine a route and to allow monitoring of the progress of the call. In simple cases, the sort of information that needs to be sent between exchanges is very similar to the basic information sent between a telephone and its exchange. Where it does differ in content is largely due to the fact that the exchanges are machines and therefore require that the signaling information is well specified and constrained. For example, a dial tone used to be specified in the United Kingdom as a "pleasant purring sound!" That is quite satisfactory for us but not so good for a machine! Between exchanges this signal is called *proceed to send* and would be indicated by a precise tone or clearly specified condition. The basic signaling protocol to set up and clear calls between exchanges is given in Table 2.5.

There is a need to go beyond this basic protocol in many circumstances, to provide such features as the following.

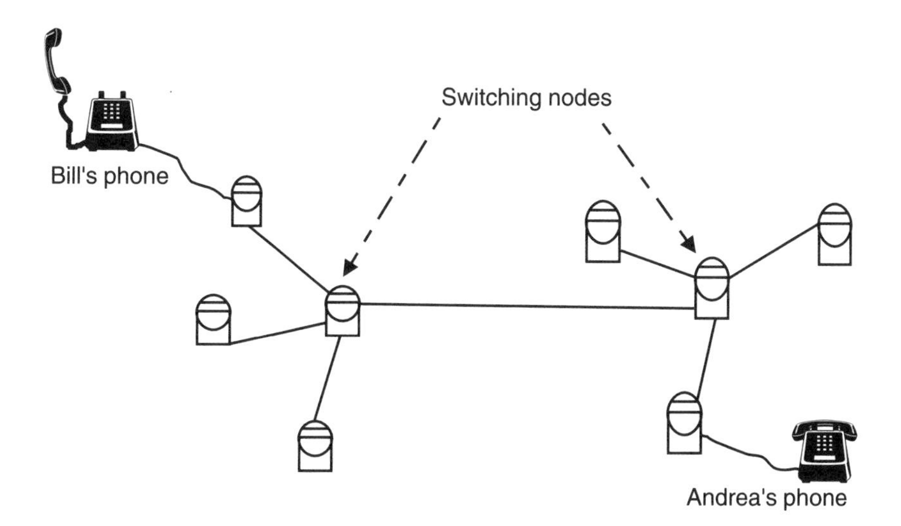

Figure 2.4 A call in set-up.

Table 2.5
Basic Signaling Between Exchanges

Originating Exchange	Terminating Exchange
Seizure	Delay dialing
	Proceed to send
Transfer of routing digits	State of called party
	Call answer
Call clear forward	Call clear back

- Billing information;
- Maintenance and diagnostic information and commands;
- Centralized network management;
- Across-network features;
- Alternative and optimum routing.

Unfortunately, there are many different signaling systems in use, some of them standard—but by far the most proprietary. And even the standard systems are likely to have national variants. In Table 2.6 the different types of signaling system are considered in overview, from simple manual signaling, which does little more than indicate that a call is required (the "generator" mentioned is the equivalent of the old hand-operated ringing generator); through to the basic, no frills systems, which carry the signals listed in Table 2.5; to sophisticated signaling systems, which are capable of dealing with most of the items listed above.

The more sophisticated approaches use common channel signaling systems—the signaling information is carried on a separate circuit to the voice, and the signaling circuit carries signaling information for a number of voice channels. For example, in the E1 multiplex introduced above, one 64-Kbps channel carries the signaling information for 30 voice channels (the remaining channel of the 32 is reserved for synchronization).

Having seen how the customer informs the telephone network where a call should go, and how the telephone exchanges then communicates that information, the next question is: How is the switching done? This topic was introduced in Chapter 1, in which a short description of the evolution of switching

Table 2.6
Overview of Signaling Systems

Signaling System Type	Use	Examples of Basic Protocols	Examples of Enhanced Protocols
Manual	Between dealer boards	Generator/generator Balanced battery	
Access (telephone)	Between a telephone and the PSTN	Loop calling plus loop disconnect Loop calling plus DTMF	ITU-T Q.931 (Integrated services digital network [ISDN])
Access (private switch)	Between a private switch (e.g., PBX) and the PSTN	Loop calling plus loop disconnect or DTMF Earth calling plus loop disconnect or DTMF Direct dialing in (or direct inward dialing)	ITU-T Q.931 (ISDN) Digital access signaling system (United Kingdom) 1TR6 (Germany) Euro ISDN
Private network	Between private switches, particularly PBXs	E&M DC5 AC15	Digital private network signaling system (UK) Corenet (Siemens) ECMA QSIG (Europe)
Public network	Between PSTN exchanges	CCITTSS6	ITU-T SS7

was given. Modern switches are all solid state, unlike the early electromechanical systems designed by Strowger and those who followed him. The focus here will be entirely on the modem switches.

There are two categories of switch: space and time. In space switching the call passes through a *crosspoint* in a switch for the duration of the call. The crosspoints that are connected are determined by the control processor according to the signaling information received. In time switching, time slots are interchanged to produce the requisite connection. If Bill and Andrea were on the same time switch and Bill was allocated time slot 3 and Andrea time slot 17, then to connect them together the time switch would receive a sample from Bill in time slot 3 but would transmit it in time slot 17 so that Andrea would receive it, and vice versa. Time switches are constructed of two digital stores. Into one the speech samples are written, sequentially, while the other controls the order in which they are read out—the connection store. The control computer manages the connections by writing to the connection store, once again according to the signaling information received.

Switches are constructed from both time and space modules to produce a near optimum design. Modern switches are generally nonblocking. This means that any inlet to the switch can be connected to any outlet regardless of how busy the switch is. There are, however, other places in the overall exchange architecture that might lead to blocking.

2.2.2 Telephone Switches

The previous section introduced the mechanism of switching and the means of controlling switching. Here, the different types of switching exchanges are introduced.

Exchanges differ according to application, though the basic architecture is almost always the same. The exchange consists of a set of line interfaces, a switch, a processor, and some common service elements (tone supplies, tone receivers, and so on) connected in a similar way to that shown in Figure 2.5.

There are a number of line interface variants, for example, a PBX might have the following:

- Normal extension;
- Analog feature phone;
- Digital feature phone;
- Attendant console;
- Normal exchange line;
- *Direct dialing in* (DDI) exchange line;
- Digital exchange line;
- Private trunk interfaces.

Line units may not have full access to the switch. It is quite common to concentrate extension line units onto switch inputs. Hence, though the central switch may be nonblocking, blocking can occur between it and the line units. This is an aspect that can vary considerably between different exchange types. Table 2.7 lists all the important exchange types, together with their area of application—CPE-based or PSTN-based.

All of these switches are computer controlled (even the dumb switch has a computer that deals with the protocol from the external control computer), but it is perhaps of more interest to examine differences rather than similarities. The PBX is probably the most general of exchanges. It covers a vast range of line sizes, features, and reliability options. A PBX may have five times more exten-

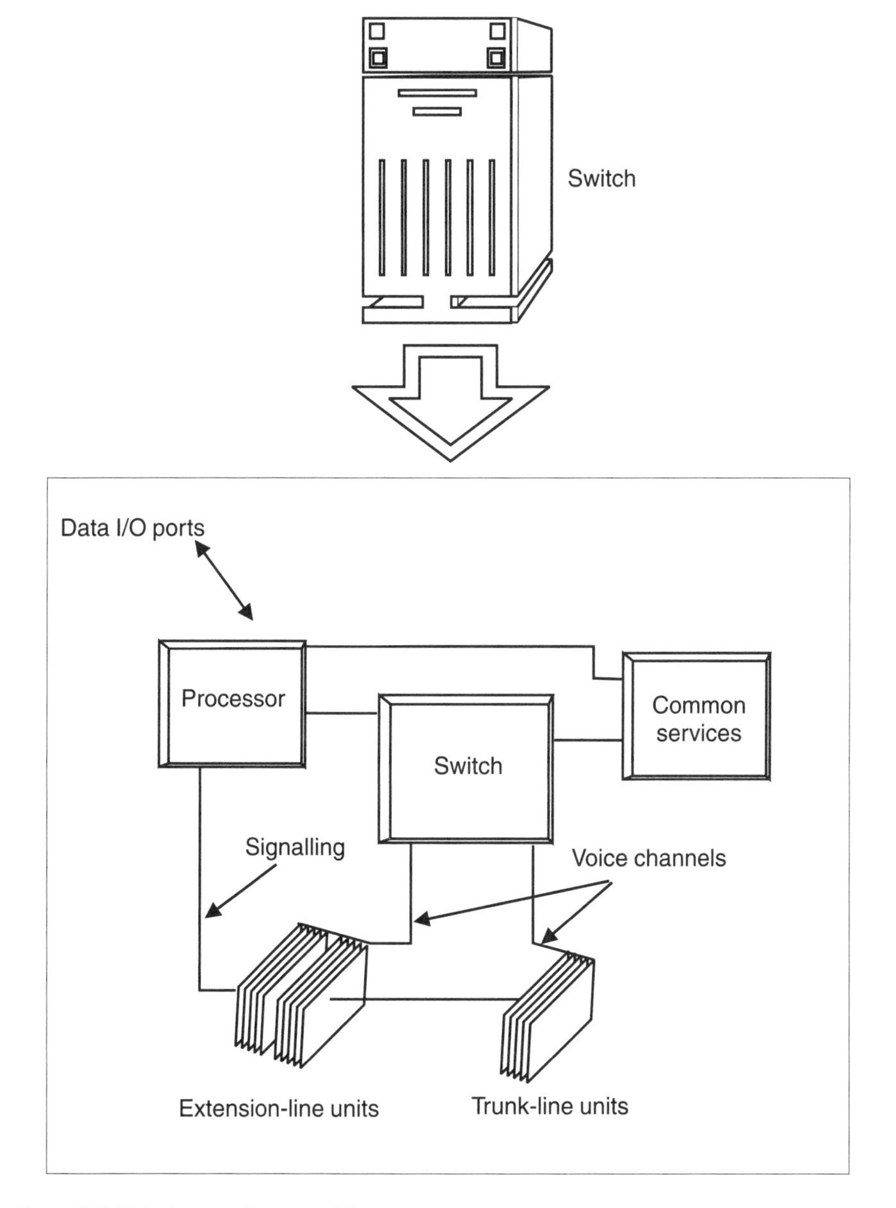

Figure 2.5 Telephone exchange architecture.

sions than exchange lines. This contrasts with the ACD, which might have twice as many exchange lines as it has agents. The ACD system therefore needs to be able to queue calls and manage the queues effectively. ACD agents are the equivalent of the PBX extension line user, except that an agent has a special

Table 2.7
Telephone Exchange Variants

Telephone Exchange	Abbreviation	Application Area	Notes
Private (automatic) branch exchange	PABX or PBX	Customer-premises equipment (CPE)	Probably the most versatile exchange
Automatic call distributor	ACD	CPE	Specialized applications in which inbound calling rate is very heavy
Key system		CPE (small end)	Uses multiline telephone sets—no attendant
Hybrid		CPE	Switches that support some combination of PBX, key or ACD system
Local exchange/central office	LE/CO	PSTN	Switch to which PSTN telephones and PBXs connect
Trunk exchange	TE	PSTN	Switches calls between local exchanges
Tandem exchange		CPE	Switches calls between PBXs in an private network
Dealer board		CPE (finance)	Simple, fast, robust key system
Dumb switch		CPE/PSTN	Specialized applications in which separate computer must have full control
Centrex		PSTN	Different software in the local exchange to create PBX-like facilities

terminal (the turret) and the agent's main task is to deal with telephone calls. The ACD differs also in the sheer amount of call information statistics that it has to deliver.

Key systems represent another extreme. Small key systems are built to a very low budget. They are highly integrated—sometimes the entire system is built on one printed circuit card! They almost always use specialized telephone sets that allow the user to continuously view the status of other users and of the exchange lines.

The local exchange or central office represents the other extreme. Highly modular in design and construction, it is made up of a very large number of different parts. Though cost is important, the guiding light here is service. Local exchanges are just not expected to fail. Reliability and performance specifications are very high. Standby power and duplication or even triplication of critical items is the norm in this environment—as is extensive monitoring and management capabilities. As a consequence, local exchanges are physically large in comparison with the CPE counterparts and PSTN designers are rather more cautious in adopting new technology. Everything that has been said about the local exchange can be said over again about the trunk exchange—performance and reliability requirements are even more exacting because the lines to which the trunk exchanges connect are much more heavily utilized than local telephone lines.

In the world of private networks, the trunk exchange is called a tandem. Mostly, tandem switching is carried out by a PBX, in addition to serving local telephone extensions. There are not many pure tandem exchanges in private networks. For that reason tandem exchanges are usually conventional PBXs configured in a special way.

Dealer boards are fairly simple devices in telephone exchange terms, though the terminals themselves can be quite complex. They are very similar to key systems. The dealer depresses a key labeled "commodities," for example, and the lamp lights immediately at the commodities position. Calls can last for less than one second so performance requirements are high, though switching complexity is generally very low.

Dumb switches form a special category in CTI and will be covered later. This leaves Centrex, which is not normally a special switch but a special function offered from a local exchange. Because this function is used to create a closed user group for a subscribing company, it seems very much like a PBX—but it is not. It is supplied and managed by the telephone company, and there can be significant advantages in this arrangement for the user—not the least of which is reduced capital expenditure.

Switches vary in hardware terms, but they also vary in software functionality. For example, an automatic call distributor system needs varied queuing algorithms to minimize the wait time for incoming callers, whereas the concept of queuing is alien to the dealer board application.

2.2.3 Telephone Traffic

This seems a rather arcane topic, but it is not really that complicated. All prospective CTI adherents should know what an erlang is and understand its connection with the American *century-call-second* (CCS).

The problem with telephone traffic is that it is fairly random. On the other hand, this difficulty can be translated into a benefit since, by presuming that it is truly random, it is possible to develop a statistical theory that has been shown to predict the behavior of telephone systems quite accurately. The formula that encapsulates this theory is termed Erlang's formula after its inventor, a Danish mathematician. The formula is not repeated here; it is available in any textbook on the subject, of which there are many [3]. It is much more revealing to graphically portray the behavior of networks as traffic increases through a graph. Figure 2.6 does just this. The upper graph shows the growth of traffic capacity with the number of trunks or lines that are available to carry calls. There is nothing very remarkable about this, except that the growth is not linear and a small number of trunks cannot support much traffic. But how is traffic measured?

Telephone traffic intensity is a measure that combines the number of calls and their duration. To make any practical sense it must be referred to a fixed time period. This period is usually the busy hour, and it will vary according to the nature of a particular business.

The American measure is the CCS. To determine the number of CCSs generated at a particular site, add all the calls and their duration (in seconds) over the busy hour and divide by 100. So, one call lasting for one hour is equivalent to 36 CCSs.

The European measure is the erlang. The amount of traffic in erlangs is often defined as the number of calls occurring during the busy hour multiplied by the average holding time of the calls. So, one call holding for one hour is the equivalent of one erlang. Hence, an erlang is the same as 36 CCSs.

Both measures therefore provide a representation of the total call duration for a group of telephones, trunks, or whatever. So, for example, if a group of 30 telephones each generates two calls in the busy hour, and the average holding time of those calls is two minutes, the traffic intensity in erlangs is simply: $30 \times 2 \times 2$, all divided by 60 to convert to hours—2 erlangs.

If the 30 telephones are connected to a PBX, and all the calls need to go to another PBX some distance away, how many trunks are needed to connect the PBXs? The easiest thing to do is provide 30 trunks. This would provide full availability—but would be prohibitively expensive. And the trunks would barely be used—in fact, each one would be in use for only four minutes in the busy hour! If fewer trunks are supplied then there is a probability that all the trunks are in use when someone wants to make a call. Erlang's formula is used to determine that probability, which is normally called the blocking probability or grade of service. Grade of service is usually expressed as calls are lost. A grade of service of 1 in 100 means one call in 100 will be lost; this is the same as a blocking probability of 0.01.

So, applying the 30-telephone example to Figure 2.6, if the required grade of service is 1 in 100, the number of trunks required is six, as determined by the upper line of the upper graph—whereas eight trunks will be required if only 1

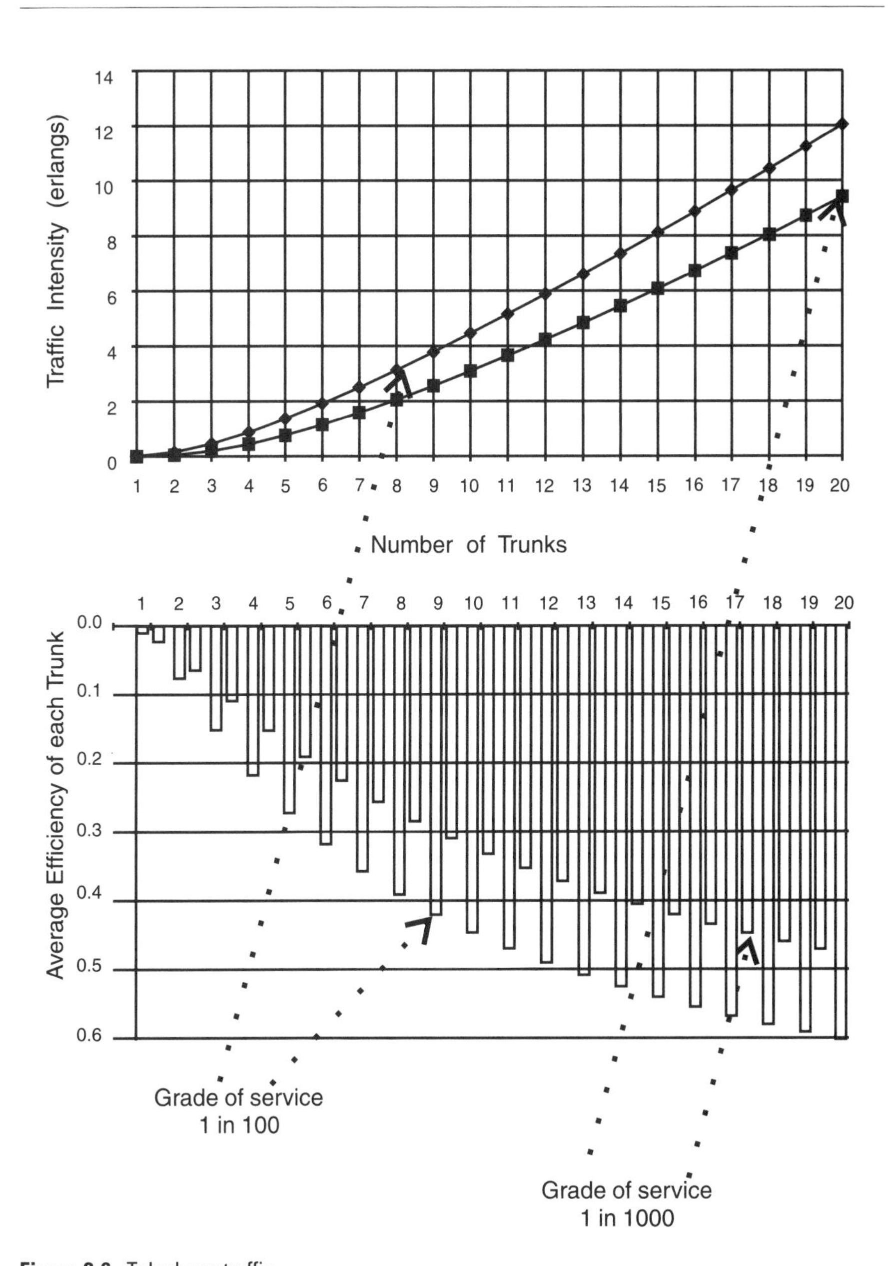

Figure 2.6 Telephone traffic.

call in 1,000 can be lost. Private networks are usually designed to a grade of service of 1 in 100.

Switches are sometimes called concentrators and some of them provide that function only. Following the previous explanation of loss and blocking, the applicability of this term should be evident. The traffic on the six or eight trunks is more concentrated than that on the extension lines.

The really important point about traffic engineering is demonstrated by the lower graph of Figure 2.6. By showing the average traffic carried by each trunk, these curves demonstrate the efficiency with which the traffic is being carried. At low trunk numbers the capacity is abysmal. Two trunks can only carry 0.15 erlangs between them. As the number of trunks increase, the efficiency increases remarkably—even for a fixed grade of service. The reason for this is fairly obvious. As the number of trunks increase, the chance of a new call finding a free trunk increases, simply because there are more gaps available between one call finishing and another starting.

Telephone traffic engineering is a discipline in its own right. Conferences are still held regularly on the topic and there are many, many textbooks written about it. However, it is sufficient here to appreciate the statistical nature of traffic and that there are tools, often in the form of spreadsheets, for dealing with it.

2.2.4 Telephony Features

The PBX that proudly boasted 168 features has already been discussed. Some features are, however, very important in CTI. In many cases, the CTI application programmer is very interested in what the switch can do in addition to setting up and clearing calls. Unfortunately, features vary in name and in operation from switch to switch. Features are often grouped into a number of user-related categories:

- Basic telephone set;
- Feature phone;
- Attendant or operator;
- Data;
- System.

All these groups are of interest in a CTI context; the problem lies in finding a method of classification that makes some sense of the confusing mass of features. To simplify things a little, broad classifications are illustrated in Figure 2.7 [4].

The classes shown in Figure 2.7 are biased towards user features. They refer to different phases of a call and are then followed by the type of feature relevant to that phase. *Incoming calls* are initially in the unanswered phase, the

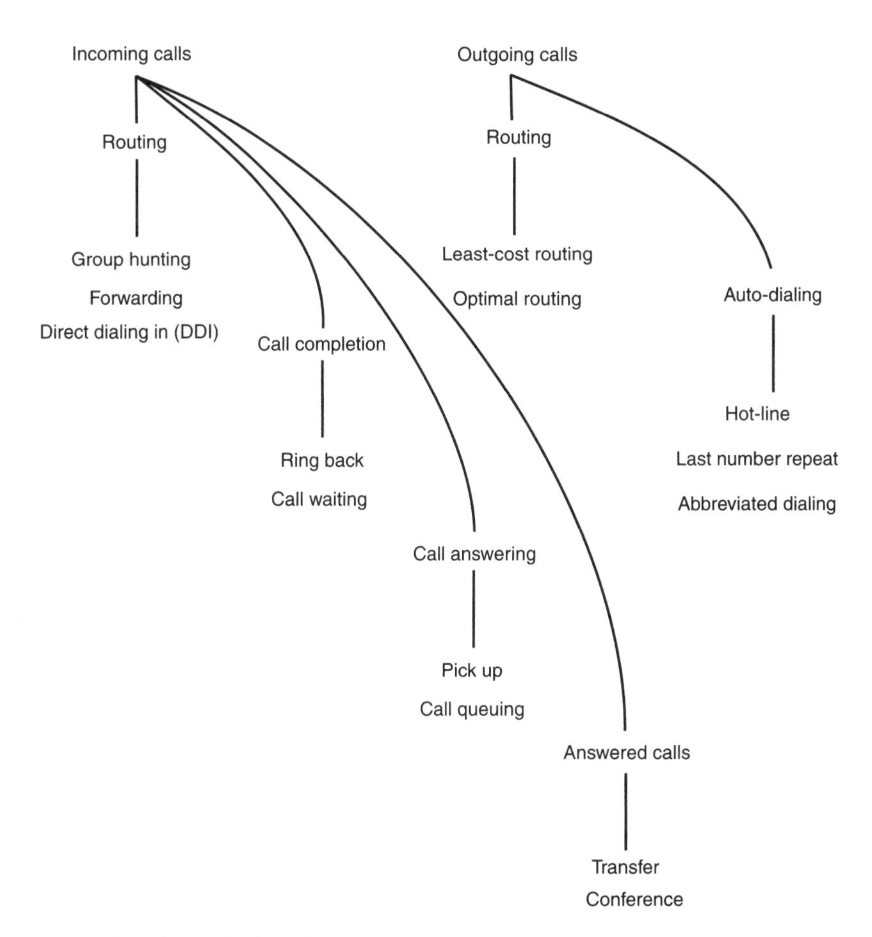

Figure 2.7 Classifying telephone features.

point at which the switch still has full control in determining the *answering point.* An incoming call may hunt over a group of telephone numbers until the first free telephone is found—group hunting. Hunt groups can be arranged in many different ways; hunting may start at the beginning of the group or at the point at which the last hunt finished.

Forwarding of calls is similar except that forwarding is applied at a single phone and is usually controlled by the phone's user. Calls can be forwarded all the time, if the line is busy, if there is no reply or, in some cases, if the line is faulty. Calls can be forwarded to the attendant or to another phone. Forwarded chains can be set up so that a call will be routed to different phones dependent on the status of the phones in the chain. For example, a call may be destined for a boss but forwarded to a secretary. The secretary might be busy on another call so the boss's call might be forwarded to the attendant, or another secretary, and so on.

DDI represents a very important feature in the world of CTI. This feature allows the incoming caller to control the routing at the terminal switch. In public network calls the dialed number usually routes the call to a particular telephone or switch. If the latter, the call will usually be dealt with by an attendant—the last stage is dealt with by human interaction. DDI automates this part of the call by allowing the last few digits to be passed from the public network to the terminating switch. These digits can then be used by the terminating exchange to route the call to a particular extension or service. In a CTI application the DDI digits may be used to identify the requirements of an incoming caller prior to answer. Traditional implementations of DDI are pulse-based and require a special interface within the local public exchange and the PBX or whatever. Modern implementations avoid this complication; the DDI information is part of the incoming call message delivered on the signaling channel.

Call completion embraces the features that allow the completion of calls that meet a busy signal or are not answered. These include ring back and call waiting. Ring back can be set in either condition: if the called party is busy, the calling party will be rung back when the called party clears; if the called party is free but does not answer, the caller will be recalled after the called party makes the next call. In either event, the called party is recalled first, and there is a possibility that the called party might commence another call while this takes place. This and any other conflicts must be dealt with sensibly as far as the caller is concerned. Call waiting applies if the called party is busy. The caller can elect to wait until the caller is free and then gain immediate connection. The called party is usually made aware that there is a call waiting and can opt to connect to the waiting call.

Call answering concerns the picking up of calls. Normally, this is simply a matter of lifting the handset. However, calls may be queued and may be answered by anyone belonging to the specified answer group (for example, a group of agents allocated to the task of answering those calls). Alternatively, a call ringing at one telephone may be picked up by entering the relevant feature code at another—this is sometimes called group pickup.

Once answered, the call may be transferred to someone more suited to dealing with it. Call *transfer* is a speech phase feature, so the feature request from the telephone must commence with a recall signal to bring the processor back into the call. This is similar to the first stage in the establishment of a *conference* call. If the call is transferred then the initiator of the transfer drops out of the call. When conferencing, that person is connected to both parties. Some switches can establish conferences of more than three parties.

As Figure 2.7 shows, the majority of features relate to incoming calls. Features that are used with outgoing calls are almost all concerned with routing. Finding the cheapest and most suitable routes via least-cost and optimal routing and making a hot-line call by simply lifting the receiver, repeating the last number called, or using short codes to achieve abbreviated dialing—these are the features that are usually available from PBXs.

Naturally, terminology is an issue here. Each supplier has its own pet name for the features that its products support. Table 2.8 provides two sets of examples: one set for the Mitel SX2000 switch [5], the other for the Nortel Meridian switch [6]. Naturally, these examples represent a small proportion of the total set of features supplied by each vendor.

2.2.5 Call Processing and Call Modeling

What is call processing? In general, it concerns the processing of signaling information to set up, clear, or redirect calls. In common usage the term is often used to identify the software within the switch that actually performs this task. Switch software, in keeping with most software systems, consists of an operating system that manages the various resources of the system, device handlers that interface to the switch hardware, and a series of applications programs, including those that deal with system management. The key application program is, of course, call processing. It has been implemented at many different times in many different ways.

To perform any control task on, or be able to interpret status messages from a switch, a CTI application programmer must know how the switch operates. In short, the computer world must have a clear model of the switch in order to ask it to do sensible things and to be able to act on the messages that it sends out. A car and its driver provide a simple analogy. The driver has a clear mental model of the way in which the car is controlled. This model is quite simple and requires little knowledge of how, for example, extra power is obtained. It is sufficient that the driver know that when the right-hand pedal is depressed the engine speeds up. The driver is also well able to interpret the feedback the car provides. This gauge for petrol, this for speed, and so on.

If the driver was shot back into the early days of motoring, he would be faced with more controls—timing advance/retard and choke levers, for example. If he had no mental model of the operation of the car's engine, he would not know what to do with them. Worse still, in those early years there were many different models to choose from! The driver might well cause the engine to stop simply because his model of its operation was incorrect.

To communicate at the functional level with a switch, there must be a model of the telephony process, and it is this model that call processing

Table 2.8
Feature Names Used by Two Suppliers

Feature Type	Controlled by	Mitel Terms	Nortel Terms
Group hunting	System	Circular hunting/terminal hunting	Group hunting
Forwarding	User*	Call forwarding busy	Call forward all calls
		Call forwarding don't answer	Call forward no answer
		Forced call forward	Call forward busy
Ring back	User	Call-back busy	Ring again
		Call-back don't answer	Ring again on no answer
Call waiting	User	Camp on	Call waiting
		Camp on retrieve	Camp on
Pick up	User/system	Call pickup—dialed	Called pickup
		Call pickup—directed	Call pickup directed
Call queuing	System	Priority queuing	ACD least call queuing
Transfer	User	Call transfer	Call transfer
		Transfer with privacy	
Conference	User	Conferencing—station controlled	Conference
			Conference no hold
Hot-line	System	Hotline	Hotline
Last number repeat	User	Last number redial	Last number redial
Abbreviated dialing	System/user	Personal speed call	Speed call

*Most user features can also be controlled by the system.

generally defines. There is little point in requesting details about a particular party to a call if the switch does not hold the necessary information. There is little point in asking the switch to do something that its call processing algorithms will not allow. Of course, different implementations have different models, but there are certain actions that are common to all. Figure 2.8 provides the most basic of call models, showing the states in which a call can exist as it moves from the on-hook condition to an established speech call between two parties. As it moves between states, an "event," such as service initiated, is generated.

Modern system design views the world in terms of objects, so there is a need to define telephony objects that are visible and manipulable. Those objects

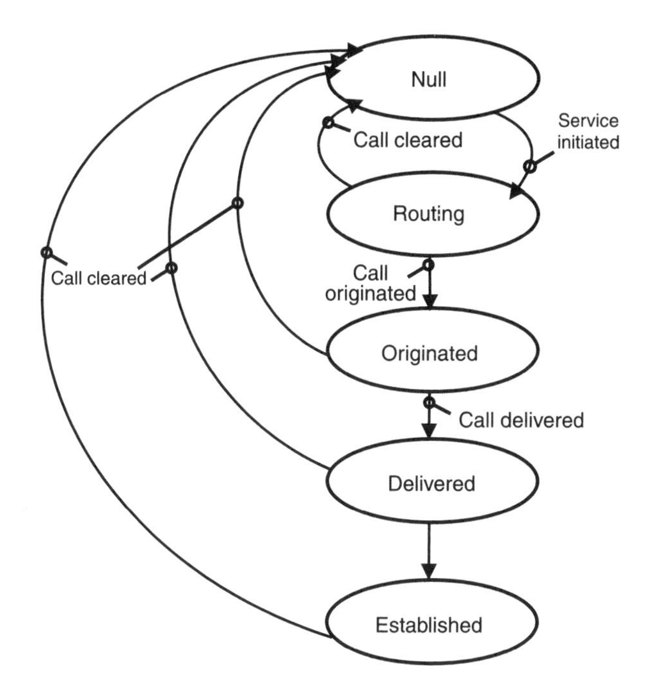

Figure 2.8 Basic call processing model.

can exist in various states, which also need to be defined. The overall state of a call is made up of the objects involved together with all of their states.

The establishment of generally acceptable models and the objects within those models is key to establishing standards for computer telephone integration. This subject will therefore be revisited in considerably more depth in later chapters.

2.2.6 Evolving Telephone Networks

The public switched telephone network (PSTN) is made up of local exchanges and trunk exchanges interconnected by a digital network of trunks. International gateways are provided, as are gateways into other PSTNs and into mobile networks. The local network that provides access to telephones (and to PBXs, etc.) is still primarily analog, although sizable PBXs and ACDs are digitally connected. The ISDN, described later in this chapter, extends digital services across the local network.

The ISDN is simply a means of extending the digital capability of the telephone network to the periphery. However, there are changes happening to the network itself that are concerned with control rather than transmission.

Modern networks have a much improved signaling system interlinking all of the exchanges. This is Signaling System 7; it provides faster call set-up, allows more intelligent signaling messages to be sent, and allows access to central databases to determine routings. This latter capability is part of the migration towards an *intelligent network* (IN). The intelligence is supplied by a central computer that allows complex routing based on a customer's particular preferences. This computer—called the *service control point*—allows the introduction of new, network-wide features without alteration of the software in all the individual telephone exchanges. Mobile telephone networks are generally IN-based from the outset, since they have to track the movement of mobiles from switch to switch.

2.3 The Computing Environment

This chapter was introduced by recounting a little story—"It's All Bits Anyway." Is the computing environment all bits now? Of course, everything in the computing world does reduce to the bit level—even if in many cases those bits are carried across analog telephone lines! However, this assumption amounts to a logical fallacy almost on the scale of:

> Flesh is made of protoplasmic cells.
>
> Marilyn Monroe was made of flesh.
>
> Marilyn Monroe was just a collection of protoplasmic cells.

It was the higher-level view of Marilyn Monroe that interested her many fans, and it is the higher level of computing that is of greatest interest to its users. In the 1970s, an understanding of computing at the bit level was essential in the production of many applications. In those days it was not at all unusual to write operating systems. In the 1990s it was becoming unusual to write code at all. Application creation, reusing objects or utilizing application generators, is carried out in a standard environment, often by reconfiguring existing application modules.

However, computers still have to interface to the outside world and to each other and bit-level compatibility becomes essential. But even this communications scene is rationalizing. Interfaces are settling onto worldwide standards and clear levels of communication are defined right up to the application layer.

This introduction to the computing environment begins by rephrasing the questions asked in introducing the telephony section directed here to the computing world:

1. What does the computer actually do?

2. What are the measures used in computing and what do they mean?
4. Why is computer networking so important?
5. How does the telephony world see the computing world?
6. What are the major trends in computing?

Once again the intention here is simply to provide sufficient background information for those who are relatively new to the computer world. The treatment here is necessarily more shallow than that given to the telephony environment. This is not to underrate the computer world. It is complex and detailed, and the books on this topic alone will fill many a shelf. On the other hand, it is a world in which general rules and concepts are sparse compared to product-specific information. Many of the concepts used in the computing world are concerned with packaging, storing, and presentation of information, together with the algorithms that allow this. The basic processor is a concept that is shared between computing and telephony and, to a large extent, taken as read. That said, the next section takes a glimpse at those basics and extracts some of the measures that apply to the computer and are therefore of key importance in CTI.

2.3.1 Computer Basics

The following sections cover the computer itself, the terminals, the software and applications, and the manner in which computers are connected together. The three major elements of computing are hardware, software, and networking.

In most applications, a computer takes input data and processes it in some way so as to provide output data—classical data processing. Process control computers do much the same thing except that the data is derived from the process machinery and the output is delivered back to it. Message storage computers do not look at the data in the messages, they simply store this data and retransmit it to somewhere else dependent on routing data. Transaction processing computers take instructions as input data and perform some operation on stored data, then provide output concerning the status of that data. It is the latter that normally are of interest in the CTI world.

The point about all this is that the basic job the computer does is always much the same. The differences lie in the performance required. The basic machinery required to perform a computing task is much the same in principle, though there can be differences that involve orders of magnitude in store size, execution time, and so forth. Figure 2.5 presented the basic architecture of a switch; Figure 2.9 presents the architecture of a computer at an even more basic level.

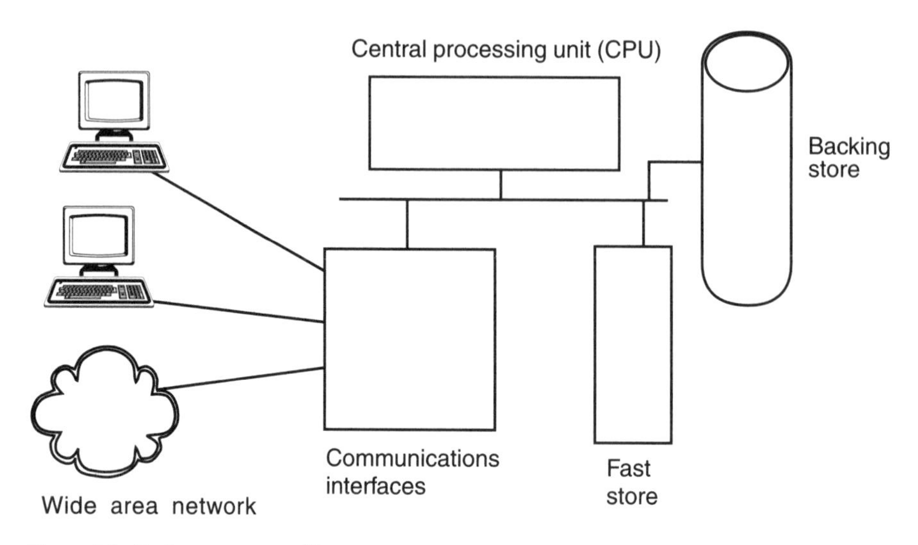

Figure 2.9 Basic computer architecture.

Simple though this picture is, it is nonetheless representative of most architectures. For those readers who wish to compare the operation of a basic computer at this level with the operation of a switch, the following paragraph may be useful.

Data is taken in from the communications interfaces either autonomously or under the control of the *central processing unit* (CPU) and transferred to the fast store (usually *random-access memory* (RAM)) or the backing store (usually a disk). The application programs usually reside on the backing store and are moved into fast store for execution. The instructions that comprise the application are then fed into the CPU to perform whatever task the application software defines. Data is read and written to the various stores or communication channels as execution proceeds. Data is usually organized into files and it is in this form that it is stored in the backing store. Most transfers of data take place across the central line in Figure 2.9, normally called the computer bus.

This is a basic view of the computer, and architectures become more complex as performance demands grow. Part or all of the modules may be duplicated to improve reliability; there may be other storage mechanisms available; multiple CPUs may be connected to the same bus; cache memories may be used to store prefetched instructions; the bus may be extended for multiple computer working; and so on. These are variants on the basic theme presented in Figure 2.9 in much the same way that there are many variants on the basic petrol engine—multiple cylinders, electronic injection, turbo chargers, and so forth.

The basic measures that apply to a computer are provided in Table 2.9.

Table 2.9
Computer Characteristics and Measures

Characteristic	Resource	Measure
Storage	Fast and backing store	MB
Performance	CPU clock speed	MHz
	Instruction execution rate	Mips
	Multiply-accumulate execution rate	Macs
	Floating point operation execution rate	Mflops
	Memory access time	Microseconds
	Latency (of disk)	Milliseconds
	Data transfer rates	Kbps
Instruction word size	CPU	Bits
Bus width	Bus	Bits
Reliability	Mean time between failures (MTBF)	Months
	Mean time to repair (MTTR)	Hours

Table 2.9 is included primarily to illustrate the point made earlier. The computer system is relatively simple to characterize in objectively measurable performance terms.

All of the items mentioned in Table 2.9 contribute to the overall suitability of a particular design to a particular application; many of them have a direct bearing on the performance of an application. The table does not include the range of interfaces provided nor the terminals that can be used; these are also important criteria in matching the machine to the application.

2.3.2 Types of Computers

Business applications are said to exist at three levels: enterprise, department, and individual. Similarly, there used to be a different computer for each of these levels: mainframe, minicomputer, and PC, or workstation.

Things do not stand still in the world of information technology generally, and certainly not in the world of computers. Soon after this pleasing logical hierarchy for computers was decided upon, the power of the PC began to soar, the cost of its memory began to drop almost as dramatically, and sensible means became available to tie these fast-growing infants into full-blown distributed computer systems.

However, technological advances do not entirely rule the business world, so the installed base still contains a good proportion of mainframes and minis. These machines are still sold into traditional markets and a new role seems to be evolving for the mini in the area of client/server computing. More than that, most of the databases that are central to large business still reside on mainframes or minis and will continue to do so for some time. At the same time, the differences between the three types of machine are being eroded by the rapid advances in technology. Those differences are described in Table 2.10.

At one time the workstation held the role of a very expensive and powerful PC and was therefore purchased for specific tasks, such as computer-aided design. However, the PC's power has risen and the workstation's price has fallen such that it is now difficult to distinguish between them. Similar trends can be seen in relation to the most recent generation of PCs and the minicomputers, even between PCs and mainframes. And so the revolution continues.

Reliability, though a problem, was not a major issue in the early days of computing. Batch jobs could be deferred and a 24-hour service was neither offered nor expected. This is in stark contrast to the world of telephony. As more and more applications were found for the computer, the reliability requirements for some of them became quite exacting. Responding to automatic telling machines in the banking world presents a good example. Applications such as this gave birth to the "nonstop" computers—specialized minicomputers and later mainframes that sported redundant CPUs, mirrored disks, and duplicated buses

Table 2.10
Historical Computer Hierarchy

System	Characteristics	Comments
Mainframe	Large, powerful, wide word size, massive backing stores	Batch operation and transaction processing; often hold businesswide customer databases; operational staff support; Air-conditioned
Minicomputer server	Moderate power, medium word size, large memory capacity	Multiuser, many different applications; normal accommodation; part-time operational staff
PC and workstation	Low power, small store, narrow word size	Single user, office, and support applications; desktop accommodation, external support

that produced MTBFs that were, and are, quite respectable, even in telephony terms.

The last computer type to be considered here is probably the Rolls Royce of the stable. These are the supercomputers—very high-speed machines with vast arrays of parallel processors to provide sufficient computing power to tackle such intractable problems as weather forecasting or complex system modeling. These are not likely to run applications that need functional integration with telephony—unless the weather forecaster needs a telephone call to let him know that the latest forecast run is complete! It is interesting to observe that what qualified as a supercomputer ten years ago is a workstation today.

2.3.3 Computer Terminals

In the early years of computing it was quite clear what a computer terminal was. It was the thing with a keyboard and a screen—the dumb terminal. Of course, the computer had many input and output interfaces, but the dumb terminal provided the user interface. Currently, the physical user interface to the computer is likely to be some form of PC. Although there are a wide variety of shapes, colors, and models of PCs, the physical user interface is much the same. All PCs have a good quality color screen, a mouse, and a keyboard. Many PCs also have sound capability through loudspeakers.

At this level, the PC can be compared with the basic telephone. It has a handset and a keypad and very little else. This physical interface it has in common with most other telephones. However, if you were to walk up to any telephone in the world, chances are that you would be able to use it to place a straightforward telephone call. This is probably not the case with the computer terminal. Before you could do anything at all, you would be likely to spend some time finding its on/off switch; then you would be asked for a password. Even if the terminal were not password-protected, it would be unlikely that you could do anything sensible with it. The applications that it has access to are not yours, and the user interface may be so alien to you that it is unusable. At least this was so. Fortunately, however, two developments have conspired to change this state of affairs.

The introduction of Windows by Microsoft has established a degree of standardization on the desktop at the user interface that goes well beyond the physical. The look and feel of the interface has become standardized—at least to some degree. Moreover, that look and feel has been extended to the applications themselves. This is to the point where it is likely that you will be able to use someone else's PC—assuming that you can gain access. Maybe you will not be able to make a phone call from the PC—but you may be able to access and use spreadsheets, word processing documents, and databases. This revolution has really taken hold in the 1990s and is being reinforced by the growing popularity of browsers. For example, the simple interface presented by the Netscape's Navigator is easy to use. It is intuitive and requires no training beyond a simple introduction to the use of the mouse and keyboard and the means of logging in. Furthermore, the browser provides a simple access to both the local machine and the local or global networks (Intranet and Internet). Thus, it seems that a reasonably consistent user interface has now arisen for the computer user. As will be apparent, that same interface can be used to make a telephone call—once again without training.

Dumb terminals are still used in the computer world—especially in the call center environment. These terminals rely, to a large extent, upon the computer to which they are connected for their operation. The most well-known terminals are contained within the IBM 3270/5250 series and the DEC VT100 range. Other forms of equipment (interactive voice response systems, for example) are connected to the IBM and DEC mainframes and minis by emulating these terminals. One of the most common emulations is based on the IBM 3270 terminal range or upon the emulation of an IBM terminal cluster controller (3174).

Two new forms of terminal began to evolve in the late 1990s. The network computer (NC), unlike the dumb terminal, does actually run applications. However, these are usually obtained from the network itself—probably in the form of Java applets. The NC does not require fixed storage since it can boot from the network and can store its long-term data in the network. The user interface offered by the network computer is, of course, the browser. Thin client terminals are also sometimes called network computers. Here the application stays on the server, and the thin client simply provides the graphical user interface. This is the modern form of the old dumb terminal. It is popular since management is minimized and remote access is simplified.

2.3.4 Computer Software

A computer is nothing without its program, and it is in the program area that the functional interface provided by CTI really exists. The role of the operating system software in managing the hardware of the computer has already been mentioned; it is presented graphically in Figure 2.10.

In all well-designed software systems the application software has no direct contact with the hardware of the computer; all communication takes place via the operating system and device drivers, as shown in Figure 2.10. Simple PCs and workstations have simple operating systems, such as DOS or the early Windows variants. They provide the user interface, file management, input/output management, the ability to run applications and other useful

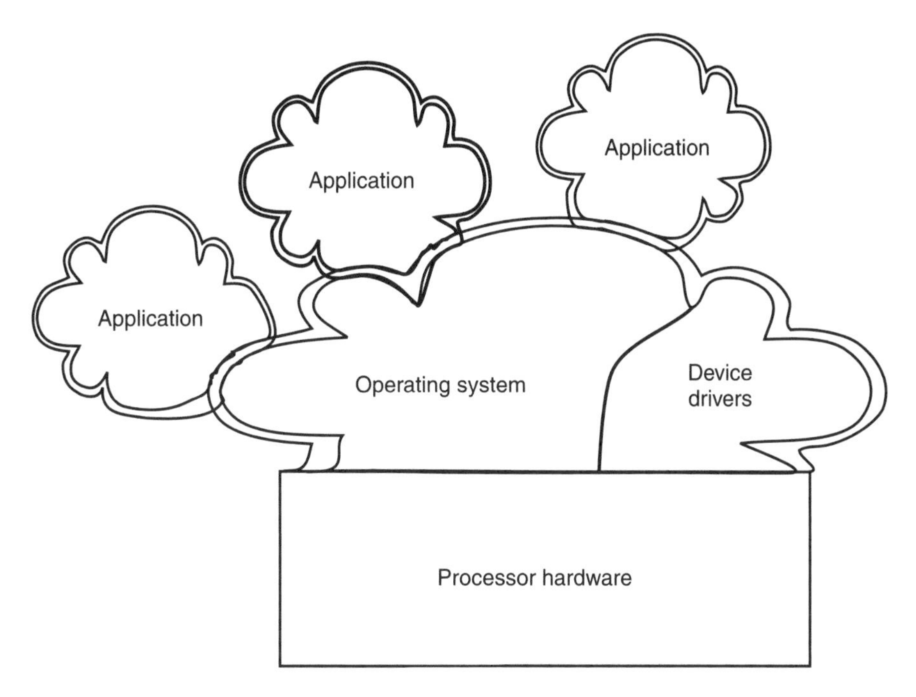

Figure 2.10 Basic software architecture.

functions for a single user engaged in just one task at a time. However, more advanced operating systems (UNIX, OS/400, Windows NT) support multiple users and multiple jobs, and here one of the key functions of the operating system is to share the available processing power between the various processes that need it. The part of the operating system that performs this task is called the scheduler. The scheduler is either executed on a regular basis—initiated by real-time clock interrupts—in which case it will then update its various time-out records—or when a process has completed its task. It is the scheduler's task to decide which of the many processes in the system should run. This is decided upon by examining the priority of processes and their state. Processes can be inactive, waiting to run, awaiting an input, or suspended. Processes that are suspended have been interrupted and are therefore waiting to run again. Those that are ready to run are in that state because they have been given an input. An input to a process is a message. The message may have arrived from another process, or from an input device, or may be a time-out message generated by the operating system and originally posted by the process itself.

Application software is traditionally written in a high-level language or a higher-level applications generator. This is compiled into code that the computer can understand or is emulated. Though there is normally only one copy of the application code in the machine, there may be many users of it. Each user has a separate and unique area of data. It is the combination of that data area and the application code that makes a process; thus there can be many different processes in many different states, all using the same code. Of course, though these many processes exist, only one process can actually run at any one time since, in most computers, there is only one CPU.

Device drivers provide the low-level interface to the terminals, disk drives, printers, and other peripherals connected to the computer. The application is isolated from the physical details of these interfaces, and its output and input channels can be switched between peripherals without changes being made to the application code.

Applications are key to businesses and so the ability to maintain and change the application program software and to protect the investment made in this software is also key. Applications can be created in various ways. Most business applications involve transaction processing on large databases. Creating the programs that perform these transactions therefore represents a major activity in the computer world. Many of the databases that are used are relational; they are formed by a collection of related tables, each of which has a number of rows that contain the individual records and a number of columns that contain the attributes. Data in one table may well appear in another, so that when it is changed in one table, the change appears in another. As far as the application is concerned, no knowledge of how the database is implemented is needed.

A *standard query language* (SQL) exists for interrogating and managing relational databases, and this language can be used to create the forms from which an application is created. At this level the language is often called a *fourth-generation language* (4GL) to differentiate it from the conventional high-level languages (Cobol, Fortran, C). Conventional languages are procedural; they are more flexible than a 4GL but are far more distant from the application.

Object orientation is a growing trend. Software objects are generalized functions that are defined in terms of inputs and outputs. Object-oriented languages such as C++ and Java are currently dominant. Java has a subset that can run on any machine with the small Java emulator (Java Virtual Machine). Java objects are therefore transportable.

Application communication within a PC has become increasingly sophisticated. Beginning with simple, user-controlled cut-and-paste operations, it has progressed to the point where it is possible to create compound documents that are automatically updated. A compound document may be produced by a word processor yet contain graphics, image, spreadsheet, voice, or video fields. Linkages can be dynamically maintained using technology such as that defined by Microsoft or the Object Management Group (ActiveX or CORBA).

Many applications are created by modifying an existing one and many application programs are written in such a way that configuring them for particular applications is relatively simple.

2.3.5 Computer Networks

Computers need to communicate for many reasons. One of the prime reasons is to access a shared database; another is to access a shared resource, such as a plotter or printer; yet another is to transfer data between users. For these and many other operations it is necessary to link computers together—to form computer networks just as telephone networks are formed by linking together telephone exchanges.

The most basic network is provided to link dumb terminals to the host. The links used are very basic and the protocols are character-based with no error recovery or other sophistication. It is interesting to observe that many PBXs offer options that allow data connections across the switch and data transmission capabilities over the extension wiring. This is a form of physical rather than functional integration between the two worlds. It has never been a popular option, partly because the early switches were blocking and therefore not suited to the task, and partly because the holding time of many data calls is orders of magnitude greater than that of voice calls so the switches were not dimensioned properly.

2.3.5.1 Wide Area Networks

Where physical integration has always been popular—and will remain so for some time to come—is in the use of the telephone network to carry computer data. Once again, the most basic application is remote access from a terminal to a host. To achieve this connection over an analog network, the character-based data and control signals from and to the terminals—which are distinctly digital—have to be transformed into a speechlike signal. This is the job of the simple modem. The modem takes digital signals from the terminal at speeds of between 75 bps and 33.6 Kbps and transmits them over the analog telephone lines within the 300 Hz to 3.4 kHz allocated to speech—and it uses some very clever modulation techniques to do so, especially at the higher speeds. Techniques are also available for 56 Kbps, but these are mainly confined to Internet service provision in one direction only.

Wide-area networking can also be supplied by fixed or switched digital connections. Circuit-switched connections are increasingly supplied by the ISDN in the public network (discussed later). There are also circuit switched public data networks in existence, mostly based on the ITU X.21 standard for synchronous operation on public data networks. Though this option has not been widely adopted, the X.21 standard does find uses in a number of fixed and switched data-link applications. Fixed digital links are provided by the telephone company (telco) or can be privately established—dependent upon local regulations. In Europe the most common bit rates available are 64 Kbps and 2 Mbps. In the United States and elsewhere the most common bit rate is 1.5 Mb. These transmission rates are derived from telephony standards (E1 and T1); higher capacity links are also available.

Other wide area networks (WANs) are packet-based, the most well-established being the X.25 networks. Most countries have at least one X.25 public network—and they are connected worldwide just as the telephone networks are. Although they are still useful, X.25 networks are too slow for some applications and are supplemented by the faster, simpler frame relay network, which is used primarily to interconnect local area networks. Where frame relay is too slow or inflexible, connectionless networks based on the switched multimegabit data service (SMDS) are sometimes used. SMDS is a cell-based technology and may be quickly displaced by ATM services, which can carry voice and data at a very wide range of speeds.

2.3.5.2 The Internet and Intranets

The most startling revolution that has occurred in computing in the 1990s is the growth in use, usability, and popularity of the Internet. The Internet is a virtual network of interconnected computers. The computers involved can adopt various guises from computing nodes, databases, or message passing nodes. Formed in 1969 with just four computers, it had an estimated 40 million users by the middle of the 1990s. Some of these use the Internet purely for the transfer of mail; others use it for text-based conferencing on specialized topics—but the major interactive use is in the access of information across the World Wide Web.

The Web is based upon the embedding of hyperlinks within documents. Selecting these links, usually achieved by a mouse click on a specific section of the screen, can cause the browser to display another document in any machine on the Internet. That document can also have links, and thus the Web is dynamically linked—the links being embedded through the use of *hypertext mark-up language* (HTML). In addition, the documents themselves can contain any media—text, image, voice, or video. Sophisticated search engines allow the user to find information of interest from the millions of documents that exist on the multitude of Web servers that can be accessed via the Web.

Communication is standardized via the TCP/IP protocol, which allows packets of information to take various routes across this virtual network.

TCP/IP also supports the three main functions used in the Internet, which are described as follows:

- *File transfer protocol* (FTP): This allows the user to explore databases on the machines and transfer files from and to them.
- *Telnet:* This allows the user to log on to a machine as a remote user.
- *Simple Mail Transfer* (SMTP): This allows the user to send messages to anyone who has an Internet address and, therefore, a mailbox on a specified computer within the network.

A common network addressing scheme—part of the IP protocol—reinforces the use of TCP/IP. This establishes unique addresses for all users of the Internet. Originally based upon a four-byte format, addresses were rapidly running out. Accordingly, the most recent scheme is based on 16-byte addresses.

Thus, the Internet is a network of networks based upon common data transfer protocols, addressing schemes, and network applications. All of this can be applied to a private data network—and it is, in intranets. Intranets are private networks that use the standards of the Internet (as recorded in the IETF's *request for comment* (RFC) documents). Intranets also use Web technology to allow users to gain access to their company's information databases through the use of a browser. The Intranet may or may not be connected to the Internet itself. If it is, this connection will usually be made through a firewall that endeavors to prevent destructive inward access.

Clearly, the dominant position of the Internet and Intranets has a major affect upon CTI. This relationship, particularly through Internet telephony, Web access to call centers, and multimedia messaging, will be explored later.

2.3.5.3 Local Area Networks

The revolution in computing that was driven by the cost effectiveness of the PC quickly spawned the need to connect PCs within an office. At first the prime use of this connection was to share peripherals between PCs—printers, for example. LAN technologies, such as Ethernet and IBM's Token Ring, evolved to fill that need. Ethernet is by far the most popular, supplying access speeds of 10 to 1,000 Mbps. Once the infrastructure of the LAN was in place and the software to support communications across it became available, more and more communications applications were found. Included in these was the use of certain machines on the LAN as servers. Servers are dedicated to a particular function and provide this function to the client PCs connected to the LAN. A typical server may hold a business database for access by PCs or may provide a common communications function—providing access to an existing mainframe computer, for example. LANs can now be interconnected by the use of bridges, switches or routers, and that interconnection can be provided across the WAN. LANs provide the backbone for distributed client/server computing, a development that has particular relevance within a CTI environment.

2.3.5.4 Computer Network Architectures

Computer networking is not just concerned with moving bits. It is concerned with the interworking of all the elements of distributed computing systems at all levels, and this is where computer networking becomes far more demanding and the protocols required far more sophisticated. Realizing this, all the major computer suppliers began defining network architectures that would allow all the computers and peripherals in their individual portfolios to interwork. The *systems network architecture* (SNA) from IBM is one example; the *digital network architecture* (DNA) from DEC is another. Both of these date from 1974.

The basic elements within SNA include network-addressable units and sessions. The former contains the so-called *logical units* (LUs), the *physical units* (PUs), and the *system services control point* (SSCP). Logical units describe a session interface. For example, LU6 defines a set of services for host-to-host communications, and LU6.2 is a subset that deals with communications between PCs, departmental computers, or both. Physical units are, as their name suggests, physical entities, such as terminals (PU1) or a mainframe (PU5). SNA is subdivided into seven protocol layers and makes use of *protocol data units* (PDUs), which in effect, define the structure of data packets that pass between users.

SNA and other network architectures were proprietary, so interworking between different suppliers' systems was not possible—except at the lowest levels. For this reason the *International Standards Organization* (ISO) began work on a supplier-independent architecture for interconnection and christened the activity *open systems interconnection* (OSI). SNA and DNA were layered architectures, defining interworking at the physical layer and gradually building upwards, layer by layer, until application interconnection is reached. OSI does the same thing and defines the now famous seven-layer model shown in Figure 2.11.

In the reference model, layers are only aware of the layers immediately above and below them and believe that they are communicating directly with their peer layer through the layer that is immediately below. So, using the example protocols on the right-hand side of Figure 2.11, if the X.400 application (this is the ITU-defined method of implementing electronic mail) sends a message to some other machine, it must package the message and envelope

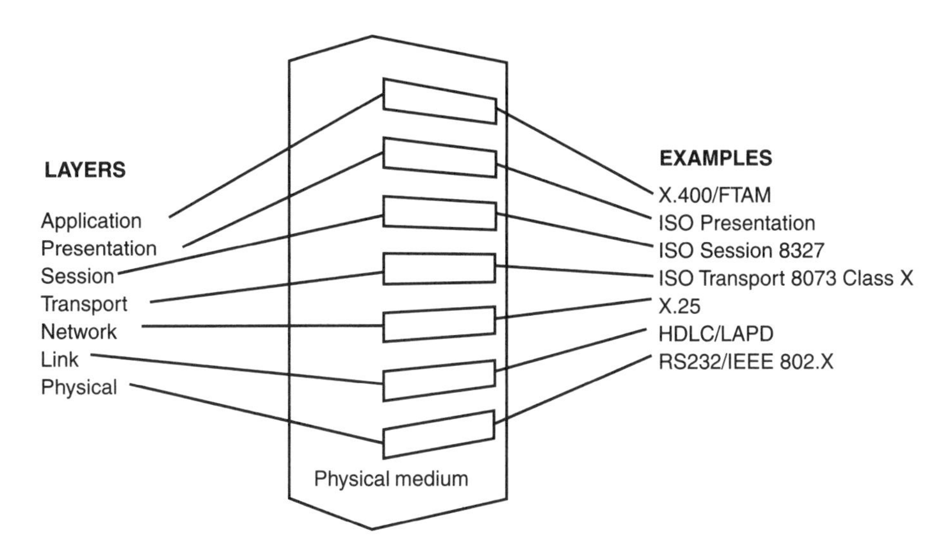

Figure 2.11 OSI reference model.

it exactly as the X.400 specification directs. It does not need to be aware that the message is carried by X.25 at layer three. Indeed, this may change as the message traverses a network; it may meet a link that uses frame relay, for example. The X.400 application is entirely isolated from this. As far as it is concerned, its communication is with the X.400 application in the computer to which the message is addressed. How it gets there should not be relevant at layer 7.

Referring again to Figure 2.11, the presentation layer provides for a common data representation between applications. The session layer deals with establishing a session between applications, re-establishing it if there is a failure, and releasing it at the end of the session. The transport layer deals with the segmentation and concatenation of the data that is transferred, flow control, and a number of other transport-related functions.

Beneath the transport layer lies the real network. A packet network is a good example of a real network and X.25 and IP definitions form example standards for the OSI model at this level. The link layer supports the network layer, providing error checking across the link between two machines and the method of organizing the frames of data for transmission. The physical level is interesting because this is where the most common standard in computing is met: this is the V.24 or RS-232 interface. There are, of course, other physical-layer standards.

This completes a very cursory overview of the OSI model. It is important as a means of providing interconnection between machines and is relevant to CTI in that context. However, for all that, it defines a complete collection of international standards at the various levels; there are industry standards that provide similar capabilities and have achieved major penetration in the computer world. *Transmission control protocol/Internet protocol* (TCP/IP) is clearly the most well known of these, providing interfaces up to the presentation layer, and a number of application layers to run above it (virtual terminal, file transfer, and simple mail transfer together with network management).

2.4 Trends in Telephony and Computing

One of the exciting, but also daunting, challenges of CTI is that its component technologies, markets, and usage are all changing at a remarkable rate. Even if the rate of change were linear in CTI's component parts, combining them produces exponential change rates. However, there is plenty of evidence that change in computing and telephony, far from being linear, is itself increasingly exponentially. In other words, CTI is changing very fast.

To put all of this into some sort of perspective, Table 2.11 provides a very coarse picture of evolution in the 1990s and over the previous three decades.

Most of these trends are driven by technology, though not all (the move from black to colored telephones was a classic example of deregulation and fashion-led need!) The main thing that Table 2.11 confirms is that change has been rapid and significant. Furthermore, the pace of change warmed through the 1980s and has reached a crescendo in the 1990s. Or is it simply that we make a précis of the past, foreshorten the future, and praise the present?

In any event, there is plenty going on from a technological vantage point. Figure 2.12 presents the major trends, one of which is the integration of the technologies of telephony, computing, voice, and image. The star bursts in Figure 2.12 represent the growth in the fundamental parameters that govern information technology.

The increase in bits affect the number of bits per word for the average processors and the density and cost of storage. The beginning of this decade marked the rise of the 32-bit supermicro and there is no reason to believe this represents a limit; the achievable word size is related to chip density and available software. Similarly, no end to the progressive increase in storage density is in sight, and the variety of storage media is increasing.

The power of processors, measured in *million instructions per second* (mips), will continue to increase, fueled partly by the increases in semi-

Table 2.11
Trends by Decade

Decade	1960s	1970s	1980s	1990s
Computing hardware	Mainframe	Mini	PC	Super micro/NC
Software	Assembler	High-level languages	Application generation	Object-oriented, reusable
Telephones	Black	Colored	Functions and mobility	Portability/ PC-based
Telephone switching	Electro-mechanical	Computer-controlled analog	Computer-controlled digital	Packet and circuit
Voice processing	Experimental	Simple announcements	Interactive via tones, early DSP chips	Interactive via speech
Telephone network	Analog (twisted-pair)	Analog (coax)	Digital/metallic	Digital/optical, integrated and intelligent, mobile
Computer network	Rare	Modem-based remote terminals	LAN and X.25	IP-based, ISDN, ATM, fast Ethernet

conductor logic speed, partly by the progressive integration of more and more functions onto a single logic chip, and partly by advances in architecture.

Bandwidth or capacity *bits per second* (bps) has not been restricted by technology for some time. Optical fiber has enormous bandwidth capability—but the availability of greater bandwidth is restricted, primarily by the lack of standard methods for interfacing to higher data rates.

Though not directly related to data processing or telephony, the *digital signal processor* (DSP) is finding more and more areas of application. It is mentioned in Table 2.11 in relation to its use in voice processing, where it performs a key role in providing high-speed processing of the digitized signals. DSPs are specialized microprocessors that have a rather different architecture from normal and are characterized by high processing speeds, very high-speed multiplication, and small, specialized instruction sets. A significant measure of the power of a DSP is the number of *multiply accumulates* (macs) it can perform per second. This is an important operation in processing digital signals and is very relevant to speech and image processing. The increasing power of the PC processor is reducing the tasks implemented on DSPs.

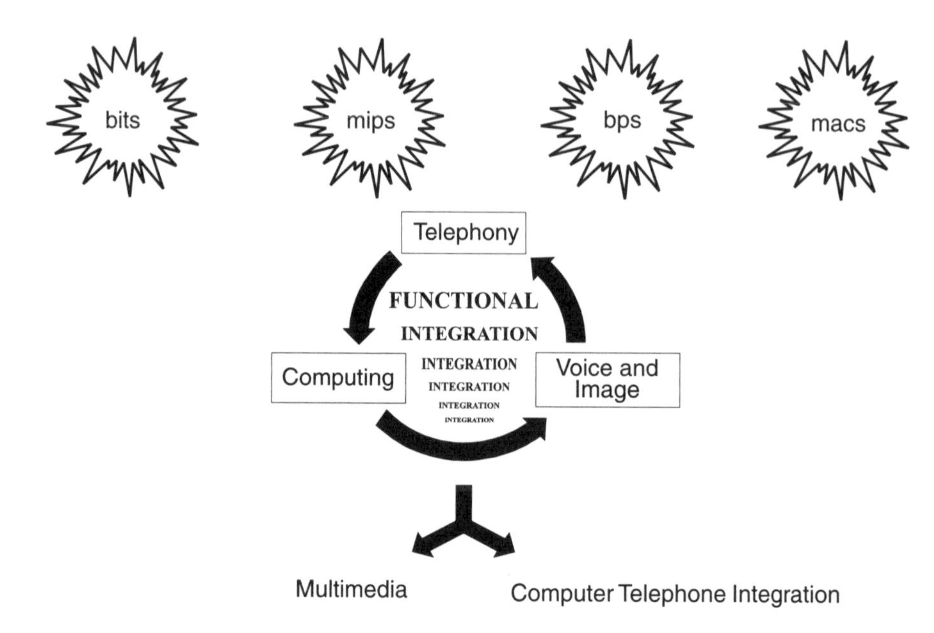

Figure 2.12 Technology trends in the 1990s.

2.5 Convergent Networks

Physical convergence at the network level is rife. The Internet carries telephony and video, frame relay carries voice, and the telephone companies offer circuit-switched solutions.

2.5.1 The ISDN

ISDN standards have been evolving since the 1980s, but it is only in the 1990s that ISDN connections have become available in numbers sufficient to stimulate commercial services across this network. ISDN can be regarded as the final stage of digitization of the telephone network—the stage that extends digital transmission of data or voice to the terminal.

The standards adopted by the ITU for ISDN define two access interfaces: basic and primary rate. The capacity provided by these are described in Table 2.12.

B and D are letters that represent the information and signaling channels, respectively. B channels are transparent as far as the user is concerned, whereas D channels carry multiplexed signaling information for a number of B channels,

Table 2.12
Capacity of ISDN Bearers

Access interface	Format	Information Capacity	Signaling Rate
Basic	2B + D	2 × 64 Kbps	1 × 16 Kbps
Primary for Europe	30B + D*	30 × 64 Kbps	1 × 64 bps
Primary for United States and others	23B + D	23 × 64 Kbps	1 × 64 Kbps

*A further 64-Kbps channel is reserved for synchronization purposes, making 32 in total.

and these can be independently switched to different terminations. The D channel can also carry packet data; this will be directed to the public packet network.

Basic access can, in principle, be provided to any telephone termination, whereas primary access is most suited to PBXs and ACDs. Two primary formats are listed in Table 2.12, reflecting the existing European 32-channel multiplexes and the 24 channels used in North America and elsewhere.

2.5.2 Asynchronous Transfer Mode Networks

Asynchronous transfer mode (ATM) is based upon the transmission and switching of very short packets [7]. It is a powerful technology that can be applied as a LAN or a WAN and combines switching and transmission. According to the ATM Forum, the cross-industry body that strives for the adoption of ATM interoperability standards, ATM does the following:

- Provides a single network for all traffic types—voice, data, and video;
- Enables the creation and expansion of new applications (e.g., multimedia to the desktop);
- Works over twisted pair, coax, and fiber optics and is usable on the existing physical network;
- Delivers its benefits to embedded networks;
- Simplifies network management—by using the same technology for all levels of the network;
- Has a long architectural lifetime; ATM has been designed to be scaleable and flexible in geographic distance, number of users, and access and trunk bandwidths.

ATM provides all of the advantages of both packet and circuit switching. It is based on a cell length of 53 bytes of which 5 bytes form a header so that the cell can be steered to its destination. The cells themselves are usually moved across the network at very high speeds, partly because the interconnections between ATM nodes are themselves high-speed and partly because the switching of ATM cells is done by hardware not software. This means that the network itself has relatively low delay. Moreover, the delay is fixed as it must be for good-quality speech transmission. This is ensured in ATM by allocating a priority to the various input streams such that a fixed bandwidth can be guaranteed for those services that require it; the remaining bandwidth is then allocated as required for the remaining inputs. In this way, ATM satisfactorily combines synchronous and asynchronous data and successfully combines voice, video, and data transmission.

ATM access rates are generally very high; current user standards are around 50 Mbps, although there are some LAN access cards operating at 25 Mbps. The approach is scaleable to any size and to any data rate. The links themselves are likely to operate at 155 Mbps or at the rates specified in the synchronous digital hierarchy of 2.4 Gbps and above.

Broadband ISDN is already well advanced in terms of standardization. B-ISDN is based on ATM, and many telcos are already beginning to build ATM backbones into their public networks. By 1997, many telcos were offering ATM network services at various bit rates. The adoption of ATM as an upgrade LAN technology has become fairly common. The attraction of ATM is that it is future-proof. Furthermore, the cost increment over an upgrade to fast Ethernet is not excessive. Meanwhile, a great deal of effort has been expended in providing solutions—LAN emulation, for example—that allow interworking with an established network. Nevertheless, the dominant solution for data-only applications is Ethernet-based. Ethernet is cheaper and easier to install. Moreover, its performance problems are largely masked by high data rates—supported by 100-Mbps and gigabit per second technologies.

Thus, at present, a viable solution is available for physically integrated voice and data networks. Interestingly enough, ATM is primarily used for data transfer only, although this is expected to change significantly as the end of the century approaches.

2.5.3 Mobile Networks

Mobile telephones use radio spectrum, and although there is plenty of bandwidth available here, it is not always usable. This stems in part from restrictions in radio component technology and from the problems encountered in using certain parts of the spectrum. Mobile telephony and mobile data transfer will therefore be more restrictive and will require efficient management of the available bandwidth for some time to come. On the other hand, mobile services are growing rapidly, particularly digital services that allow almost unrestricted geographical movement, such as *global system for mobile communication* (GSM), and radio LANs that allow simple interconnection of devices within an office.

2.6 Multimedia

This leads to a final word on the subject of multimedia. This chapter covers trends that are specific to telephony and computing; however, one of the exciting trends that has a particular relevance to CTI is the integration of voice, video, image, and data on the desktop. This is multimedia, and the driving force that CTI provides to the roll-out of communicating multimedia systems is addressed in two future chapters, including the last chapter in this book, which looks towards a computer telephone integrated future.

References

[1] Richards, D. L., *Telecommunications by Speech,* Butterworths, 1973.

[2] Coleman, A. et al., *Subjective Performance Evaluation of the RPE-LTP Codec for the Pan European Cellular Digital Mobile Radio System,* Globecom, 1989.

[3] Bear, D., *Principles of Telecommunication Traffic Engineering,* Peter Peregrinus, 1980.

[4] Walters, Robert E., *Voice Information Systems,* NCC Blackwell, 1991.

[5] *SX2000 Product Specification,* Mitel, 1987.

[6] *Meridian 1 Consultant Handbook,* Northern Telecom, 1992.

3

Integration Technology

3.1 Don't Mention Technology

Some years ago I traveled to Canada with an avowed antitechnologist. Having carefully avoided any subject that might remotely touch upon technological matters, we spent much of the journey in a somewhat uncomfortable silence. Upon arrival in Canada, we began a series of meetings that culminated in dinner with the people we were visiting at a luxurious and, no doubt, expensive restaurant in the older part of the city. Immediately after we were seated my colleague made a short speech.

"It is very nice to visit this splendid country, and we wish to thank you for an interesting day. But now it is evening, so we relax. From now on the conversation must not touch upon any technical matters. Anyone who so much as mentions bytes, microprocessors, UNIX, or any related subjects must leave. Now, please enjoy your meal."

This was followed by some querulous laughter, much exchange of looks, and then everyone began to concentrate very intensely, and silently, on the menu. Fortunately, the taking of orders and the serving of wine broke that silence. Conversation began to flow and, as instructed, everyone studiously avoided talking about technology—which would mean talking about work—which was the one thing that we all had in common.

Later in the meal my colleague had to absent himself for a little while—a call of nature. During his absence, rather like schoolchildren babbling away while the teacher has left the class, we indulged in a brief orgy of "technical talk." I seem to remember that we spoke of superconducting, neural-network-based, optical computing—this being the most depraved area of technology that we could think of at the time!

In time my colleague returned and the conversation settled back into more everyday matters. You may wonder why this person had so much control over the content of our discussions. Suffice to say that he paid the bill for that very enjoyable evening meal. There is no such thing as a free lunch!

3.2 Introducing Small CTI Technology

What my erstwhile colleague would really like about small CTI is that there is not a lot of technology. So, why does this book include a chapter on technology and what is it all about?

Naturally, much depends upon your definition of technology. My colleague regarded everything that he did not understand as technology and people whom he did not understand as technologists. Anything new brings with it a new vocabulary or, worse still, a redefinition of current vocabulary. Terms like *third party* and *first party, discrete* and *merged,* are not new, but they are used in a particular sense in CTI. Defining *objects* is by no means new, but the telephony objects that a computer sees and can manipulate in CTI are new in the sense that they have not been viewed in this way before; they have therefore not been named in the way that they will be here. It is said that "the beginning of wisdom is to call things by their right name [1]." In some areas of CTI, that beginning is still awaited.

Small CTI is not a technology, though many people call it so. It is, exactly as its name suggests, an integration. When a company produces an alarm clock that is integrated with a radio and then calls it a *clock radio,* no one regards this as a new technology. It is simply an application of existing electronic technology, the physical integration of an electronic clock with a radio. The clock radio provides an interesting analogy in that, just like CTI, the integration here is also functional. Producing a clock radio is not just a matter of putting the two elements into the same case. The clock turns the radio on and off and, simple though the action may seem, this is a functional integration.

The underlying technology in CTI is also electronic—primarily digital electronics. However, in this case the integration is not at this basic level; the integration is between the technologies of telephony and computing, which is why Chapter 2 introduced the technologies of those two different worlds.

The technology of integration is the means by which the computer and switch are linked, the mechanisms used to communicate across those links, and the programming interfaces and tools that enable the creation of integrated applications. These are the areas that are explored in this chapter. The technology of automation is discussed in Chapter 4.

3.3 Integration Architecture

Chapter 1 introduced the taxonomy of CTI, breaking the topic down into first-party CTI and third-party CTI. These are the basic architectures of CTI, and it is beneficial to list them here once more—together with their subdivisions—as a precursor to a more detailed discussion.

- First-party CTI:
  - Basic;
  - Enhanced.
- Third-party CTI:
  - Compeer;
  - Dependent;
  - Primary.

3.4 First-Party Architecture

Figure 3.1 illustrates first-party CTI architecture. The implementation may be basic or enhanced—this will be determined by the signaling system supported by the switch. The important thing to remember about this form of integration is that it is based on intercepting the line signaling that would normally pass between the telephone and the switch and vice versa. It is therefore provided on a single-line basis. Merged first party is shown within the dotted lines in Figure 3.1. This is primarily a packaging issue. The "guts" of the telephone become part of the PC.

The components of a first party system are

- Application programming interface;
- Application program;
- Switch;
- Computer;
- Telephone;
- Terminal;
- Attachment interface;
- Signaling intercept;
- Line signaling.

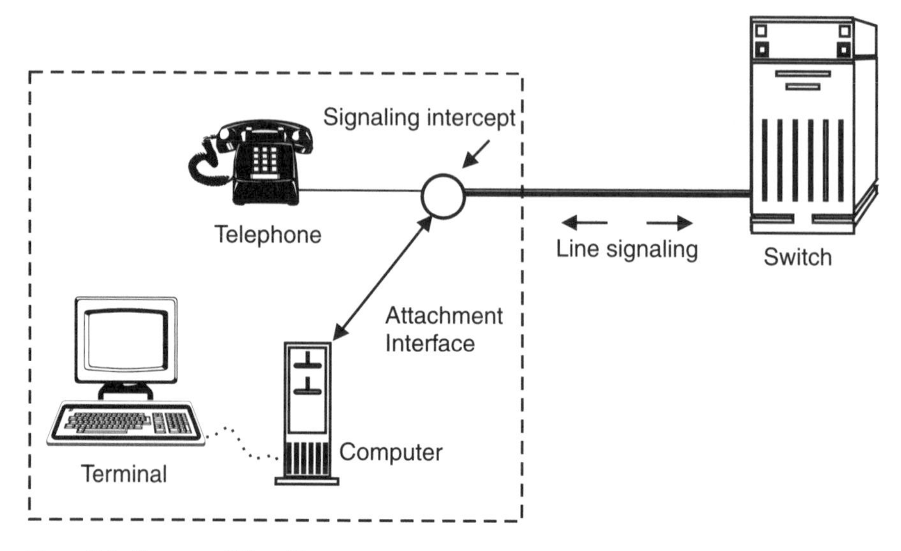

Figure 3.1 First-party CTI architecture.

Each component deserves separate consideration.

3.4.1 The Application Program and API

These items will be covered extensively later. Suffice to say here that in first-party CTI, the application copes only with one telephone line so, to a large extent, the application software is simpler than that required for third-party CTI.

3.4.2 The Switch

The switch can be any telephone system. Because first-party CTI generates and receives the same signaling that the telephone would transmit or receive, the switch is usually not aware of the connection.

3.4.3 The Computer

The computer will normally be a PC of some sort—though this is not necessarily so. First-party CTI can be implemented on a computer that deals with many lines and many terminals. However, even when this is so, there will be as many connections to that computer as there are lines.

3.4.4 The Telephone

Types of telephones were discussed in Chapter 2. The feature phone and key phone belongs to first-party enhanced and the conventional phone to first-party basic. Everything depends upon the nature of the connection to the PBX. Feature phones require enhanced signaling capabilities and are generally proprietary. First-party basic architecture is therefore appealing; it can be applied to any switch, since almost all switches provide an interface for conventional phones. But there is a danger in taking that approach. Many users already have feature key phones of various types. They probably like them. In any event, they can be very unhappy if, to gain the benefit of CTI, they lose facilities from the feature phone.

Furthermore, the feature phone signaling system is much more extensive than the POTS capability of basic first party. In first-party enhanced, the extra facilities are made available to the computer. Key system signaling particularly provides status information on the exchange lines and on other users. Hence, the computer knows far more and can display the information it has on the screen. This is particularly relevant for workgroup applications. The screen can show the current status of all other users—via a telephone icon attached to the user's name. This increases the power of first-party CTI enormously.

In merged first-party, the telephone is usually reduced to a single handset or microphone and speaker. The other telephone parts are absorbed into the computer (usually as a card) or are provided by a sound card.

3.4.5 The Terminal

In most cases the computer in first-party CTI is a PC, so the terminal is simply a PC monitor, keyboard, mouse, and any other relevant peripherals. There are a number of instances in which special terminals have been designed for merged first-party CTI.

3.4.6 The Attachment Interface

In first-party CTI the attachment interface (AI) shown in Figure 3.1 can be very simple. Physically, it can be a conventional serial interface (RS-232) or the parallel interface, USB, etc. The control protocol that allows the set-up and clear-down of basic calls may be the AT set used for modem autodialing (often called the Hayes command set). The AT set (AT stands for attention) is quite comprehensive; it does in fact provide functionality considerably in excess of that required for simple autodialing [2]. It allows for the storing of telephone numbers, speed control, flow control, and so on. Basic call control is supplied by the dialing command that, in common with all other AT commands, uses a very simple protocol. Commands are commenced with AT and terminated by a "Return" or "Enter" character. To call my telephone number, a computer would need to send the following command to a modem:

++ATD01865208930

Where "D" is the Dial command, there are a number of parameters that can also be embedded within this command—for example, the use of the "T" or "P" parameters to indicate tone or pulse dialing.

Feedback is also very simple. The modem will respond with one of the following:

- CONNECT (Speed);
- BUSY;
- NO ANSWER;
- NO CARRIER.

International standards do exist for this form of control. The CCITT V.25 bis recommendation defines the protocol. It requires that commands are sent in this general format:

COMXX;YY—YYZST{Return}{Line Feed}

Not all of the fields have to be present. Those shown are defined as the following.

- COM: A three-letter mnemonic;
- XX: Two digits that specify the location of a stored number;
- YY—YY: Up to 18 digits for the telephone number;
- ZST: Three qualifiers that are used only in responses.

Examples of commands and responses from this simple protocol are listed as follows.

- CRN: Call number;
- CRS: Call stored number;
- CIC: Connect incoming call;

- INC: Incoming call;
- CFI: Call failure indication.

So, for example, a call to my number would be initiated by the computer as

CRN01865208930

And the response might be

CFIDT

The call failed because no *dial tone* (DT) was detected.

These are examples of very basic protocols that can be used on the link for simple call control. Both protocols are based on ASCII characters passed across an RS-232 connection. There are other possibilities. In the final analysis the link and the protocol are often determined by the signaling intercept, which is the next component to be described.

3.4.7 The Signaling Intercept

Looking again at Figure 3.1, the signaling intercept is the point at which the computer can:

- Observe the signals passing between the telephone and switch;
- Insert signals to be sent to the switch;
- Divert signals from the switch to the telephone itself.

Figure 3.3 provides a more detailed view of the signaling intercept and its functions. Note, however, that implementations vary considerably. In particular, in the simplest implementations, the signaling intercept may merely allow the computer to monitor the line. The letters "T" and "S" in the figure label signals from the telephone and the switch, respectively.

The implementation described by Figure 3.3 provides facilities for the telephone to be isolated from the switch and for the computer to insert signals into the line, if required. It can either monitor the telephone and the switch independently or observe the signals that pass between the two of them. This is probably the most sophisticated form of intercept. Because it can obtain access to the line directly, it can insert signaling information *and also* play voice

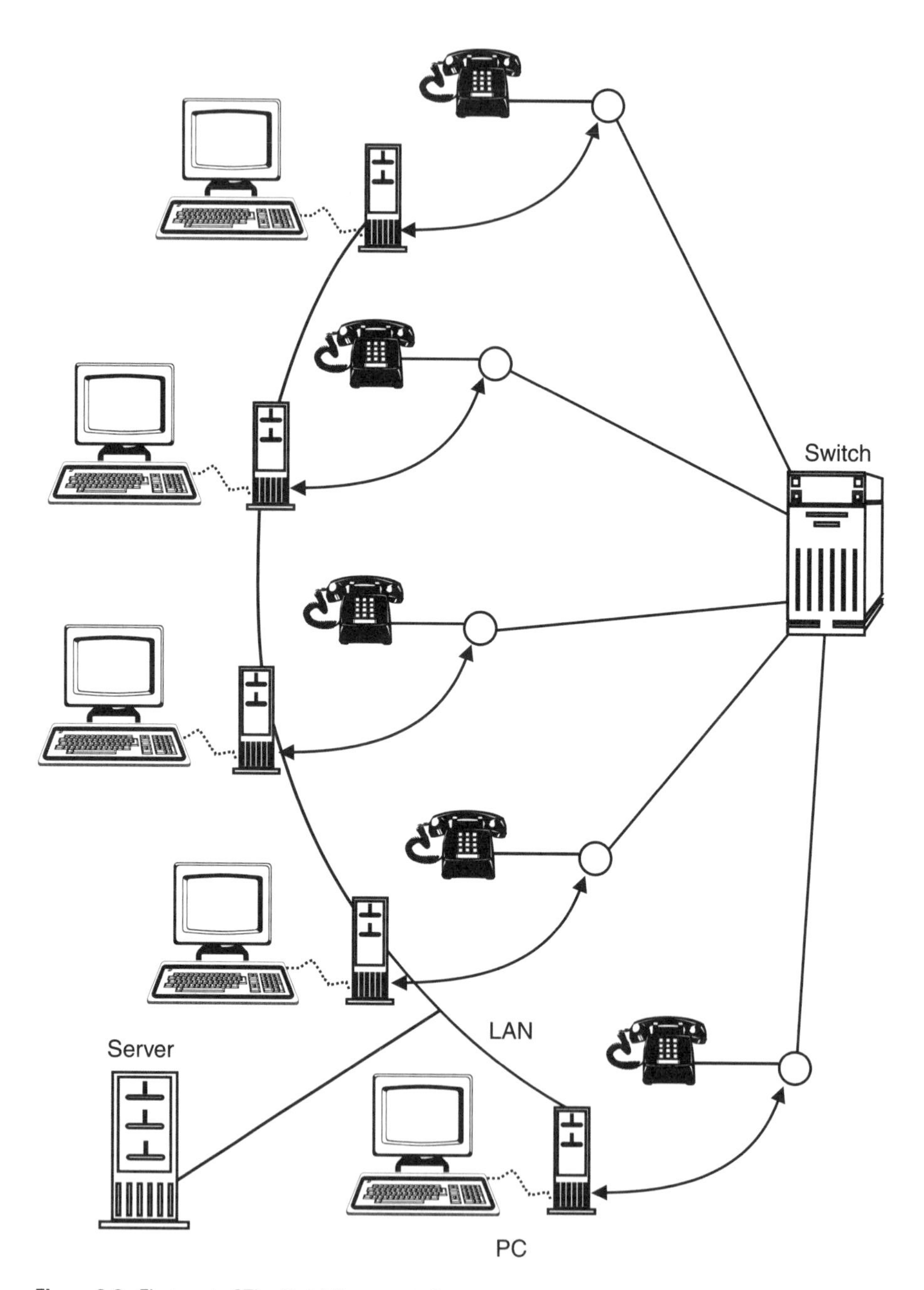

Figure 3.2 First-party CTI with LAN connected computers.

announcements, if these are available. Similarly, it can connect the line to a voice recording mechanism and so automate call processing if the telephone user is absent. Simpler implementations have data-only access.

Here are a few of the ways in which a signaling intercept can be provided:

- PC cards;
- Data feature phone;
- Stand-alone units;
- Complete systems.

Each of these will now be examined in this role in a little more detail in Sections 3.4.7.1–3.4.7.4.

3.4.7.1 PC Cards as Signaling Intercepts

Here, the PC absorbs the intercept as a plug-in card. Probably, the first card that I used as a first-party intercept was the Callbox XM. It was very basic, offering little more than outdialing. There are, of course, a number of PC modem cards that offer this function—though the support software is not so sophisticated. The Callbox card allowed a PC user to dial from any number on the PC screen. It did not monitor the progress of the call; it simply connected the line to a monitor loudspeaker and left that to the user. It was little more than an autodialer but, for all that, offered functionality that approaches the definition of first-party CTI. It is interesting to note that it achieved a degree of functional integration that was entirely application-independent. Callbox XM obtained its information from the memory area of the PC containing the screen image. The application was activated by a "hot key," which froze the program that was running—a database manager, for example—and searched the screen image for anything that resembled a telephone number. The user was then asked if this is the number to be called and, if the answer was yes, the card pulsed out the number. Of course, it would quite happily pick up the height of Mount Everest (29,078 feet) if that happened to look like a telephone number! Maybe this is the price of application independence. The Callbox XM card did not provide the switches shown in Figure 3.3. It could, however, send tone digits to line; it did this by bridging the line rather than breaking it.

Cards like Startalker (similar to Bigmouth in the United States) are much closer to the signaling-intercept functionality shown in Figure 3.3. More than this, they proceed to the next stage of functionality—the ability to play voice messages to line and to convert received speech into digital form and thus into files for storage. Some of the cards go yet a step further. Not only can they recognize incoming tone digits, they can also recognize speech as a command input. Examples of the more advanced cards include Best Data's ACF 5000 and Computer Telephony Ltd's ST32.

In general, PC cards form an excellent solution to the provision of signaling intercepts—especially the single-line variants. The level of integration is good; the link shown in Figure 3.3 is contained within the PC. Communication between the computer and signaling intercept (PC and card) is across the PC bus. Most cards provide some form of application programming interface that presents the programmer with a comprehensible set of command and status messages that translate to and from the line signals. Following the introduction of Microsoft TAPI, most suppliers accepted this as an interface.

On the whole, a PC card provides a physically neat solution. To install the Callbox card, you plug it into the PC in the normal way, transfer the telephone jack from its wall socket into the socket on the card, and plug the jack on the end of the cord connected to the card into the telephone wall socket. Only basic telephones can be used, of course. Feature phones are not compatible with PC cards designed for the general market. However, some switch suppliers are now supplying PC cards of their own that do allow access to the proprietary enhanced signaling that is utilized by the suppliers' feature phone range. And some PC card manufacturers are designing cards that are compatible with some of the more widely used feature phone interfaces—mainly to provide integration for voice messaging systems.

One of the most important developments in first-party CTI is the availability of intercepts for key systems. All key phone interfaces and protocols are proprietary, though key systems usually provide a POTS interface capability to allow the connection of conventional telephones, modems, etc. This interface costs more than the key phone interface and, not surprisingly, provides less functionality. Key phone signaling is message-based. It is carried on a separate signaling channel and provides all of the functions available from a POTS phone plus the ability to manipulate multiple lines and observe the activity of other key phone users. The keyphone interface lets the user see more of what is going on and allows the user to do more. The interface is usually digital and is superior to the POTS interface in all respects, except that it is proprietary.

Citel, a United Kingdom-based company, began to penetrate the barrier set by proprietary key system interfaces in 1995. It contracted with most of the major key system suppliers to provide a screenphone solution (card, software, and handset) with the understanding that the suppliers would release details of their interfaces. Citel acts as an OEM in the supply of screenphones. The key system suppliers specify the user interface, and these vary considerably from one supplier to another. In the simplest examples, the screen looks very much like the keyphone that it replaces. In more imaginative implementations, an entirely new layout is used. Such layouts incorporate all of the functionality of the keyphone and extend the interface into a form more suited to the PC screen. In one implementation, other telephones in the user's key group are shown as icons on a mini screen. They can be moved and manipulated as graphical objects and, therefore, grouped in any way that suits the user. Calls can be dragged from an incoming call mini screen to any of these telephones, and calling another member of the group is achieved with a mouse click on the relevant telephone icon.

The Citel product is sold in a box that is badged by the key system supplier and distributed through the normal key system channels. The box contains a handset, the intercept card, and the software on a disk.

Although single-line PC cards are fairly cheap, the expenditure is on a per line basis and therefore can be compared directly with the cost of the PC, the telephone, or both. In this context the PC card solution loses some, but by no means all of its attractions.

3.4.7.2 Phones as Signaling Intercepts

Here the phone absorbs the intercept. However, by far, the majority of phones are hermetic. They present no interface to the outside world beyond their connection to the line. Yet first-party enhanced CTI requires access to the facilities that exist within the phone's signaling system.

There are many ways of providing data access via data phones. These provide a data path to the switch in some way—often digitally but sometimes using data over analog voice or simply a modem. However, this is not important in the context of signaling intercepts. What is important is that data phones support a data connector, usually RS-232-compatible, sometimes X.21-based. The data phone may also be a feature phone or it may connect to a POTS line.

In very simple applications the data call is set up via the telephone keypad. These are of no practical value to first-party CTI. The phones that are of interest here are those that allow access to signaling information. When this is made available (via an RS-232 port, for example), direct connection can be made to a PC or any other computer—at no extra cost. No cost, that is, apart from the software driver needed within the PC to interface with the call control protocol supported by the phone. This will probably be an AT or V.25 bis protocol, as described above. It may therefore be rather limited in scope—suitable for autodialing applications but not for more sophisticated transactions. What is needed for this level of functionality is an extended AT command protocol, access to D channel signaling in an ISDN environment, or full access to feature phone signaling. Connecting via a phone data port does not, of course, provide access to the voice channel—this is managed by the telephone itself. However, some specialized data phones also provide this level of access.

Examples of data phones abound, although they are by no means the most popular of telephone variants. Most PBX suppliers provide a feature phone with data capabilities. These feature phones are usually digital. For example, Nortel's Meridian Business Telephone, working with the Meridian 1 PBX, has an optional communications adapter that provides data and control capabilities. This is a proprietary phone—as are most PBX feature phones.

Many PBXs do offer a standard ISDN interface on the extension side, and there exists a whole range of ISDN digital telephones that can be used with these PBXs or on a basic rate direct line. Many of these digital telephones have a data option—usually supplying an X.21 interface that gives access to call control.

Some analog telephones have in-built data capability, and some have been designed with first-party CTI in mind. One such telephone is the *Special Plain Ordinary Telephone* (SPOT), a phone with an in-built modem and serial port connection that is produced by the UK company SDX. The first SPOT simply supported call set-up from the screen. However, a more advanced SPOT version also provides voice messaging, data modem, and fax capability—all of this without the need to plug a special card into a PC. The phone works with any analog line, and applications can be written to use Hayes AT commands or Microsoft TAPI.

USB telephones can also supply the first-party line intercept—and take advantage of the USB's ability to carry digital speech. Mitel produces a USB telephone, the Mitel Personal Assistant.

3.4.7.3 Stand-Alone Units as Signaling Intercepts

There is no practical reason why the circled signaling intercept of Figure 3.3 should not be provided as a separate unit—with its own power source and connections to the telephone, switch, and computer. For first-party basic CTI, all the connections are standard. The company that produced the Callbox XM PC card also produced such a device and found many applications for it. Dubbed Callbox 2, it has the functionality implied by Figure 3.3. This device has extended telephony facilities, including the provision of a headset and loudspeaker/microphone for hands-free operation. It performs outdialing, digit detection, and call progress monitoring within the box. The attachment interface to the computer is, of course, RS-232. The protocol offered is an extended Hayes AT command set. The basic protocol was described in Section 3.4.6. Table 3.1 provides a few examples of instructions that the computer might send to this device. Terse, isn't it!

Table 3.1
Examples of the Commands Issued by the Computer to Callbox 2 [3]

Command	Meaning
ATS32 = 1 V1 [E1	Set to Earth loop recall method; enable verbal results codes; set to headset operation
ATQ3 D9,0600714478	Go on line and dial 9, wait for secondary dial tone, then continue with the rest of the number; monitor line for outcome, then send result code as appropriate
ATH0	Hang up

One or two points in Table 3.1 might require clarification. S32, in the first command simply happens to be the register that contains the configuration information for the dial and recall method to be used. S32=1 sets it to pulse dial and earth loop recall. The reference to "verbal results code" indicates that responses should be given in full rather than as a number—"Ringtone" rather than "18," for example. There are 20 status messages (results codes). In the second command, Q3 indicates that enhanced responses are required. The rest is self-evident, I think. In the last command, "H" does not stand for hang up; it stands for *hook state.* The "0" following the "H" indicates *hang up.*

Installation of this form of signaling intercept is very straightforward. If the original telephone is not to be retained, you unplug it, plug in Callbox 2 instead, then connect Callbox 2 to the computer and mains supply. If the original phone is to be retained, then it is plugged into Callbox. In either case, you use the relevant CTI software. There are many products similar to Callbox; some of them connect to the parallel port of a PC.

3.4.7.4 Complete Systems as Signaling Intercepts

The use of an entire system as a signaling intercept may seem an odd concept, but it has been used and it does have some merits. The history of CTI was covered in Chapter 1, and it included a reference to the Mezza system. The Mezza architecture made the computer itself act as the signaling intercept. The computer was special in that it dealt with voice and data. Called an *information manager,* its role as a multiline signaling intercept is demonstrated in Figure 3.4. Here, each of the analog switch connections is transformed to digital as they traverse the information manager. Within the information manager, the role of computer and signaling intercept are physically integrated.

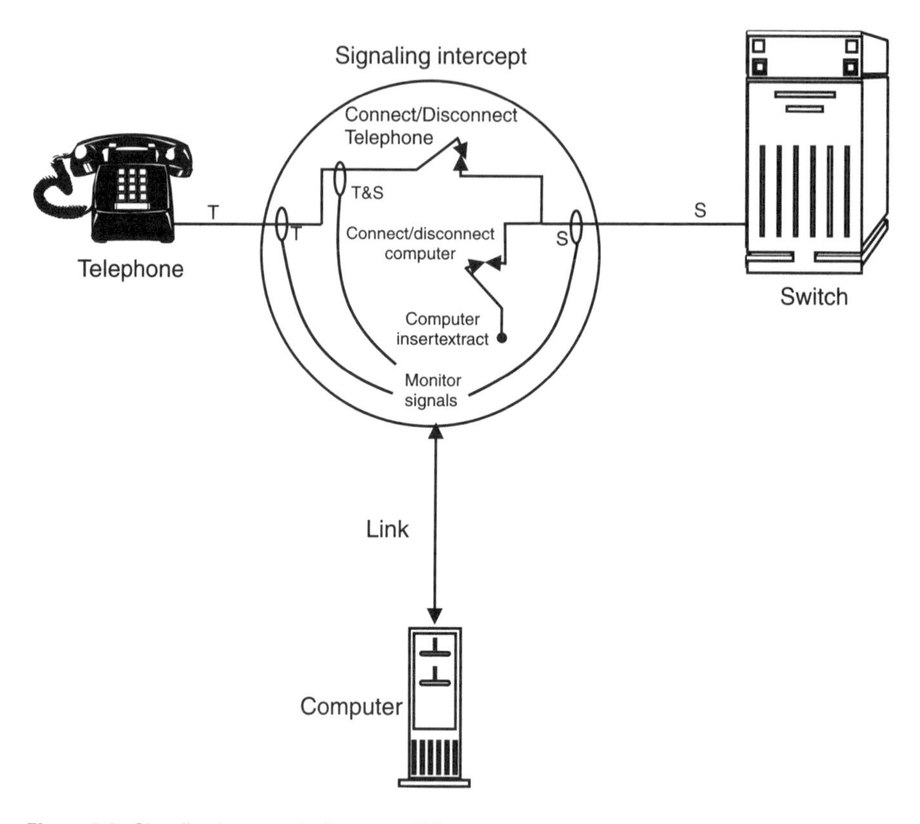

Figure 3.3 Signaling intercept in first-party CTI.

The Mezza approach provides complete control over the lines that pass through it, so anything that can be achieved via a normal telephone with regard to telephony functions could be controlled by the information manager. This should be advantageous, but the architecture proved difficult to change and the provision of special terminals proved prohibitively expensive.

3.4.8 Line Signaling

Conventional line signaling has been described in Chapter 2. There are not many signals involved. To review, a normal telephone produces the following signals:

- Off hook/on hook;
- Digits;
- Recall.

The POTS interface from the switch produces the following:

- Ringing current;
- Supervisory tones (dial, ring, engaged, unobtainable, equipment engaged, etc.);
- Reversals to indicate called-party answer (sometimes).

These are the line-signaling conditions that the signaling intercept has to produce and detect in first-party basic. Though these signals differ around the world, the differences are not major and most signaling intercepts can be programmed to cope with the variations. Major variations that do occur concern the recall signal (earth or timed line break) and the supervisory signals. Anyone who makes a significant number of international calls can describe how bewildering the tones of other countries can be. Over time, the POTS interface improves. It is now common for a POTS interface to provide an indication of calling line identity (CLI)—where known. CLI is often sent via modem like signals prior to ringing or with a call waiting indication.

First-party enhanced CTI is a different matter altogether. Often based on the line signaling provided for feature key phones, each switch supplier has its own signaling system. Often these are not published, which means that only the switch supplier can provide the signaling intercept. In fairness to the switch suppliers, this is not a situation of their own making—standards for this part of the signaling world do not exist. It is not an area to which international standards bodies have directed any effort.

Fortunately, with the advent of ISDN, that is beginning to change. The ITU has provided a specification of the signaling system to be used in the D channel of ISDN lines. This signaling system is called *digital subscriber signaling system No. 1* and, as its name implies, this system caters to signaling from ISDN-compatible terminals. The recommendation number for this standard is Q.931, and it represents a standard basis for enhanced signaling to the telephone and hence a firm basis for first-party enhanced CTI. Q.931 is an ISO-conformant layer 3 specification for basic call control; however, it does allow for the inclusion of supplementary services, and other standards bodies are enhancing it in this way—the *European Telecommunications Standards Institute* (ETSI) has already published a superset of the protocol called Euro ISDN (ETS 300/102).

Q.931 is a message-based signaling system. Table 3.2 indicates the type of messages it defines. Note that this protocol allows for user information to be transferred. It therefore provides an alternative route for the supply of computer-related information (the current database record the caller is view-

Table 3.2
Examples of Q.931 Messages

Message Set Type	Examples of Messages
Call establishment	Alerting
	Call proceeding
	Connect
	Progress
	Setup
Call information phase	Resume
	Suspend
	User information
Call clearing	Disconnect
	Release
Miscellaneous	Congestion
	Facility
	Information
	Status

ing, for example). What may not be clear from Table 3.2 is that this information can be conveyed before the call is answered. This can therefore be used to provide a communication channel between PCs that are involved in the same call.

3.5 Networked Computing and First Party

If the first-party computer is a PC, it may belong to a network of similar devices connected by a LAN and sharing network servers that provide common services to the community of PCs. Figure 3.2 provides an example of this more complex arrangement—not an uncommon one by any means, since the majority of office-based PCs are networked. (Note that the server shown in Figure 3.2 does not perform any telephony-related functions. Later in this chapter the topic of client/server CTI is introduced, and then the server *is* intimately involved with telephony. However, this is a form of third-party CTI and should not be confused with the discussions that follow.)

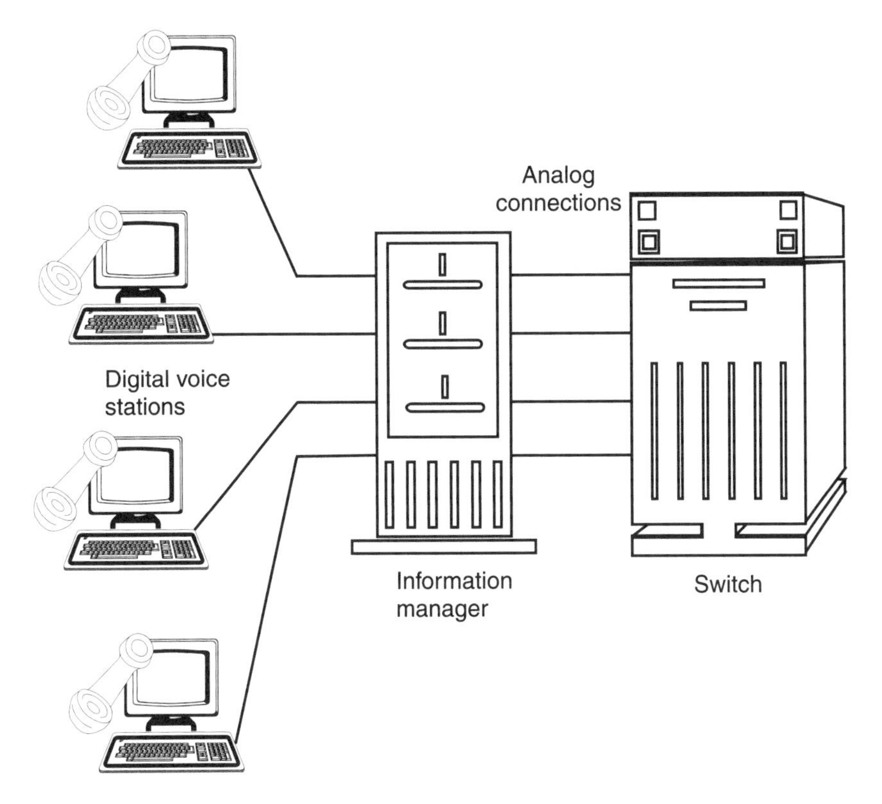

Figure 3.4 Multiline signaling intercept.

Of course, the networking of PCs and the provision of a server, as shown in Figure 3.4, may have no bearing at all upon CTI—each PC may deal with first-party CTI in an entirely independent manner. But, because the PCs are networked they may, of course, communicate information about telephony—simply by sending messages to one another across the LAN. Whether or not they do so depends upon the applications that exist in each PC. If the applications do provide a networked CTI capability, many things are possible. Consider the following scene.

A call arrives at the PC of Hugh Daglish. The screen brightens and a flashing telephone icon appears in one corner, together with an audible tone. Hugh moves the screen cursor to the telephone icon and double-clicks his mouse. A window appears, providing the following information:

Incoming call on line 1
Calling Party — Not Known
Call Type — Incoming Trunk Call from PSTN
Associated Data — None
Auto Answer ❑ Ignore ❑ Minimize ❑
Transfer ❑ Park ❑ Features ❑

Hugh lifts the telephone receiver and answers the call; the first line of the window changes to "Call in Progress." The call is a help desk inquiry. Hugh obtains the customer account number, clicks on "Minimize" to clear the telephone window from the screen and enters the account number into the help desk management program. Data on the client appears on the screen.

Hugh says, "Now let me see. You are Harvest Foods and you have a Mark 7 control system. How can I help?"

"We can't control the temperature on the fifth floor. It is over 30° in there and people are getting pretty irate. We've tried all the usual things but nothing helps. Soon we will have to close down that floor and relocate the staff."

Hugh types some of the details into his PC, then selects an inventory of Harvest Foods' equipment.

"There's a master controller in room 304 on that floor. Have you reset that?"

"Yes, of course. We've reset it from the central console and we've even walked over to reset it manually. No effect at all!"

Hugh clicks on the telephone icon. The telephone window appears.

"I'm going to talk to Jim Clow. He knows your set-up and is an expert on Mark 7 controllers. Just hold on for a minute."

Hugh clicks the transfer button. A directory appears. He clicks Jim Clow's entry and the window changes to reflect the changed status. Meanwhile, Jim's screen flashes and he immediately clicks on the telephone icon. This is what he sees.

Incoming Call on Line 1
Calling Party — Hugh Daglish
Call Type — Inquiry. Caller also has Incoming Trunk Call
from PSTN
Associated Data — Data Record for . . . . . . Harvest Foods
Auto Answer ❑ Ignore ❑ Minimize ❑
Transfer ❑ Park ❑ Features ❑

Jim answers the call, and the data on Hugh's screen also appears on his screen. They briefly discuss Harvest Foods' problem and decide that Jim should take over. Jim then clicks on "Features." He selects "Pick-up Held" and begins to talk to the caller. Hugh is now out of the call.

This is a classical CTI feature and one that is usually associated with third-party implementations. How is it achieved in first-party CTI?

From the computing angle, what happened is not particularly difficult to achieve—although it does raise some interesting questions with regard to data security. When Hugh initiated the inquiry call to Jim, a message was sent across the LAN that provided a complete picture of Hugh's current status—in telephony and computing terms. Hugh is actually accessing the Harvest Foods data from the central server. The message sent to Jim contains enough information to allow Jim's PC to select the same data record when he answers the call.

That explanation covers the database transfer. Now what about the telephony side of the transaction? Remember that the only link the computers have to the switch is via normal line signaling. When Hugh initiated the inquiry call to Jim, Hugh's computer caused a recall signal to be generated. This causes the incoming caller to be placed in the held state, that is, not connected to any other party, simply waiting while the party to which they were connected does something else. Hugh's PC then generates the necessary digits to indicate that Jim's phone should be rung. In first-party basic, Jim's PC will receive very little information about the call. At best it may be able to discriminate between local and trunk calls. So how then did the PC know that the call was from Hugh and that Hugh had another call held, as shown on Jim's screen?

The answer is simple. All of the telephony-related data is carried in the message originally sent to Jim's PC—across the LAN. The switch is not involved at all in the transfer of this data!

Now, at the time that Jim and Hugh were talking, the customer was held, probably listening to music! This is a normal inquiry call, but it was initiated by Hugh. Usually, if the inquiry is such that Hugh wants to transfer the call to Jim, Hugh would replace his telephone and Jim would be automatically connected to the caller by the switch. But note that here it was Jim who requested "Pick-up Held." This is something that cannot normally be done in telephony; the held call belongs to Hugh. So, what is happening? Once again the LAN is used to send a message from Jim's PC to Hugh's PC. This message causes Hugh's telephone line to be cleared.

This example has been considered in some detail to demonstrate the potential of first-party architecture. It is a potential that aligns well with trends in computing—toward personal work stations, distributed computing, and so on. It does, however, require the existence of effective distributed software—

effective from the compatibility, manageability, and performance viewpoints, particularly the latter two. An alternative method of producing similar capability across a group of PC users connected to LAN will be considered in a later section on third-party CTI.

3.6 Third-Party CTI Architecture

Third-party CTI in its most basic form is shown in Figure 3.5. There are many installations that possess this simple architecture. The components of third-party CTI are very similar to those of first-party CTI, but in Figure 3.5 more emphasis is placed on the CTI link between the computer and the switch. In third-party CTI this is often called an intelligent link. This link usually operates at a high level, carrying messages between the processors of the switch and the computer. For this reason, and because the link is so key to the implementation of third-party CTI, Figure 3.5 provides a more detailed illustration of the CTI link and its communication stacks.

In third-party CTI the only physical component that is added to the computing and telephony system is the link, hence its importance in this context. These are the components of a third-party CTI configuration:

- Application programming interface;
- Application programs;

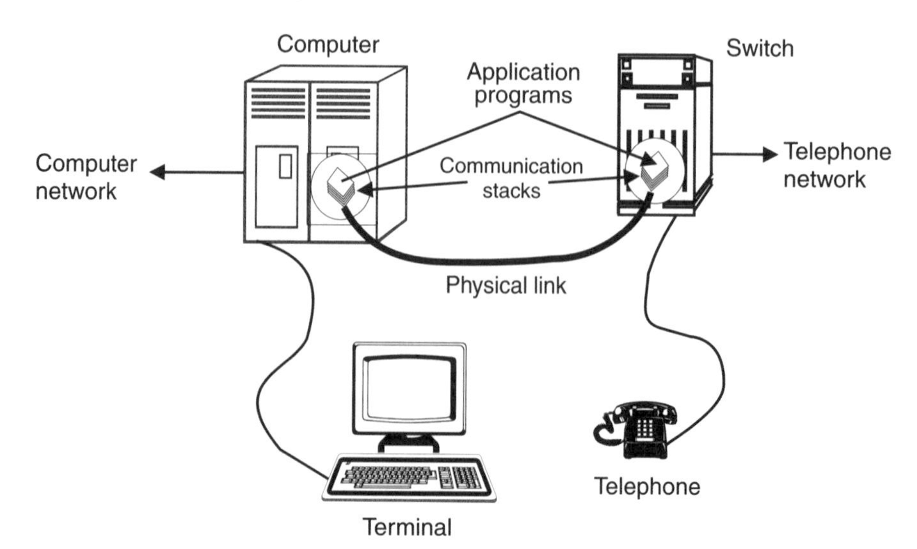

Figure 3.5 Third-party CTI architecture.

- Switch;
- Computer;
- Telephone;
- Terminal;
- CTI link.

The following sections study each of these components in sufficient detail to fully introduce this architecture. Then, some of the implementations available are examined.

3.6.1 The Application Programs and API

These items will be extensively covered later. Sufficient to point out here that, in contrast to first-party CTI, in third-party CTI there are at least two application programs—one in the switch and the other in the computer—as shown in Figure 3.5. Thus, to a greater or lesser extent, the switch is aware of the existence of the computer and its connection to it. Naturally, this awareness is greatest for third-party dependent CTI, in which the switch needs the computer for its operation, and least for third-party-primary CTI, in which existing interfaces are utilized.

3.6.2 The Switch

Any of the switches listed in Table 2.7 can be used—with the proviso that they have a suitable link or the interfaces to enable third-party-primary CTI . Those most commonly used are the PBX and *automatic call distributor* (ACD). However, where Centrex is a common alternative to a private switch, there are a number of central office–based solutions.

Physical linkage is not usually a problem, but the switch software does generally have to be upgraded to provide CTI capability. This means that, even where a particular switch model is known to support CTI, a particular installation may not. The CTI software is usually offered as an option.

Many switches are networked. In telephony this means that they are linked in some way—usually by fixed transmission paths. At its simplest, networking provides alternative methods of connection over that provided by the PSTN—a private network. However, where advanced private network signaling systems are used, the private network offers functional advantages. Telephony facilities, such as transfer and conference, can be made available across the network and, in the most comprehensive networks, a collection of switches can appear as one switch. This raises a question: where full telephony networking is provided, is the CTI capability usable networkwide? This question is addressed in Section 3.9 on distributed third party.

3.6.3 The Computer

It is conventional to present third-party CTI in the form shown in Figure 3.5, and many installations can be just like this. However, the computer shown is by implication a mainframe or mini, whereas the norm in computing is client/server operation. Figure 3.6 provides a picture that is representative of many installations. The telephones have been omitted for clarity.

Note that Figure 3.6 shows the switch connected to the server. This is simply an example; the CTI link may well be physically connected to the LAN (if the switch has a LAN interface) or it may be connected to one of the PCs. In any event, this does present a rather different architecture to that suggested by the simple approach taken in Figure 3.5. However, some reflection will show it to be a simple extension of the centralized approach of Figure 3.5. Figure 3.6 embodies the client/server approach, in which the telephony server extends telephony functionality over the LAN to the PC-based clients. It must not be confused with switch networking, which was discussed earlier. In client/server CTI, there are many computers controlling a single switch, whereas in switch networking, there are a number of switches controlled by a single computer. The question here is: Can any of the computers send CTI messages to the switch? The corollary of that question is: How does the switch direct its responses to the correct computer? These issues are discussed under "CTI Protocols."

3.6.4 The Telephone

Any telephone supported by the switch can be used with third-party CTI. There are no physical restrictions what-so-ever here; any limitations that might occur will be traceable to the implementation of the CTI software in the switch.

3.6.5 The Terminal

Any terminal can be used. There are no restrictions here either—except those imposed by the computer and its software. Naturally, the most common terminal is a PC.

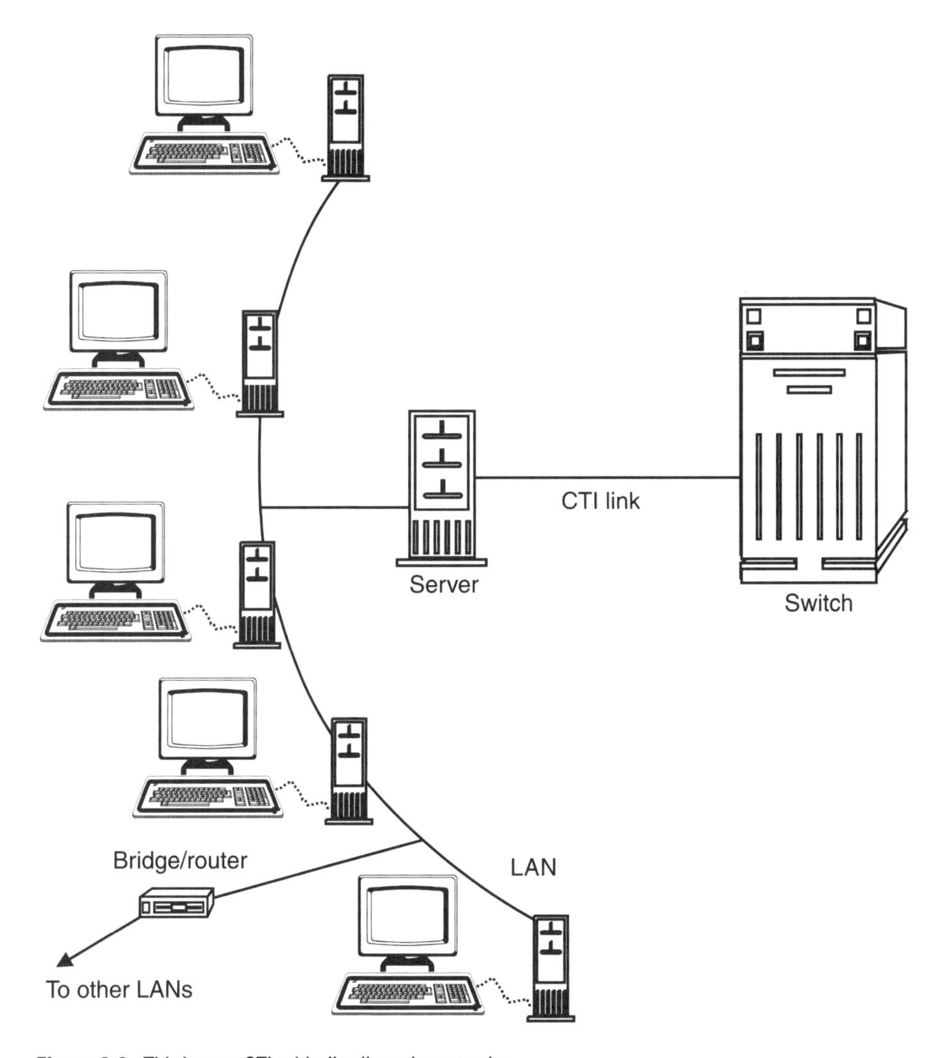

Figure 3.6 Third-party CTI with distributed computing.

3.6.6 The CTI Link

The link has three distinct components: the physical link, the communications stack, and the protocol.

3.6.6.1 The Physical Link

Often no special connections are required to provide a third-party CTI link. Most switches and computers have serial interfaces and these can be used to provide the link. Hence, the physical connection was often made via a bulk-standard RS-232-type cable. Some switches connect directly to a LAN and therefore provide Ethernet coaxial connection, for example. For small installations, the requirements of this link are not especially exacting. It will be shown later that data rates of around 9.6 Kbps are usually quite adequate, and most protocols provide for error detection, so error free performance is not essential.

The switch and computer are often collocated so a simple cable connection is all that is required. If they are not, all the usual means of remotely connecting terminals can be used—from base-band modems to a channel within a data network. If the switch is PSTN-based, long links are obviously required. Some Centrex-based implementations use an ISDN channel for the link.

3.6.6.2 The Communications Stack

The communication stack provides the basis for error-free communication of messages between the computer and switch application programs.

Most CTI implementations use a layered architecture, and most use standard interfaces at the lower layers, even though the high-level protocol is itself proprietary. At the physical layer, V.24 (RS-232) was the most common specification but was supplanted by Ethernet. HDLC is common at the link layer and X.25 is sometimes used at layer 3 although IP is the most popular. Above this, it is difficult to generalize, though many of the available protocols have null layers between layer 3 and 7—no transport, presentation, or session layer interfaces. However, TCP is dominant here. Examples of implementations will be given later.

There is a requirement to observe the performance, test, and fault-find on the link. Some protocols do provide maintenance facilities, usually at the lower layers of the communications stack.

3.6.6.3 The Protocol

This is the key to "small CTI" implementations, although it becomes an internal matter in merged third party. The protocol defines the language used at the computer and telephony subsystems in creating an integrated system. It is of sufficient importance that Section 3.7 is devoted to this topic.

3.7 CTI Protocols Under the Microscope

Without a CTI link, third party compeer CTI is not possible. Since this is the leading form of integration in the call center, a closer look at the link protocol is essential.

3.7.1 What is a Protocol?

First of all, what is a protocol? When the first North American Indian hit upon the idea of communication by smoke, he couldn't rush over to the fire and send a birthday greeting to his cousin on the other side of the valley. Even if his cousin happened to be looking in the right direction at the time, he would just see puffs of smoke. It would require a remarkable level of intuition to guess that the puffs had anything to do with his birthday! What happened, of course, was that the two Indians held a meeting prior to the birthday and agreed on a coding scheme for the puffs. What they did was specify a protocol, one puff for "Happy," two for "Birthday," and so on. These are the high-level messages. The cousins also needed some parsing rules, such as defining the interword puff-pause to be longer than one minute, within-word puff-pauses to be less than 30 seconds, and so on. They then identified the need for sign-on and sign-off conventions and all the rules to enable communication by smoke. Very few people know that this meeting led ultimately to the establishment of ANSI (American National Smoke Institute). With knowledge of this protocol the cousins could communicate—but only with each other. Larger meetings were necessary to promulgate the puff protocol to other members of the tribe, some of whom wished to change the puff-pause length, while others wanted to add messages. At this point, cooperation with ETSI (European Telegraph and Smoke Institute) might well have been suggested.

A protocol is a set of rules used for communicating between processors and specifying the message set that can be sent. The message set must be specified in sufficient detail to ensure the following:

- Its scope will allow the application programmers to implement the required applications;
- The messages have meaning to those who are to use them—application programmers;
- The messages can be encoded and decoded by the two processors.

So, for example, if there is a requirement to "make a call between directory number 2133 and 2224, calling 2133 first, then 2224 when 2133 is answered," the message set must support a message that could be called "Make a call." This message will have a specific bit pattern allocated to it and a prescribed method of specifying the two directory numbers, together with an indication of which is to be called first.

Third-party CTI protocols therefore establish the set of rules and the messages that allow bidirectional communication between the switch application software and the computer-application software.

3.7.2 Models

In designing a protocol that has sufficient scope to allow programmers to implement the required applications, it is essential to establish a meaningful and accurate model of the way in which the computer and the switch do things. In Chapter 2, a very simple analogy was used to explain this concept—the car driver and his model of the way the car works. There is another way of demonstrating the idea. Imagine that your brain is transplanted into the body of a whale. The objects you then control, flippers, fluke, and so on, are going to be rather different from arms and legs. I think it would be very hard indeed to become accustomed to the loss of a neck in particular! However, you would need to build up a model of the whale's body—probably by experimentation. In CTI there is little room for experimentation.

To communicate at the functional level, the computer must have a model of the telephony process and the switch must have a model of the computing process. There is little point in requesting details on a particular party to a call if the switch does not hold the necessary information. One way of viewing third-party CTI, which helps in identifying the protocol requirements, is to equate the computer with a telephone operator or attendant and the switch with a computer terminal user.

3.7.2.1 Switching Model

The telephone attendant receives calls, listens to instructions, and routes calls accordingly. To do this the attendant uses a database of some sort (the telephone directory or personal knowledge). The attendant also monitors the calls that are being set up. If the called party is busy, the attendant might intrude on that party to offer the new call or, if the called party fails to answer, might offer an alternative. Generally, in a normal two-party call, the attendant is a third party, providing supervision and control during call setup and leaving the call to attend to others once it has been established. In CTI the computer plays a very similar role, the major difference being that, because of its speed, it can deal with a far greater number of simultaneous calls. To perform the telephony task, the attendant needs to understand what the switch does—not the technical detail of how calls are switched or how the software deals with dial-pulse collection—but an appreciation of what each key on the switchboard does, the order in which keys must be pressed, the meaning of the different lamps, and so on. Often the attendant receives training on the operation of the keyboard and through that course builds a mental model of the operation of the switch. Ideally, it is a model that is sufficiently detailed to allow the performance of the switchboard operator's task.

Models generally consist of a set of objects, the states that the objects can exist in, and the rules for manipulating them. What are the telephony objects that the attendant is aware of? They are the people who ring in or are called, the calls that are set up, and the connections between the calls and the people. Similarly, the computer needs a model of switch operation. It will not be exactly the same as the attendant's, but it will be similar. The objects that the computer "sees" and can manipulate will be some variant of the list in Table 3.3. The table lists the attributes and examples of what can be done to them. Note that connections represent the relationship between a call and a device. Connections have states (e.g., ringing), and the state of a call is the sum of the connection states. Table 3.3 is derived from the *European Computer Manufacturers Association* (ECMA) standard protocol [4], which is covered in detail in Chapter 8.

3.7.2.2 Computing Model

What does the switch "see" when it looks at the computer? In the CTI world this is not a topic that gets much airing since most protocols are computer-to-switch-oriented. The computing environment is a very different world from that presented above, though the basics of modeling still apply. The computer presents objects that can be manipulated by the switch; the switch must know what they are like and what it can do to them.

Using the analogy established earlier, the computer will appear to the switch much as it will to a terminal user. The terminal user normally establishes a session with a computer to extract data, perhaps running programs against that data to convert it to a suitable form. Similarly, the switch may well contact the computer to obtain routing information for a particular call. The data returned will depend on characteristics of the caller: the time of day, the stage reached in a particular task, and so on. That is, data is extracted and then

Table 3.3
Switching Model Objects and Parameters

Object	Attributes	Sample Operations
Device	Identifier (e.g., directory number)	Monitor
Connection	Identifier, state	Hold, clear
Call	Identifier, state	Make, clear

modified by running other relevant programs in much the same manner as described for the terminal user.

The objects that the switch may be able to manipulate are presented in Table 3.4. However, most protocols supply only minimal functions in the switch to computer direction—usually confined to route request and statistic recording functions. Furthermore, these functions tend to use the switching model objects of Table 3.3, leaving the computer to translate these into those of Table 3.4.

3.7.3 General Requirements of a CTI Protocol

The attendant and terminal user viewpoints form a basis for surveying the general requirements of CTI protocols. At one stage, connections between computers and switches were termed command and status links. It became less usual for the term command to be used, partly because it can start alarm bells ringing in the regulatory ear and, more importantly, because it is not an accurate description of what happens. In practice, the computer sends a message to the switch that contains a command, for example, "make a call." The switch endeavors to carry out the command but only after it checks the validity of the message. There are two levels of checking: syntactic and operational. Syntax checking simply tests the format and content of the message. For example, it may ask the following questions:

- Is the command code a valid one, that is, is it on the list of available commands?
- Are the parameters of the message within the range allowed for those parameters?

Table 3.4
Computing Model Objects and Parameters

Object	Attributes	Sample Operations
Application program	Identifier, state	Execute
File directory	Identifier	Search
File	Unique identifier within directory	Open/read . . .
Database record	Static or dynamic identifiers	Read/write
Database field	Static or dynamic identifiers	Read/write
Terminal	Device identifier/terminal type	Read/write/monitor

- If certain parameters have to be present, whereas others are optional, are the compulsory ones present?

If the request passes syntax checking, the switch will acknowledge with a response message, then attempt to perform the request. If the message is "Make a call," the switch will attempt to establish a call to a specified directory number. One of its first actions will be to determine whether the directory number represents an internal extension or an external telephone. Assuming the former, it will then translate the directory number to an equipment number; this tells the switch processor where the equipment for this extension is. It may be that the user of the specified directory number has left the company—the number will be disabled, so an equipment number is not available. The switch cannot therefore set up a call to this number and will inform the computer of this. There are numerous other reasons why the "make a call" request may fail during its operation. Another example is that the originating directory number may be barred from making outgoing calls; once again the request will be rejected and the computer informed.

The point is that the command sent by the computer cannot make the switch operate on messages that are incorrectly formatted or make the switch perform operations that are impossible or illegal. In this sense, the message sent is a request, not a command. There is, however, one class of CTI for which the term "command" is correct and that is in third-party dependent CTI. The protocols used by dumb switches are command-based simply because they are dumb—by design.

Most CTI protocols are request/response-based; any message sent is acknowledged by the receiving system. The acknowledgment response is sent after the syntax checking has been carried out. On the basis of the syntax check, a positive or negative response is returned. A positive response indicates nothing about the operation of the request; it simply indicates that the request has been received and is a valid request. Further messages will be received to trace the operational success or otherwise of the original request—these are the status messages.

There are a number of different message types linking the actions of the computer and switch. Sections 3.7.3.1 and 3.7.3.2 provide a generalized introduction to these.

3.7.3.1 Functional Messages

Request messages. Requests from the computer to the switch are often similar to the functions of buttons on a telephone or attendant switchboard. Examples of request messages are the following:

- Answer a call;
- Hold a call;
- Make a new call;
- Toggle between two calls;
- Join two parties together.

A list of parameters containing the information that allows the request to be performed will be attached to each request. For example, if the request is "Make a new call," the parameter list must contain the directory number of the party to be called. If the request is to join two parties together, the identity of each party must be transmitted. These parameters supply data for use by the switch in performing the request; others contain values that modify the operation of the request. For example, the "Make a new call" request might have a parameter that indicates which of the two directory numbers should be called first. Yet another might indicate the action to be taken should the called party be busy.

Requests made from the switch to the computer will be similar to instructions that a terminal user might send to the computer. For example:

- Provide a route;
- Run a program;
- Get an item of data;
- Change an item of data;
- Store an item of data.

These too will need to be accompanied by a list of parameters. For example, the computer must be told all that is known about a call in order that it can give routing information (e.g., CLI or DNIS), or it must be told which program to run or which data item to read.

Status messages. Status messages let the switch or computer know what is happening. There are two classes of status information: the snapshot and the continuous monitor. As its name suggests, the snapshot provides an instantaneous report of the status of a specified object. The computer may, for example, wish to sample the state of a telephone line prior to setting up a call to it.

The continuous monitor provides very similar information to the snapshot, but it provides this information over the whole period that the monitor is enabled. That is, any change of state of a monitored object will result in the generation of a status message.

When starting a monitor, it is often possible to specify a filter as a parameter. The filter will be used to disregard event messages that are of no interest. If, for example, a call center application is only using call-based data selection so that it can pop up the details of incoming callers, then a monitor filter might be set up as follows: Any new call originating from a particular incoming trunk group is reported, though all the changes of state involved in the set-up and clear-down of the calls are not.

Status messages from snapshot requests are transmitted in response to a request; status messages resulting from the establishment of monitors arrive asynchronously.

3.7.3.2 Housekeeping Messages

These are the messages that have no direct bearing on the CTI applications. The housekeeping messages included in existing protocols are limited in extent and number. Some protocols include services to establish associations, provide call information, obtain the system status of the computer or switch, and control telephone feature settings. There is a need for housekeeping messages within a CTI protocol and, as management requirements become more demanding, the range of housekeeping messages grows.

3.7.3.3 Association and Networking

Association is concerned with binding a particular computer and switch session together. Taxicab dispatchers provide an analogy. They communicate with a large number of taxi drivers over a shared radio channel. Dispatchers hand out jobs to the drivers who are nearest to the customer and provide routing information, fare estimation, and so forth. These dispatchers are really rather clever in the way they time-share themselves between multiple callers who want cabs and multiple cab drivers out on the road. They do it by labeling each taxi driver—by name or number—and appending each conversation that they have with the driver with that label. Similarly, the taxi driver announces himself by name or number when calling in for information.

In much the same way, a computer application needs to establish an association with the switch before it begins sending requests—and that association will ensure that the responses returned by the switch get back to it. Association needs to be established, even in the situation in which there is just one switch and computer application; the computer can, of course, have a number of separate applications that require CTI capability.

The description of the switch and the computer as third-party components touched on the problems raised by networking. There are two related issues here: the physical transportation of the messages over the networks and the business of association that has just been discussed.

In a properly layered protocol, there should be no inherent difficulties in remotely situating a computer and switch that support a CTI application. Indeed, it is quite likely that the computer-based application program is distributed over a network and that the CTI messages may travel over one or more networks. Furthermore, there is no inherent reason why a number of computers should not communicate with a single switch or vice versa, or why multiple computer systems should not communicate with multiple switches. This does require an association between switches and computers such that messages relating to a particular call arrive at the correct switch and status messages from a particular switch can be associated with the computer system that requested them. This is not particularly dissimilar to the general problems faced in computer networking and similar solutions apply. The options are discussed in a later section.

3.7.4 Proprietary CTI Protocols

The importance of the switch/computer protocol has already been stressed. In an ideal world, one protocol would be designed, agreed on by all and sundry, then implemented by all the switch and computer suppliers. Although there are a number of standards available (see Chapter 8), the real world is not standardized. In Chapter 1, the long and active history of CTI was introduced. The detritus of all that activity is a plethora of proprietary protocols and a bewildering compatibility list of computers and switches. Table 3.5 collects together some of the proprietary protocols that are in use at the time of writing. These form a representative selection covering third-party compeer protocols on various switches, together with one example of third-party dependent CTI on a dumb switch. The detailed methods used to implement a particular protocol are often closely guarded by suppliers—so much for openness! The switches range from PBXs to ACDs to local exchanges. Note that many of the PBXs also offer ACD functionality. This is normally an option—as is the CTI link itself. Where known, the table includes the physical link and communications stack used. Naturally, there is no table of first-party protocols—these are determined by the signaling system. The Harris 20/20, however, in addition to offering its HIL, also offers the Workstation Integration Link (WIL). This is a specialized method of implementing first-party enhanced CTI. Note that, at the time of compilation, the protocols that are known to use the standards described in Chapter 8 are Application Link, CSTA link, Sopho IS link, Callbridge, and CompuCALL.

3.7.5 Protocol Converters

Table 3.5 shows only a sample of the available protocols, but it easily demonstrates that there is already a large incompatibility problem within the CTI world. Protocol standards may be an answer to this problem, but the standards came too late to make much of an impact. Meanwhile, the computer world looks out upon a bewildering selection of protocols—some specialized, some generic, some layered with OSI conformance at the lower layers, others proprietary from the lowest layers up. The only characteristic that these various protocols seem to have in common is their mutual incompatibility.

One way to deal with this protocol jungle is to convert one protocol into another. This can be done for all sorts of secondary reasons—though the fundamental motivation is always to provide compatibility. The position of a protocol converter in CTI is demonstrated in Figure 3.7. The types of conversion that are undertaken are

- Standard to proprietary;
- Compeer to primary.

Sections 3.7.5.1 and 3.7.5.2 consider these in more detail and provide some examples.

3.7.5.1 Standard to Proprietary Conversion

Though there are not too many of these around at present, this is an area of protocol conversion that may become very important. Given that a switch supplier may have established an installed base of product that supports the company's proprietary protocol, the motivation to implement a standard protocol as well—or worse still, to upgrade the installed base to a standard—is slim. On the other hand, some customers may be very keen to adopt a standard, perhaps as part of an open strategy of equipment procurement. Developing a protocol converter that converts the proprietary protocol to the standard and vice versa seems a perfect solution. It provides a point of flexibility at which upgrades within either protocol can be implemented within a contained and a dedicated environment, isolated from the switch and the computer.

Perfect solution though it may seem, protocol conversions of this kind do require a certain degree of compatibility between the two protocols. Protocol converters can also provide a performance bottleneck and a support complication.

Table 3.5
CTI Link Summary

Supplier	Protocol	Abbreviation	Physical	Comms Protocol	Switch Type	Switch Name
Alcatel	CSTA link		Ethernet	TCP/IP	PBX	4400
Aspect	Application bridge		RS232, Ethernet, token ring	TCP/IP LU6.2	ACD	Callcenter ACD
Lucent	Adjunct switch applications interface	ASAI	BRI, Ethernet	TCP/IP	PBX ACD	Definity
AT&T	Processor switch applications interface	PSAI			LE	5ESS
Ericsson	Application link		RS422	TCP/IP X.25	PBX ACD	MD110/ ACP1000
GPT	Integrated call control link	iCCL	Ethernet	TCP/IP	PBX ACD	iSDX
Harris	Host interface link	HIL	RS232		PBX ACD	20/20
Mitel	Host command interface	HCI			PBX	SX2000
NEC	Open applications interface	OAI			PBX	NEAX 2400
NT	Meridian link		RS232, Ethernet	X.25 LU6.2	PBX ACD	Meridian 1
NT	CompuCALL		BRI	X.25	LE	DMS 100
Philips	Sopho iS link		BRI	Q.931 (user to user)	PBX	Sopho iS3000
Rockwell	Transaction link		RS232	TCP/IP X.25 LU6.2	ACD	Spectrum
Rolm	CallBridge				PBX	9751 CBX
SDX	Synergy link		RS232, Ethernet	TCP/IP SPX/IPX	Key PBX ACD	Index
Siemens	CallBridge application connection link		RS232, BRI		PBX	Hicom 300
STS	Hi-stream		RS232		ACD	Supercall 2000
Summa Four	Host control link	HCL			Dumb	SDS-1000

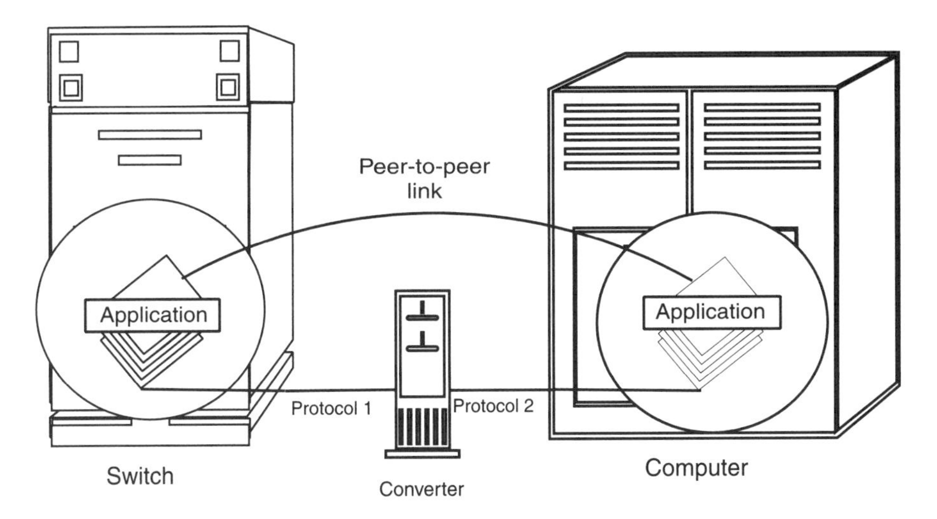

Figure 3.7 Protocol conversion.

3.7.5.2 Compeer to Primary Conversion

Switches often bristle with interfaces, each of which provide some subset of a compeer-level protocol. In some cases it is possible to collect a number of these interfaces together at a protocol converter and provide a compeer interface (standard or proprietary) to the computer. This is a very pragmatic approach. Often the attendant console provides a lot of information on the progress of calls, and some PBX implementations provide external access to the console or its interface card. Call information logging ports can provide information on the progress of calls. Often the information is not supplied until call completion, which is not very useful; however, some switches provide a message at the commencement of a call and another at the end. Some compeer-to-primary converters obtain the current status of a line simply by calling it.

3.8 Primary, Dependent, and Merged Solutions

Third-party primary solutions rely upon the use of existing switch interfaces rather than a special CTI link. Some examples are provided in the section on protocol converters. However, there are other primary solutions that evolve conveniently into dependent solutions.

Consider the case of Siskins and Sons, a company that has been supplying air conditioners for the last 25 years. Business is stable, and so is the company. It has about 100 employees in the head office—and they all spend quite a lot of time on the phone. John Siskins, Jr., who has just taken over the running of the company from his father, is bent on change but shares the old man's caution about spending money. John's friend, who works for the local telco, has convinced him that he really needs CTI for the 20–30 people who are most on the phone—buyers and the like. The problem is the switch. It is an old one, but John, Sr. found it quite adequate and so does John, Jr. Unfortunately, however, the switch does not have a CTI link, and the telco cannot or will not supply one. Another company says that it can provide an upgrade, but the cost is astronomical. It might be cheaper to buy a new switch.

John's friend introduces him to a company that has a solution. The company explains to him that he has a legacy PBX and that it has a server that will front-end it. Further, the company explains that it will interface to the telco digitally and that Siskins and Sons can then have customers dial individuals in the company directly. Also, everyone connected to the company LAN will be able to make calls from their PCs. In addition, they will get screen pops on incoming calls and a host of other things. John hardly understood any of this but asked the cost. Since it was less than half the cost quoted for a CTI link, John gave the go ahead.

Subsequently, John told an architect friend about this decision. She has a real estate company that had three different offices in different towns. Though her company used the same switch as John's, she had upgraded to digital interfacing some time ago. Hence, she already had direct dialing to the desk—but she was very keen on making calls from the PC. She was also keen on the screen popping facility since the company did a lot of repeat business with corporate clients. Accordingly, John gave her the name of the company that had upgraded his legacy system, and she was pleasantly surprised to learn that since her firm was already digital, the price to upgrade her three switches was only twice the price John had paid. She contacted John's first company just to check. It declined to quote, saying that the CTI link for her switch had been discontinued.

What is going on here? Figure 3.8, which depicts John's arrangement, might help to explain. In the figure, the interfaces between the front-end server and the legacy switch are simple extension interfaces. Calls arrive digitally, and their DDI/DID numbers are used by the server to route them through to the relevant extension—and to pop up a screen at the relevant PC. When a call is made from a PC, the server calls the related phone and then makes the second half of the call.

The server contains interface cards to the digital telephone network and to analog ports on John's switch. The server is itself a switch. The implementation in the real estate company is similar—except that the switch is bypassed by the server. The server is connected digitally to the switch and benefits from an advanced signaling system (DPNSS or QSIG). To the legacy switch, the server

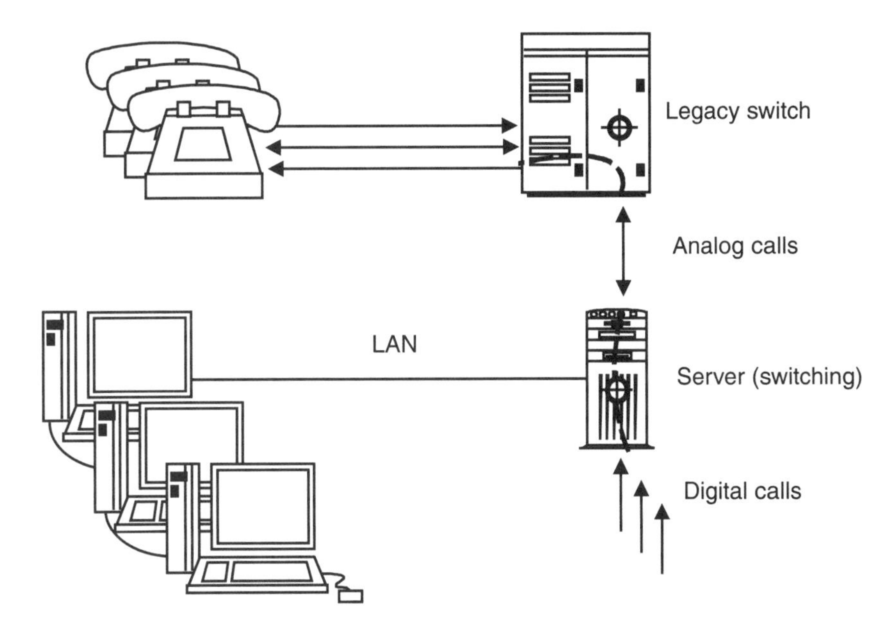

Figure 3.8 Third party primary—front-end solution.

appears to be another PBX. It is reconfigured to route incoming calls to the server, which then routes them back again—directing them to the specified phone and popping the screen at the phone's related PC. Outgoing calls from the PCs are handled in much the same way that John's are.

Is this a sensible arrangement? For legacy PBXs it can be. However, once the proportion of users rises above the halfway mark, it might make more sense to do all of the switching in the server. This then suggests a migration path from legacy switch to merged third party.

Merged third party is intrinsically dependent in its operation. In most implementations, the processor that controls the server's LAN-related applications is also in charge of call processing. The concept of a link has gone entirely—although the programs may be quite separate, they can communicate directly within the merged CT-server. That is not to say that all third-party dependent solutions are merged. On the contrary, there are a number of dumb switches that are used in CTI based systems; an example from Summa Four is included in Table 3.5. Naturally, the protocol for these switches is different to that for a compeer link, but the differences are in the details. After all, it is possible for the computer to take over much of the role of call processing even in a third-party compeer arrangement. The Siemens Hicom provides one partial example of this. At one stage, the Hicom's processor was not considered sufficiently powerful to support ACD operation. Siemens therefore decided to implement the ACD functionality within an attached processor that controlled the Hicom across the CTI link.

3.9 Networked Switching and Third Party

In a previous section, the interaction of a LAN-connected computer system with first-party CTI was examined. The example demonstrated that the two can interact beneficially—such that first party can achieve some of the functions that are normally the prerogative of third-party CTI. Nevertheless, third party is the preferred implementation if functionality is the major decider.

It may be clear that the LAN connection of computers is not an important issue in the third-party CTI world. Most implementations are client server-based, and the CTI server assumes the responsibility for distributing CTI capability to the clients. How that is done is a software issue, and the options will be considered later. Functionally, the outcome is the same for these options. The clients can, unless specifically limited, have the same capability as the server. The true limit on functionality then becomes the CTI link and its protocol. But what if there is a network of switches?

Paul Siskins' real estate friend had three separate locations, each with the same switch. Assuming that a CTI link upgrade was available for the switch and that it was cheap, should the link be made at all of the sites, at just one, or at two? One question that immediately arises is this: Are the sites linked from a data point of view? In other words, is there a corporate data network? In most companies, it is possible to log on at any site and to obtain data from any site This is also the case for John's friend. As a result, screen pops can be delivered to any PC in the corporation, regardless of where the call enters. Besides the architectural choices that will be now be discussed, there are addressing implications. Within a switch, the telephones connected to it and the calls that are in progress have local numbers. When the switch is networked to other switches, the telephones must have network numbers. When CTI is networked, the calls also need network numbers. A conference call between users at each of the three real estate offices is a single network call. Yet it may have commenced as a call between two phones on the same switch.

The lack of network numbering has been a major problem in implementing multisite CTI. The problem is particularly apparent where network-based voice and data call association is required. Transferring a call across the network is simple; tying it together with the associated data is not—unless the call is tagged with a network-wide address.

Some of the options for implementing the real estate CTI network are listed as follows.

1. A single LAN-based CTI server can be provided.
2. One CTI server can be connected to all switches.
3. Each switch can have a CTI server.

This chapter ends by considering these individually. It should illuminate some of the factors to be considered in enterprise-wide CTI. There is no clear solution. The choice depends upon the importance given to reliability versus performance and cost.

3.9.1 Single Server Connected to the LAN

This implies that all CTI messages must arrive at the central server via the LAN. Since there are three different towns involved in the real estate company example, then messages to and from two of the switches may well arrive across WAN connections that link the three LAN segments. This is not good news from the point of view of cost or performance. Will the server know which switch it is dealing with? If the messages sent to the server are normal LAN frames then it should be able to ascertain the source of CTI messages by examining the sending addresses. Similarly, it can direct messages to particular switches because it knows their IP or MAC addresses.

This approach does simplify the placement of the server, which can be installed anywhere that is convenient. However, it does assume that all switches can connect to the LAN, that their CTI links are LAN-conformant, and that they can send and receive frames just as any other device does that is connected to the LAN.

Message delay, in this case, is a composite of WAN and LAN delays. Furthermore, it is not a reliable solution—unless any other server can take over the CTI server role on failure. It will require some mechanism for tracking calls between the PBXs if voice and data call association is required.

3.9.2 Single Server Connected to All Switches

This is similar to the first approach, but it demands a multi-CTI link server. The CTI links are connected directly, which means that performance can be controlled. However, CTI link connection does have to be provided, and this can be costly for remote links. The expense, in fact, may well dictate the physical location of the server. However, this solution does not dictate the physical nature of the CTI links, and it can therefore cope with switches from different suppliers, provided that the relevant drivers are supplied.

Addressing is simplified to some extent, since the single server automatically knows which messages come from each switch and decides which message goes to each switch. However, this does not obviate the need for network-wide call addressing.

Because it is a single server solution, there are reliability concerns; however, the single server does have the advantage that it can reroute calls to lightly loaded switches across the network.

3.9.3 Server per Switch

This solution requires that the real estate company buy three servers rather than one (although the switches can be of different types, and a multi-CTI link capability is not required). On the positive side, however, this method minimizes LAN traffic and is inherently reliable. Since it is a distributed solution, having one server per switch complicates management and voice and data call association. Furthermore, there is no obvious location for CTI-based network routing to reside.

References

[1] Treffert, Donald A., *Extraordinary People: An Explanation of the Savant Syndrome,* Bantam Press, 1989.

[2] British Telecom Datacomms, *DM424X Modem User Guide,* Issue 6.

[4] *Callbox Two Application Programmer's Specification Document,* 8th ed., April 1992.

[5] *Services for Computer-Supported Telecommunications Applications,* ECMA 179, 1992.

4

Media Processing Technology

Media processing technologies were barely mentioned in Chapter 3. That was quite deliberate. Chapter 3 was concerned with call control; this chapter deals with media control. In call control, you examine a call's origin and destination. In media control, you look inside the call at what it's carrying. Since we do both it's no wonder the two things get confused.

The media of interest are voice, video, image, graphics, and text. Media processing technologies provide functions that can be extremely beneficial in many CTI environments. The functions provided are complementary to integration but not essential, yet the increased savings that can be made by fully automating some part or all of a business application, via the use of voice processing in particular, is such that a chapter on this topic is essential.

4.1 What are Voice Processing Systems?

The term *voice processing* can be applied to a lot of products. In the widest sense, anything from a toy that uses a synthetic voice to tell you it loves you to a computer running an application to convert spoken English into text are voice processing systems. However, the largest market for voice processing systems is supported by telecommunications and it is this environment that is of interest here.

There are many different ways of categorizing voice processing systems. The usual categories into which these systems are placed are the following [1]:

- Voice messaging systems;

- Voice response units;
- Interactive voice response systems;
- Audiotex systems.

These four types of systems allow information to be obtained or telephone calls to be completed through interaction with a machine. There is a certain blurring of the categories because the general trend is towards converging the various systems onto single, adaptable platforms. These platforms are capable of performing most voice processing functions when suitably programmed and provided with suitable peripherals. However, the categories are still useful from both a technical and market viewpoint, and each type of voice processing system will be explored in further detail.

4.1.1 Voice Messaging

Voice messaging systems can be connected to the telephone system in many different ways, as shown in Figure 4.1.

Originally, the use of voice messaging systems via a centralized bureau was the most prevalent used. This was quickly overtaken by voice messaging at customers' premises—usually behind a PBX, as shown on the left of Figure 4.1. Meanwhile, network-based voice messaging, both for residential telephone answering and mobile telephone back-up became more and more prevalent. Computer-based solutions are the norm as voice messaging moves to PC-based solutions. All of these solutions are relevant to CTI, though the prime areas of

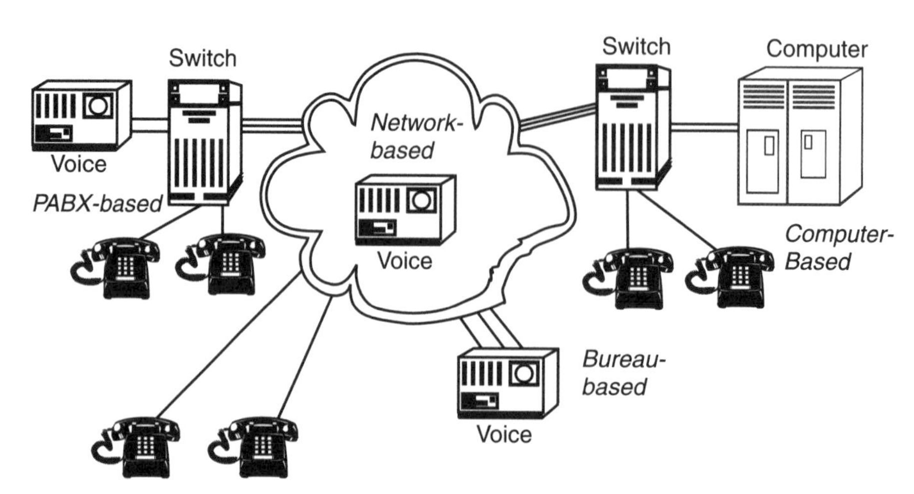

Figure 4.1 Placement of voice messaging systems within the telephone network.

activity that are of particular relevance are customer-premises-based—particularly in overflow and transaction completion.

Voice messaging can be conveniently subdivided into three categories:

- Telephone answering;
- Autoattendant;
- Voice mail.

In many cases, voice messaging systems are capable of providing all three functions on one platform. The three functions form a useful classification in that telephone answering tends to concentrate on incoming calls destined for an individual, voice mail tends to concentrate on responding to or producing outgoing messages, and the autoattendant function deals with incoming calls for groups of people. The functions are described in more detail in Sections 4.1.1.1–4.1.1.3.

4.1.1.1 Telephone Answering

The telephone answering market is dominated by simple telephone answering and recording machines. Though telephone answering machines are relatively simple products, it is interesting to observe that some voice messaging machines cannot perform the answering function as successfully. In particular, these unintegrated systems cannot directly associate an incoming call with the person for whom it is destined and cannot provide an indication that a message has arrived. In an unintegrated system, placing a call to the extension of Julie Lancashire of Artech House would be answered along the following lines:

"Hello, this is the Artech House answering service. Please enter the mailbox number of the person you want to speak to."

Of course, this is doubly annoying. Not only is the called party not there, but you are being asked to identify the person twice! Once, when the call was made, and again, to leave a message. And if you do leave a message, there is no guarantee that Julie will get it because there is no means of informing her that a message has arrived.

Such problems are cured by integrating the answering system with the PBX, and this is an essential step in providing acceptable telephone answering solutions. The subject of integration will be covered later in this chapter.

Telephone answering is important in the CTI context because, if it is well integrated into a call center, it can provide significant benefits by acting as an overflow route when all the agents are busy. Also, if integrated with the LAN, it can become part of a unified messaging system—capable of dealing with multimedia and mail.

4.1.1.2 Autoattendant

The autoattendant is simply a replacement for a switchboard operator. It can be used all of the time, at peak periods, or outside normal office hours. The autoattendant has to deal with both users who possess tone telephones and users who do not. Some systems can respond to the spoken word or to digits from a rotary dial phone. Autoattendant users can either key the extension directly or request a specific department. Some autoattendants offer a directory function for users who know the name but not the number of the required person.

In some cases the autoattendant monitors the status of the called extension. If the extension is busy or there is no reply, the call is recovered by the autoattendant and further options are given for completing the call. In particular, a caller can choose to deposit a voice message for the owner of the extension, since the called party's identity has already been determined. This demonstrates the synergy that exists with the telephone answering subset of voice messaging.

4.1.1.3 Voice Mail

Voice mail is probably the most functionally complex of the voice messaging family, certainly when judged by the range of features it offers. It is often described by analogy to electronic mail, and there are many similarities. One major difference is, of course, that the terminal required for voice mail, the telephone, is used by all.

A key term that is often used in relation to voice mail systems is *mailbox*. A mailbox belongs to a particular user and each mailbox has a unique address, the mailbox number. This number is often, though not always, associated with the user's telephone number. The mailbox contains either incoming voice messages awaiting attention or voice messages that have been listened to and then saved. The mailbox may contain other messages—fax, for example.

Associated with each mailbox are details particular to the owner, such as spoken greetings, spoken name, the facilities that can be used, the maximum size of the mailbox, and a telephone number to which messages should be delivered.

In practice, the majority of users tend to record messages, send them to one or a number of other mailboxes, play messages, forward them to someone else, or reply directly to the sender. These are the so-called *record, send, play, forward, reply* (RSPFR) functions. Systems do possess a great many other features, but it is the RSPFR set that really gets exercised. The RSPFR functions can be likened to the basic functions needed to make and receive calls in normal telephony—and there exists, of course, a multitude of supplementary features in voice mail, just as in telephony. Voice mail systems usually offer a spoken menu of options at each stage of operation; these are called prompts. Regular users tend to ignore the menus, and most voice mail systems are designed to allow the prompts to be overridden.

All the evidence suggests that, unless there is some significant motivation, users will not regularly call voice mail systems to ascertain whether messages have been left for them. The most widely used solution to this problem is a message-waiting lamp at the telephone itself. However, intensive users find that the lamp is more or less permanently lit and therefore becomes redundant. These users do have sufficient motivation to call in to the system at regular intervals—it has probably become their prime method of communication. There are, of course, a number of alternatives to the lamp to indicate that a message is waiting, including pager activation, stutter dial tone and forced delivery. One of the most important in a CTI context is the use of the LAN or Internet as a delivery channel.

4.1.2 Voice Response Units

Voice response units (VRUs) are, in many respects, the most basic examples of the voice processing system—no need for integration here. The VRU simply connects to the telephone system or network as a telephone does. In contrast to the interactive voice response system that follows, it is self-contained. The voice processing application lives within the VRU. It can support various functions, listed as follows:

- Simple announcements and/or music-on-hold;
- Voice information store;
- Voice routing;
- Voice forms.

The first might well be used in the call center to service the customers waiting in queue. The second is often called audiotex. In this form, it provides a dial-up information store. The service is often accessed via premium rate lines, and this, together with its use to dispense dubious entertainment services, has given audiotex a bad name. The information that is played can often be chosen from a menu via the use of the keypad, etc.

Voice routing has a clear overlap with the auto-attendant function described above. It is commonly used in the call center to screen calls. Callers are asked to key a number for the department required, or to speak the name of the department. The VRU will then place the call on hold and generate a call to the requested department. The call can then be transferred in the normal way.

Voice forms represent a specific voice service. Here the VRU asks questions, and the caller answers by pushing appropriate keys. In the call center, a completed call might be transferred from an agent to a VRU. "Was your call to our center dealt with satisfactorily?" the system might announce, "Press one for yes or two for no." Other questions can be asked, and in this way the caller completes a voice form. In some cases, the VRU may offer transcription facilities so that responses can be converted to text.

4.1.3 Interactive Voice Response Systems

Interactive voice response (IVR) systems are quite distinct from voice messaging systems, and VRUs. They can often be the most important category of voice processing system in a CTI environment. An IVR system can be regarded as something that transforms the telephone into a computer terminal. Alternatively, the IVR system can be envisaged as a device that allows a computer to communicate directly with a telephone user—in some senses, the two views are equivalent. With regard to communicating with computer systems, the main advantages the telephone has over a terminal are that it is:

- Cheap;
- Available everywhere;
- Easy to use.

Of these three, the last point does require some qualification. Basic telephones are easy to use, but they are really only useful for simple transactions. Once the amount of data the computer is feeding back to the user becomes significant, or the amount of data required of the user becomes significant, the terminal becomes the superior—if not the only practical—solution. Nevertheless, there are many, many simple transactions for which the telephone provides a usable alternative.

Figure 4.2 indicates how an IVR system sits between a host computer and a telephone user. The computer itself is not usually aware that it is associated with an IVR.

Often the computer in Figure 4.2 will be an existing installation originally designed to perform some transaction-based applications. It usually serves a group of terminals or is connected to a computer network to which terminals are connected. Most applications of interest will be based on multiuser database access—that is, a conventional transaction processing environment. In these cir-

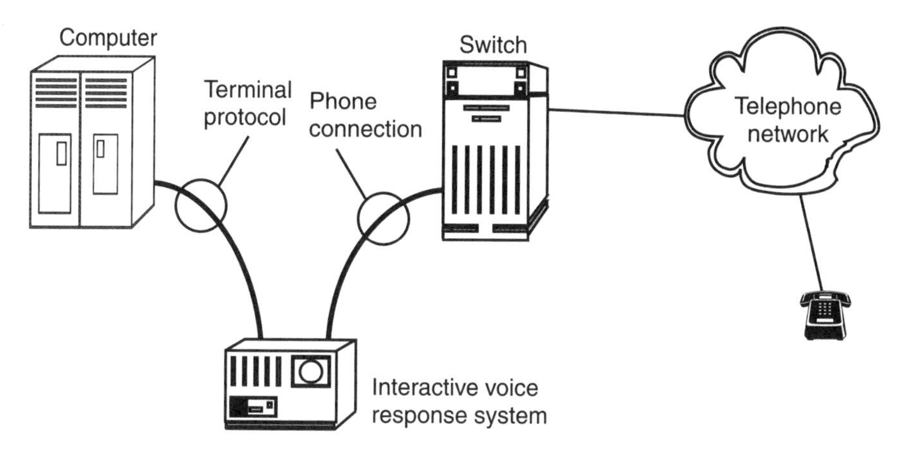

Figure 4.2 Placement of an IVR system.

cumstances the physical connection to the computer system and the protocol utilized between the terminal and the computer are usually predetermined. This means that in most implementations the IVR system has to emulate an existing terminal. However, in more advanced architectures, the computer may already provide an application-to-application-level interface for communicating with other intelligent machines, in which case this may well prove the most suitable connection point for the IVR system.

The task of the IVR system is primarily that of a translator. Commands from the telephone are received by it, usually in the form of tones but increasingly by voice. These are then translated into commands for the computer. Responses from the computer have to be translated into verbal responses for onward transmission to the telephone.

The application of an IVR system to automate telephone banking is a common example of its use. Here the caller is connected to the bank's computer via the telephone and the IVR system. After passing the necessary security checks, a user can do most of the things that one needs to do to manage a bank account—with the one obvious exception that cash cannot be transmitted over the telephone line—yet! The user can, for example:

- Request a checkbook;
- Determine how much there is in the account;
- Transfer funds between accounts;
- Obtain a mini statement.

In each case the interactive voice system prompts the user to indicate the service required and transmits the request, with any necessary data, to the bank's computer. The computer's response is then translated into speech.

The link between the IVR system and the computer is usually a standard input/output channel, as already described. It does not carry telephonic information and should not be confused with the CTI links described in Chapter 3.

When an IVR system is acting as a translator for *incoming* calls only, it is quite clearly not involved in call processing. Its involvement is limited to the speech phase of the telephone call. However, the IVR system is connected to the switch in exactly the same manner that a telephone is. It is therefore capable of doing whatever a telephone is capable of doing. In this sense it is capable, at least in principle, of providing the basis of first-party CTI. The IVR system can become a signaling intercept—as described in the last chapter—but a signaling intercept for what? There is no telephone associated with the IVR system; the only way a person can become involved with a call is by call transfer. Call transfer can be initiated in the usual way: the IVR system sends a recall signal to the switch and then sends the digits to specify the recipient of the call—precisely as a telephone would. But who or what initiates the transfer? Normally, in first-party CTI, this would be the user or the user's application. There is no user, as such, associated with the IVR system, hence the decision to transfer the call must lie with the *application.* The question then becomes: Does that application reside in the IVR system or in the computer? If the latter, how are the telephony commands and responses communicated between the two systems? If the former, how much of the business application needs to be moved into the IVR system? These points will be examined in more detail later.

IVR systems perform useful functions both independently from and together with a linkage between the telephone system and the computer system. Interactive voice systems do have their place in the world of integration. Where IVR really works well is as a complementary capability to third-party CTI. It is here that real benefits are to be found.

4.1.3.1 IVR and Integration

Since there is some confusion concerning the role of IVR within the CTI world, it is worthwhile examining the role and capabilities of these systems in further detail. In Chapter 1 the generic functions of integration were introduced. These are the following:

- Screen-based telephony (SCB);
- Call-based data selection (CBDS);
- Application-controlled routing for incoming calls (ACRI);

- Application-controlled routing for outgoing calls (ACRO);
- Voice and data call association (VDCA);
- Data transfer (DT);
- Coordinated call monitoring (CCM).

It is worthwhile to examine each of these to determine whether or not IVR can provide them. In particular, can they be provided without third-party CTI linkage?

IVR and Screen-Based Telephony

Screen-based telephony is not very meaningful in an IVR context. After all, the purpose of an IVR system is to replace the terminal with a telephone, whereas the objective of screen-based telephony is to allow control and monitoring of call set-up, manipulation and clear, from a terminal. Also, in the IVR world, the telephones are not directly associated with an IVR port. As shown in Figure 4.2, the calls are switched to a port on the system as required, through a local switch or through the public network.

Nevertheless, it may be possible to achieve screen-based telephony with an IVR system. Imagine that a suitable application existed within the computer—a terminal user sees a telephony window on the screen and makes a call. It is possible that messages could be sent to the IVR system instructing it to set up the call. It could then translate the messages into those required to set up a call from its telephone ports (assuming it is capable of doing that). Progress reports on the call set-up could be detected by the IVR system and sent as messages to the computer, which can then update the telephone window. Of course, to do that, a protocol would be needed between the IVR system and the computer—a protocol that carries telephony command and status messages—a protocol similar, if not identical to the CTI protocols introduced in Chapter 3!

Having done that, what purpose does the IVR system serve? It simply acts as a first-party intercept to the telephone port on the switch. The problem then is: How is the call associated with the telephone of the person to whom the terminal belongs? This is simple if each telephone is connected directly to the IVR system—but they are not, they are connected to the switch. So the computer would have to generate another series of messages to cause the IVR system to transfer the call to the relevant telephone.

However, this is all getting out of hand. This is exactly what CTI linkage does well—and perhaps there are already a few too many protocols without inventing another set for the connection of IVR systems to computers. Furthermore, if screen-based telephony is created in this way, the feedback and facilities will always be limited to those that can be provided by first-party basic CTI.

So, in conclusion, it is possible to provide screen-based telephony via an IVR system—but only in a restrictive way and only by creating CTI linkage between the IVR and the computer. This is something that a number of IVR suppliers have done.

IVR and Call-Based Data Selection

Calls that terminate on the IVR system can, in principle, cause the relevant data for that call to be displayed on a terminal. There are two problems here. First, not all calls terminate on the IVR system. Second, there are the usual difficulties that are met when the computer and switch are not tightly coupled: How to associate the terminal with the correct telephone and how to transfer the call to that telephone?

The latter difficulties are surmountable but not easily or effectively. The problem of monitoring all incoming calls can be solved by arranging the system such that all calls *do* terminate on the IVR system—but this solution might not be at all suitable for the type of business being transacted and might well be expensive.

So, once again, it is just about possible to provide this function without a CTI limk, but it is not particularly practical. However, it will be shown soon that IVR can have a key role in implementing this function in cooperation with third-party CTI.

IVR and Application-Controlled Routing for Incoming Calls

Incoming calls that are answered by the IVR system could be redirected to another number based on information from the computer application. This would demand that a suitable message set exist between the IVR system and the computer to transfer the requested number and receive the modified routing. The IVR system would then have to transfer the call, as required.

Provision of application-controlled routing would therefore require that a protocol similar to the CTI protocol be provided between the IVR system and the computer *and* that all calls that are to be monitored are answered by the IVR system. Because application-controlled routing will usually require that all calls be monitored, the IVR system would require a lot of ports. Another problem here is that in answering and then transferring the calls, the IVR system adds significant delay to call completion—especially if the IVR needs to ask questions of the incoming caller. There is a possible source of confusion here. A VRU can ask a caller for routing information as explained above. Here, the application lies in the VRU, whereas in application-controlled routing the application is considered to be in a separate computer system. Naturally, if the computer and VRU are merged, such distinctions become academic.

IVR and Application-Controlled Routing for Outgoing Calls

Almost everything written so far about the routing of incoming calls applies here—except that the interactive voice response system has to monitor all outgoing calls if they are being generated automatically. This does not seem practical. It is a simple matter for the switch to do this—and to refer to the computer if a CTI link exists. To refer the call to an IVR system, it would need a similar link to that system! However, IVR systems do have a strong role in the provision of call progress monitoring information, particularly where supervisory tones provide the main basis for determining call progress. This is, of course, a single line function—it will be reintroduced when power dialing applications are considered.

IVR and Voice and Data Call Association

Voice and data call association could be the easiest of the functions to achieve with IVR, but only when the voice call is being transferred from the IVR system. In this case the system can "whisper" the account code of the incoming caller to the transferee, who can then input that number into the terminal and get the caller's details on the screen prior to the transfer of the call. This system is hardly automatic, but it is practical.

In the call center, for example, it is not possible to identify which agent answers a particular call transferred to an agent group without the use of a CTI link. Whisper mode is then the only option.

IVR and Data Transfer

Data transfer is a nonstarter. Using an IVR system to transfer data is a little like using a modem to carry speech. However, this is the rarest of the generic functions and perhaps, therefore, the least important in this analysis.

IVR and Call Monitoring

The main challenge in call monitoring lies in getting call-related information from the switch to the computer—the interactive response system has no role to play here. It also has little to add with regard to call information data, since the switch is already entirely aware of how long calls are connected to its ports for and, therefore, can accurately measure the use of IVR ports. It does, however, have its own statistics to add to those collected by the switch and computer.

4.1.3.2 IVR Complements Integration

The examination of IVR as a substitute or alternative to third-party CTI has proven quite damning, but that should not detract from its normal role in interfacing a telephone to a computer or in supplementing the effectiveness in a CTI system. The following dialogue between a certain Mr. Beusch, who wants to book a holiday, and the agents who interact with him may help to demonstrate the latter. Mr. Beusch rings the travel agent and is answered by a friendly agent named Francine.

Francine: "Hello, this is Premier Travel. My name is Francine. How can I help you?"

Mr. Beusch: "Uh, hello. I want to book a holiday."

Francine: "Certainly, sir. Could I have your name, please?"

Mr. Beusch: "My name is Beusch. René Beusch."

Francine: "Yes and how are you spelling that, sir?"

Mr. Beusch: "Beusch—B E U S C H."

Francine: "Did you say B U E S C H?"

Mr. Beusch (patiently): "No, Beusch—B E U S C H."

Francine: "Oh, sorry, Mr. Beusch. I've got it now. Unusual name, isn't it? Now where is it that you would like to go and when?"

Mr. Beusch: "I would like to go to Turkey in October."

Francine: "Oh, I'm terribly sorry. I don't cover Turkey. Would you mind if I transfer you to my colleague, Tracy?"

Mr. Beusch: "No, please do."

The transfer is arranged, and the dialogue continues.

Tracy: "Hello, this is Premier Travel. My name is Tracy. How can I help you?"

Mr. Beusch: "I wish to book a holiday in Turkey . . ."

Tracy: "Oh yes, I deal with that. Could I have your name, please?"

Mr. Beusch: "I have already told the other lady that it is Beusch, René Beusch."

Tracy: "Is that Bush spelled B U S H?"

I won't go on. Everyone has met this interchange—in reality or in Chapter 1 of this book. Suffice to say that Mr. Beusch slammed down the receiver when he was asked again for the dates on which he needed to travel. He decided that Premier Travel wasn't for him. His second choice failed to answer his call, but his third choice, Tertiary Travel, did. The company had automated its office, so Mr. Beusch's call was answered by an IVR system.

IVR: "Hello. Thank you for calling Tertiary Travel. If you wish to obtain information on the holidays that we offer, say 'information.' If you want to book a holiday, say 'booking.' If you want to speak to an agent, say 'agent' or simply wait."

Mr. Beusch: "Booking, please."

IVR: "Please give me your postal code."

Mr. Beusch: "IP13 6HX."

IVR: "Now please give me your name."

Mr. Beusch: "Beusch, René Beusch."

IVR: "Do you know the reference number of the holiday you wish to book?"

Mr. Beusch: "No."

IVR: "Can you now tell me the country you wish to visit?"

Mr. Beusch: "Turkey."

IVR: "Do you wish to go to Tanganyika?"

Mr. Beusch: "No."

IVR: "Do you wish to go to Turkey?"

Mr. Beusch: "Yes."

IVR: "We have holidays based in Istanbul and Izmir. If you wish to go to Istanbul, please say 'yes' now."

Mr. Beusch: "Yes."

IVR: "Please say the month in which you wish to go to Istanbul."

Mr. Beusch: "October."

IVR: "The only package holiday available for that month starts on Monday the fifteenth and is for two weeks. Do you want to book this holiday?"

Mr. Beusch: "No, I only wish to go to Turkey for one week."

IVR: "Please hold the line. One of our agents will deal with your request in just a moment."

Mr. Beusch holds on. The IVR indicates to the computer that a problem has occurred. An agent is contacted and presented with Mr. Beusch's spoken name and a record of what has been established so far. The agent indicates to the computer that she wishes to be connected to Mr. Beusch.

Agent: "Hello, Mr. Beusch. My name is Wendy. I understand that you want to book a holiday in Istanbul, but the two-week special from October the fifteenth isn't suitable. How can I help?"

Mr. Beusch: "Well, I only wish to go for a week. I have to be back by the twenty-third. I have to attend a wedding on that day."

Wendy: "I'll check for you, but I think that it might be possible to return on the twenty-second. You would still have to pay for the two-week holiday, but the difference between a two-week holiday and a one-week one isn't that much—just under; £60, in fact."

Mr. Beusch: "Well, I'll do that. I'll take the two-week one as long as I can fly home on the twenty-second."

Wendy: "Good, just hold on while I check the flights and, if it's all okay, I will get you back to our order taker. How do you spell your name, by the way?"

Mr. Beusch finally spells his name. Then, once Wendy checks the return flight, he is returned to the IVR system to complete his order. So, his unusual problems were dealt with by a person—a person who knew his name and the business he had transacted so far. This example uses speech recognition and CTI to complete a relatively complex transaction. (How long does it take you to book a holiday through your local travel agent?) It may seem a little drawn out—that is in part due to the use of IVR as a front end and in part due to the nature of the transaction. The point is to demonstrate that CTI provides the means by which data collected by an IVR can be passed to an agent when difficulties occur, and also to demonstrate the role of the IVR in interfacing the telephone to the computer.

4.2 Voice Processing Technology

Much of the technology of voice processing is shared with telephone switching. The areas that differentiate it are the following.

- Voice compression;
- Voice recognition and verification;
- Voice synthesis.

None of these technologies are essential to voice processing systems. Indeed, all of the systems described above can function quite well with pulse or tone detection and simple recorded voice announcements. However, the availability of good implementations of the technologies listed is doing much to extend the market, and these technologies can all be important ingredients within a CTI-based voice processing system.

All three technologies are advancing in quality and cost as time progresses. It would be invidious and possibly dangerous to provide a clear picture of the state of each technology here. Suffice to say that at the end of the 1990s, all three technologies are commonly used in practical voice processing systems.

Voice compression is claimed to produce reasonable quality speech at 2.4 Kbps—the international standards currently specify techniques that produce 5.3 Kbps speech. The techniques used are the following:

- Pulse code modulation (PCM) at 64 Kbps;
- Adaptive differential pulse code modulation (ADPCM) at 32 Kbps or lower;
- Low delay-code excited linear predictive (LD-CELP) at below 16 Kbps.

Telephone networks generally carry PCM digital speech at 64 Kbps. The other techniques are somewhat noisier than standard PCM—but the difference in quality will be barely noticeable to most users. There are a number of proprietary and national compression algorithms available, all of which can be useful in reducing the amount of memory required to store voice and the capacity required to transmit digitally encoded messages. Compression is of particular interest in the world of digital cellular where 6.5 Kbps speech is now in use. The actual quality of each of these techniques can only be assessed via extensive subjective experimentation, and their use can present compatibility problems when voice files need to be shared between different systems.

Speech recognition and verification are now well-established in the world of practical voice processing systems. Recognition is of key importance in Europe, where the penetration of tone-based telephones is, on average, low and varies from country to country. Table 4.1 provides an estimate of the penetration. However, note that no one actually knows just how many DTMF telephones are in use at any one time. The telephone company knows how many lines are *capable* of supporting tone signaling. This has no direct relation to the number of telephones that can send tone signaling, since most lines can support tone or pulse. Moreover, the proportion of the phones varies according to the nature of the social group surveyed.

Wide vocabulary, speaker-independent applications are now widespread across the telephone system. Some of these are beginning to incorporate artificial intelligence techniques to improve speech understanding. The break-

Table 4.1
Penetration of DTMF Telephones Across Europe [2]

Country	Telephones with DTMF
Switzerland	66%
Belgium, Netherlands	85%
Denmark, Norway, Sweden	61%
France	69%
Italy, Germany, Spain	48%
United Kingdom	40%
Germany	34%

through of the early 1990s was the use of subword modeling. This enabled the introduction of flexible vocabularies to the speaker-independent scene.

Speech verification is used to confirm that a speaker is who he or she claims to be. It is a more difficult technology to implement successfully, and it is not so important in a CTI environment, although it can provide a useful security check in some environments.

Speech synthesis, particularly the production of speech from text, has been under development for many years. The output is certainly comprehensible but still lacks the quality required for general application. However, where the impression made upon the user is not of prime importance, text-to-speech technology provides compact storage and fully flexible phase creation. It is also useful for converting e-mail to speech for delivery over the telephone line.

4.3 Fax and Image

A fax machine carries out three functions when it sends a fax to line. First, it scans a document to convert the image on the paper to a digital form. The scanned image is represented as an array of pixels, each of which represents a small single spot on the paper. (A fax representation of a normal sheet of paper would contain about one quarter of a million pixels.) Each pixel is encoded as black or white. The scanning process feeds directly into a coding stage, where strings of white or black pixels are detected and converted into appropriate codes. The codes are short for the most likely sequences and long for the least likely sequences. In this way the *bit map* produced by scanning can be significantly reduced in size before the fax is transmitted. The output of the coding stage can be regarded as a compressed digital representation of the original image. In group 3 fax, this representation is then passed through a suitable modem stage to prepare it for transmission over a telephone line. Received faxes pass through the reverse of this process. Group 4 fax is purely digital and is therefore suited to transmission over the ISDN. It uses a more powerful compression scheme and can carry color images.

4.3.1 Fax Processing

The important thing to note is that the fax machine produces an encoded bit image of the original document in much the same way that a voice messaging system produces a compressed digital version of a spoken message. Both outputs can be regarded as, and treated as, computer files. Thus, faxes can be stored in voice mailboxes and voice and image files can be managed by a conventional computer.

The simple fax machine is a point-to-point device—an image scanned in your office can be made to appear at some other fax machine that is connected to a telephone line elsewhere. However, the fax message can be processed just as voice messages are. More sophisticated fax systems offer useful services, such as the following:

- Fax distribution;
- Fax overflow;
- Fax selection and response.

Many of these services can be integrated into voice messaging and voice response systems. A physical integration is taking place here that adds to the usefulness of these platforms in a CTI environment.

Many faxes do not go anywhere near a fax machine; they are generated in a PC and delivered to a PC. They are produced by converting documents within the PC into a compressed image and sending this through a fax card. They are carried across the PSTN and/or the Internet and enter the PC through a fax card. Within the PC, they are stored and displayed as images—or can be converted to text using optical character recognition software.

4.3.2 Document Image Processing

Here the paper images are scanned into the image system and are then available as image files. Image and fax are much the same thing, and anything that can be said about one can be implemented for the other. In a CTI context, image systems provide the computer with another input and output mechanism and introduce the need for very large backing stores.

Scanned documents form the basis of workflow in some call centers. All incoming mail is scanned and compressed (usually using group 4 fax techniques). The mail is then manually or automatically indexed so that it can be stored, recovered, and processed. In some call centers, scanned letters are popped onto the agent's screen automatically when the letter writer calls in. The writer's calling line identity is used to identify the customer record and then pull out any recent correspondence. The correspondence can be annotated by voice or text and entered into the general workflow process within the company as a whole.

Once the concept and practicalities of scanning in "real documents"—the ones that arrive via the external post—are accepted, there are many possible applications that combine voice and image at the desktop and may include CTI operations. Take the example of a typical office scene: documents arrive in the morning post and are scanned into the system. One of those documents might be a resumé of a possible employee. The document is delivered to someone's screen for consideration. That person thinks the resumé is interesting and sends a voice message to all in the company who might be interested, providing the document reference. Recipients can then call up the resumé at their screen without the need to create a paper copy. CTI allows them to contact applicants, once again directly from the screen. The linkage between the voice processing system and the image system will allow voice annotation of working documents and the linkage to the computer will enable the storage and retrieval of such composite documents.

4.4 Interworking

Interworking is key to the implementation of integrated systems, this section examines the interface options.

4.4.1 The Integration Interfaces

Figure 4.3 suggests a number of interfaces that can exist between the voice or image processing system, the switch, and the computer. Most, if not all of these interfaces will be provided physically over the LAN.

4.4.1.1 Voice System to Switch

The normal connection between a voice system and a switch is identical to that used between a telephone and a switch—though voice systems are, of course, multiline. These interfaces can be analog or digital.

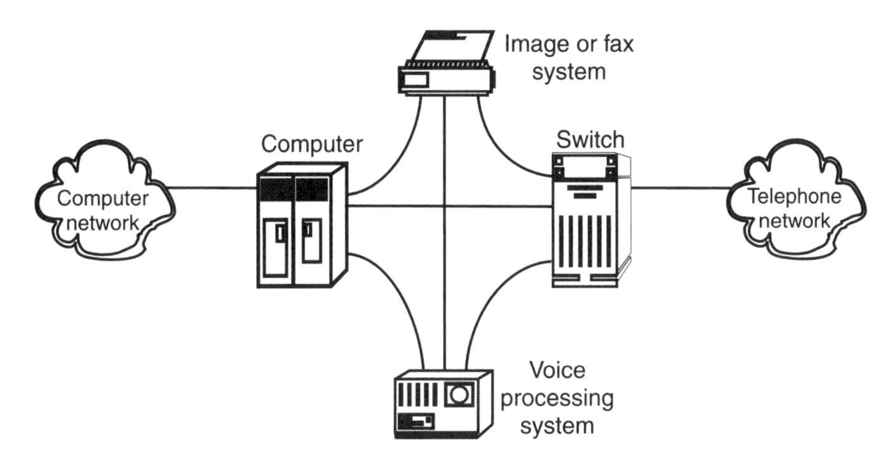

Figure 4.3 Possible integration interfaces.

However, there is a need for a closer connection, especially where voice messaging is concerned. The functions required include the following:

- Presenting the number that was dialed;
- Message waiting indication;
- Message delivery.

The many forms of implementing these integrations are sure proof of the inventiveness of man! Naturally, the CTI link can form one these—some protocols do provide voice processing functions that go well beyond those listed.

4.4.1.2 Voice System to Computer

The classical application here is interactive voice response, and the configuration that has already been described in some detail. Often the computer is unaware that a voice system is connected, since the voice system simply emulates a terminal. However, there are a number of other integration opportunities here, including the integration of voice and text mail. The interfaces tend to be proprietary and closed, though some terminal emulations are standard. However, some suppliers provide open interfaces for the control of voice messaging systems and interactive voice response machines. There are also some relevant Internet standards.

4.4.1.3 Computer to Fax System

Messages can be sent by fax directly from the computer to remote users. These interfaces address the flow of information and text files from the computer to the fax server. In many cases, the fax system has become a fax server—the linkage between this and other servers/PCs is then the LAN.

4.4.1.4 Fax to Voice System

This connection is a good example of the integration of fax and voice. The voice system interacts with a caller to ascertain what information is required. A description or reference number is then transferred to the fax system, which sends the required information to the user's fax machine. The form of this interface may be a simple data link or a more intimate connection. The function may in fact be handled by the computer, making this interface internal. The function is usually called fax response.

4.4.1.5 Fax to Switch

This a straightforward telephone-type connection that gives the fax machine access to the telephone network.

4.4.1.6 Switch to Computer

This is the CTI link and is covered adequately elsewhere.

4.4.2 Unified Messaging

Clearly, the interfaces described above lead to the possibility of delivering voice, fax, and, e-mail to the PC via one interface. Also, with the use of suitable speech technology, all three can be presented at the telephone.

The benefits of such an arrangement were fully explored in my first book, *Voice Information Systems* [1]. The LAN can be used to deliver message waiting indications to the relevant terminal, to facilitate message retrieval, and to provide integration with e-mail and fax mail. Similarly, the LAN can be used to supply e-mail to the voice system where it can be connected to voice ports for collection via the telephone.

Some call this integrated messaging; others call it unified messaging. Whether a particular implementation is integrated or unified messaging causes many arguments, and there are many definitions for both. The difference for the user is often unclear. The idea is that you can manage all of your messages from one or another of two interfaces—the phone or the PC. This can be achieved by middleware that brings together voice, fax, and e-mail at the same user interfaces. If that works, does it matter whether the solution is integrated or unified? I believe that it does.

One definition of unified messaging insists that all the different mail messages must be saved in the same store, use a single directory, and have a common administration point. One example of where a single store is beneficial is that of listening to an e-mail via text-to-speech at the telephone. This is quite practical for integrated or unified messaging—but what happens if the telephone user deletes the message? This is no problem if there is one store, but a lot of work is required if there are multiple stores. It is also quite clear that those responsible for managing the system will look favorably upon a single administration point, and single directory.

One example of a unified system is the Unified Messenger from Octel, a company subsequently taken over by Lucent. The user interface to this product is shown in Figure 4.4. Like other suppliers, Octel does as little as possible in developing this product. It is Windows-based and hence uses Microsoft's Exchange Server as the message store and Exchange Client as the user interface. All that is added to the latter is a telephone icon to distinguish a voice message. Clicking on the voice message brings up Octel's player interface, which can be configured to play messages through the sound card or the telephone. It can also be used to record a voice annotation, which can then dragged and dropped into a Word document, if desired.

The Octel architecture relies on a unified messaging server that connects to the PBX through PC voice cards. It also connects to the LAN to allow the compressed messages to be sent, or recovered, from the Windows Exchange

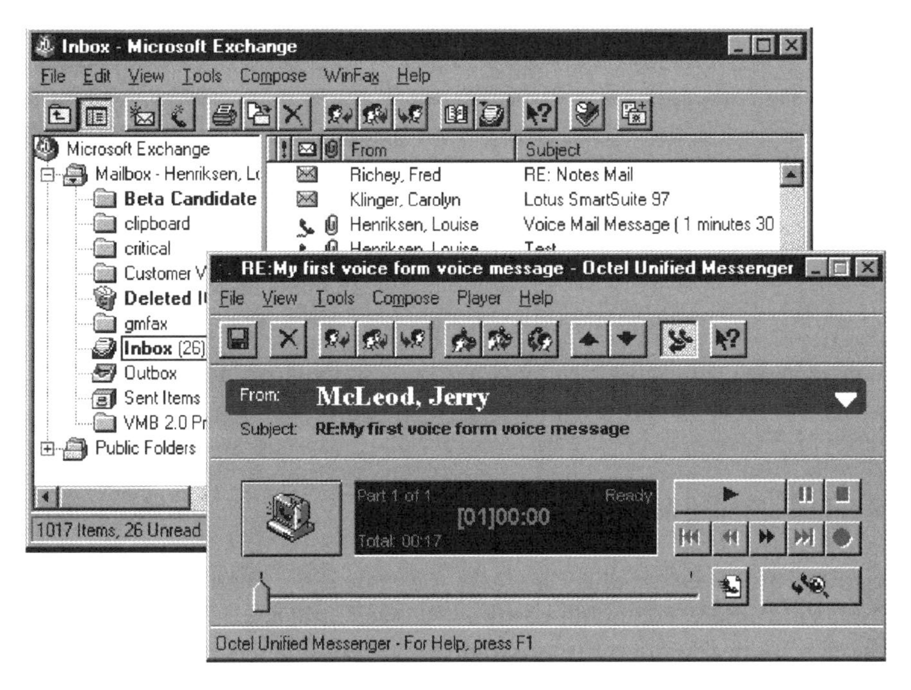

Figure 4.4 Unified messaging user interface. *Source:* Octel

Server. Naturally, this means that messages can be stored anywhere in the enterprise—and recovered from any PC. Multiple messaging servers provide resilience and scalability. The voice servers can continue to take messages in a reduced functionality mode even if the LAN fails.

4.5 Multimedia

Most of the media found in offices today has already been addressed by the description of voice processing and fax and image processing. Multimedia systems need to control all of these from the same terminal. However, the term *multimedia* usually encompasses the use of video material, in addition to voice and still image.

Many multimedia applications are stand-alone applications. The multimedia terminals do not need to communicate and do not connect to the telephone system. In these applications, CTI has no role.

Where a multimedia application does extend to the use of voice telephony—or, indeed, video telephony—then the integration of telephony with the computer becomes as important as it is in single-media communications. CTI is almost intrinsically multimedia-based, in that most applications use both text and voice. The extension into the media of fixed and moving image is not a big step, it is a natural one. It is an area in which the association of data (multimedia data, in this case) with a call can be very important indeed.

4.6 Video Processing

The earliest approximation to video is, I believe, represented by the "What the Butler Saw" machines that were once popular as fairground entertainment. These machines were self-operated. The user placed a coin into the machine and peered through a slot at a title card. The card announced the content of the show. ("What the Butler Saw" was not the only program on offer.) The user then wound a small handle on the side of the machine, which flipped through a series of pictures, and—just like the cartoons—this appeared to animate the program; the pictures appeared to move! The subject was often an extremely overdressed young lady who hastily discarded some garments and then jumped into bed. She was usually still decently clad as the shutter fell and the show ended—this was an era where eroticism was more in the mind than in the eye.

Video processing is still based on the same principle [3]. However, here the moving images are electronic, and the repetition rate is sufficient to make the image realistic.

The video signal is similar to the voice signal in many ways. In audio processing, sound is converted into an electrical analog, and in video processing, light intensity and color is converted into an electrical analog. This is the baseband video signal. One complete scan of an image is called a frame, and, in much the same way as facsimile, the number of pixels in the frame determines the basic quality or graininess of each image. In the full video baseband signal, each pixel is described in terms of the intensity of its color components.

It can easily be shown that the bandwidth of the analog video signal for a good quality signal is in excess of 5 MHz. Thus, the video signal does need a lot of bandwidth, it requires over 1,000 times more than that needed for telephone speech. Converting analog video to a digital form requires the same basic process as that used to convert the analog speech signal into a digital one. The resultant bit rate of the video signal is very large indeed; the standard CCIR-601 encoding produces bit rates of up to 270 Mbps. Given that, in CTI, we feel the need to compress digital speech from its paltry 64 Kbps to 16 Kbps or less, the video signal certainly presents a challenge. Fortunately, plenty of activity has been focused on this seemingly impossible task—primarily to provide reasonable quality videoconferencing links at a reasonable cost.

Speech compression depends on the fact that there is a lot of redundancy in the encoded signal that can be removed without overly distorting the speech itself. Video compression is based on the same approach, and fortunately there is a significant level of redundancy in the video signal. There are various techniques in use—often at the same time. First, the quality of the signal can be reduced by enlarging the pixel size; in the extreme, this produces a grainy picture, but it may be acceptable if the picture that you see is small. Second, transmitting only the pixels that are changed for each frame can substantially reduce the bit rate of the signal. Third, it may be possible to transmit information that describes the shape of the elements that make up the picture rather than the image itself. These, and other techniques, can be used to produce a much reduced bit rate digital video signal.

It is essential that a standard approach is used for compression. Luckily, there are many standards to choose from! Some are deliberately designed for efficient communication of digitized video, whereas others are primarily designed for efficient storage or delivery. Some of the major techniques are listed as follows:

- MPEG1 is defined by ISO and the ITU. It compresses to 1.2–2.0 Mbps, includes a specification for compressed audio, and is used for low-quality "movie" applications.

- MPEG2 is also from ISO and the ITU-T, but its quality is much better than that of existing analog TV. It compresses to 4–60 Mbps.
- H.261 is part of the ITU's group of standards (H.320) for narrowband videophone applications across the ISDN. It compresses to rates between 64 Kbps and 2 Mbps.

The ITU has developed complete standards for a number of networks besides ISDN, as shown in Table 4.2.

Possibly the area of greatest interest in the CTI area is the transport of video (and voice) across packet networks that exhibit variable delay. Such networks include most LANs and, of course, IP networks. H.323 is of most relevance here. It references H.263 video codecs for use at less than 64 Kbps and G.723.1 audio codecs for use at 6.4 and 5.3 Kbps. This is the standard favored by industry leaders such as Microsoft for use over the Internet.

The most common use of video processing technology for communication is in videoconferencing. However, there are other applications, such as the following:

- Videotelephony;
- Video mail;
- Video database.

Videotelephony is quite simply the addition of vision to conventional telephony. Videoconferencing is the use of similar technology to allow more than two people to see and hear each other—although the term is also applied

Table 4.2
ITU-T Video Standards

Network	Standard
Visual telephone terminals over the ISDN	H.320
Visual telephone terminals over ATM	H.321
Visual telephone terminals over guaranteed quality of service LANs	H.322
Visual telephone systems and equipment for LANs that provide a nonguaranteed quality of service	H323
Visual telephone terminals over PSTN	H.324
Visual telephone terminals over mobile radio	H.324/M

to videotelephony. Videoconferencing terminals are usually full-screen implementations that allow the simultaneous viewing of three or four people. Most products are based on the H.320 standard, taking full advantage of the selective bit rate capability of that standard (selective in 64 Kbps increments). Video mail is an obvious extension of voice mail, but only makes sense when videotelephony reaches critical mass. Video databases are accessed from video servers. *Video* (or *media*) *server* is a term given to devices that act as video juke boxes. Video-on-demand usually is supplied across the telephone line or a cable TV connection. Video download needs a high-speed connection since the rate is likely to be in excess of 1.5 Mbps. Using the LAN for this is feasible, and some products provide a means of reserving bit rate for video.

Video communication has been given a fillip by the Internet—as have many technologies. It is a small but growing part of CTI.

References

[1] Walters, Robert E., *Voice Information Systems,* NCC Blackwell, 1991.

[2] Schema, *Voice Processing in Europe,* 1994.

[3] Walters, Rob, *Computer-Mediated Communications: Multimedia Applications,* Norwood, MA: Artech House, 1995.

5

Application Elements and Application Creation

Once upon a time, technology dominated small CTI. In those days, all eyes focused on CTI links. One man, at a now forgotten CTI conference, was so skeptical that he asked to walk the wire. He did not believe that a switch could be connected to a computer and insisted on seeing the link itself. Now CTI is all about the benefits obtained by providing functional integration and automation. It is this that allows the business application to be carried out more effectively and more efficiently.

These areas of functional integration must show themselves in some way within CTI and automation applications. The question is, can they be categorized? Are there general functions that can be realized via CTI and that form the conceptual building blocks of CTI applications? Chapter 1 indicated that this was possible. The generic functions that were identified there will now be described in further detail.

5.1 Generic Functions of CTI

The functions introduced in Chapter 1 are listed here, together with the abbreviations that will be used for them in future references.

CTI Function	*Abbreviation*
Screen-based telephony	SCB
Call-based data selection	CBDS

Application-controlled routing for incoming calls	ACRI
Application-controlled routing for outgoing calls	ACRO
Voice and data call association	VDCA
Data transport	DT
Coordinated call monitoring	CCM
Interactive voice response	IVR
Messaging exchange	ME

5.1.1 Screen-Based Telephony

SCB is precisely what its name suggests. Telephone calls can be set up from a screen, keyboard, and mouse, and call-progress events will be written onto the screen. This has been called "mouse telephony." Figure 5.1 provides an example of screen-based telephony.

5.1.2 Call-Based Data Selection

CBDS uses information about a call to select the most relevant data from the associated computer for display on the screen. This feature is often referred to as *screen popping.* The call-related information can be derived from any known aspect of an incoming call. Examples include the following:

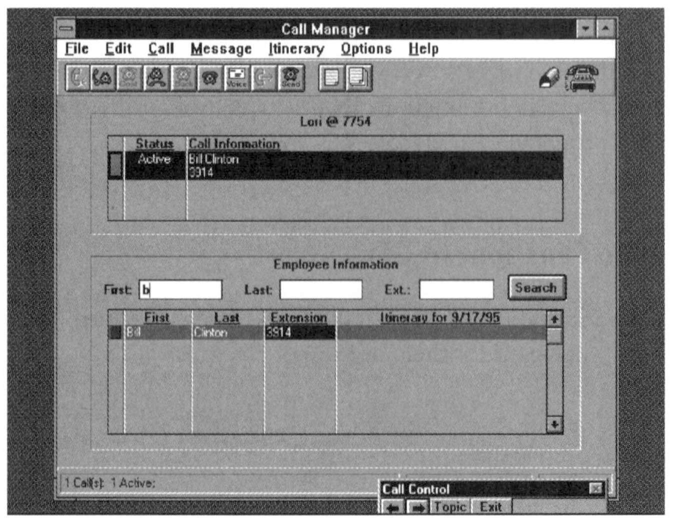

Source: Q.Sys

Figure 5.1 SCB in action.

1. *Calling line identity* (CLI);
2. Calling line location (area code);
3. Incoming trunk number;
4. Dialed number;
5. Call type;
6. Previous answer points;
7. Public or private status;
8. Queuing time.

The first item, CLI, is often the most useful, but it is not always available. Where it is, it may well be used to identify the caller by name by referring the CLI to a customer database. If the number is found in the database and the caller is an existing customer, any data on previous transactions can be written to the screen prior to call answer. The agent might then say, "Hello, Mr. Johnston. Thank you for calling. I hope you are enjoying driving the Mazda MX-6 you purchased from us in August." However, experience shows that it is best to ask if the caller is Mr. Johnson before leading with personal information. It could be Mrs. Johnston calling, and it is possible that Mr. Johnston had not told her about the MX-6!

The fourth item is sometimes called *dialed number information service* (DNIS), especially when the information is delivered from the ISDN. DNIS allows the computer to determine what number the calling party originally dialed or keyed to get to the current answering point. This information may, for example, indicate that the caller is interested in a particular product—say, motorcycle insurance policies. The relevant data on motorcycle insurance can then be displayed, prior to call answer, if necessary.

Item 6, previous answer points, allows the names of people who have already handled the call to be displayed—together with information they have collected so far.

Thus, there are many aspects of incoming call information that are known and can be used to select the data presented on the recipient's screen.

5.1.3 Application-Controlled Routing

Incoming and outgoing calls can be controlled by the computer application. Both types will be described by reference to Figure 5. 2. For incoming, the call is shown entering the switch from the PSTN. The call is trapped in some way by the switch, and reference is made to the computer for routing information. The

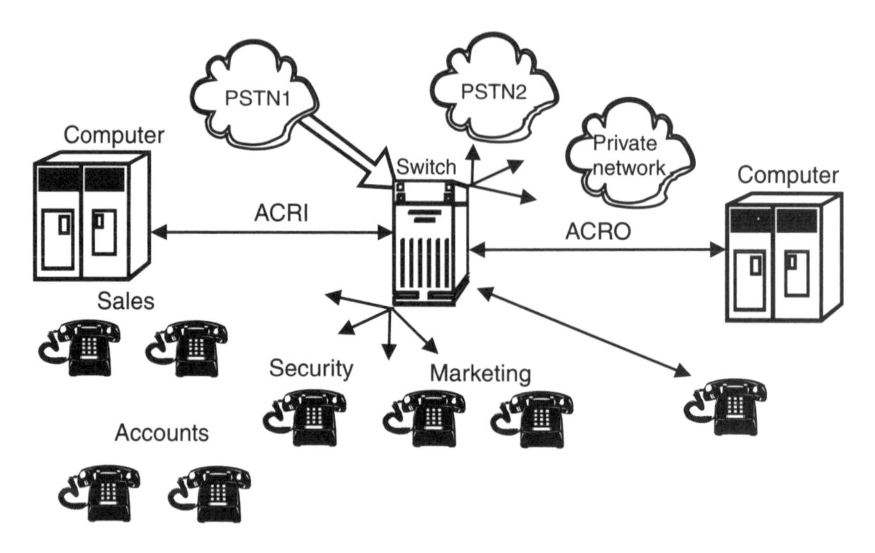

Figure 5.2 Application-controlled routing of incoming and outgoing calls.

computer application may use many sources of information to determine the best route or answering point for this call, for example:

- The CLI and DNIS from the caller;
- The time of day;
- The skill level of the people available to answer;
- The current activities of people in different departments;
- Information about the incoming call (see above);
- The route taken by previous calls.

Having determined the favored route, the application conveys this to the switch and the call proceeds. So, in Figure 5.2 the call may have been for the sales team. However, the application may know that the only person in sales at this time is already on the telephone. Since it is after normal office hours and the call is not from a customer known to the application, the application routes it to security.

Application-controlled routing of incoming calls is very useful in the call center. It can be used at a macro level to get the call to the right group of agents. It may be used at the micro level to get the call to the agent who usually deals with the caller.

The telephone on the right in Figure 5.2 may be used to make an outgoing call. Once again the call is referred to a computer application—but this time for outgoing routing control. The application examines various information sources before deciding how and where to route the call. In the example, the call can be routed over PSTN1 or PSTN2 or the private network. The decision may be made on the basis of cost, time of day, how busy various routes are, the priority of the outgoing caller, and so on.

ACRO is often used to automatically establish outgoing calls. Here the telephone user or agent is not involved with the call until it is established. This is the basis for power and predictive dialing, which are examined in more detail in Chapter 6. Here, it is the application that originates the call—usually from a list of target numbers that are provided to it.

5.1.4 Voice and Data Call Association

This is the magic of CTI referred to in Chapter 1. An active call is being dealt with by an agent whose job it is to sell insurance policies. As the call progresses, relevant files are opened, data concerning the caller is displayed, and new data about the caller's insurance is entered. Then a problem arises; the caller wants to know something about the insurance policy that the agent is unable to explain. The call is transferred to a supervisor and *the data follows the call.* The supervisor knows exactly what the original agent has done. All the necessary details are displayed to let the supervisor carry on with the transaction.

Another way of looking at this is to pose the question, why can't you do all the things with a computer session that you do with a telephone call? Why shouldn't the data be conferenced when you bring a third party into a telephone call? After all, the conference will probably concern some information that is on file. Why shouldn't the data screen that is transferred to the supervisor be returned if the call is returned, and so on? It is worthwhile thinking through each of the more common telephone facilities to assess their relevance to complementary data facilities.

5.1.5 Data Transport

The linkage between the switch and the computer is predominantly used to transport command and status messages. It will, however, be evident that it can also be used to transport data between the two machines. Whether the link is an efficient data transport mechanism or not depends on many factors, including the link traffic and data transfer rate. Nevertheless, there are applications that use this function as follows:

- As a means of transferring data direct from the telephone keypad or attached input device to the computer application;
- As a means of displaying computer application data at the telephone—on a screen, if the telephone has one, or on any other suitable output mechanism.

This is probably the least commonly used of the generic functions—possibly because there are other ways of achieving the same thing, and because very few of the existing third-party compeer protocols support it.

5.1.6 Coordinated Call Monitoring

One of the greatest benefits of CTI is the visibility it can provide into the way business is really done. Day-to-day call statistics are generally of little interest to managers. They really need to know what people are doing with their time, not where their calls are going. CTI, because it binds together the telephone and business applications, is able to collect combined information on what is going on—and the computer is well qualified to process that data into meaningful reports.

One example of CCM that originates from a U.S.-based call center demonstrates the benefits of this function. Some call center agents were doing a lot more business than others, and it was not at all clear why. CCM and subsequent analysis cleared up the mystery. The agents were all dealing with incoming calls. The more successful agents were able to identify whether a call was likely to lead to a high-value order in under 15 seconds; if it was not, the agent simply hung up! If the caller rang again, it would usually be a different agent who picked up the call. Thus, there were two sets of agents, the nice and the not so nice, and the not-so-nice agents were getting richer at the expense of the nice ones. Cross-referencing computer use and telephony events allowed the call center manager to detect, and correct, this practice.

5.1.7 Integrated Voice Response

IVR is both a system and a function. Its use as a system is covered in some detail in Chapter 4. As indicated there, the function of the IVR system is primarily that of a translator placed between the telephone system and the computer system. Commands from the telephone are received by IVR, usually in the form of tones but increasingly by voice. These are then translated into commands for the computer. Responses from the computer have to be translated to verbal responses for onward transmission to the telephone.

In the context of a basic function of CTI, IVR covers many subfunctions that allow the creation of voice systems. These range from the playing of prompts to the emulation of legacy terminals.

5.1.8 Messaging Exchange

As is the case for IVR, the messaging exchange function is an umbrella term covering a number of closely related facilities that form the basis of integrated messaging applications. These facilities include message arrival indication and message handling; the latter embraces message storage delivery and generation.

Where a voice messaging system is linked to the computer world, the computer can be informed by messaging exchange that a message has arrived and is given details of it (who sent it, the priority, the length, etc.).

When a message is retrieved from the computer system, delivery can take place via the telephone system to an associated telephone or to a sound card within the computer. Delivery may involve some conversion—text-to-speech for example.

Messages for transmission can be recorded via the telephone and voice messaging system with the computer controlling the operation, or via the computer for transmission to the voice messaging system as a file. Messaging exchange enables all of the above operations.

5.2 Application Generation

These generic functions are the conceptual building blocks of integrated applications. Though they provide an excellent means of examining the functional makeup of CTI applications, they are at too high a level to be used to implement applications. Application generation can take place at many levels, and this will now be examined. The generic functions can then be used to indicate the functionality of some of the application tools that are available. However, before the solutions are examined, it might be worthwhile to illustrate the problem. The following story provides an excellent example of the problems involved in implementing CTI applications. The story begins when the hired consultant Quantrill, offers his findings to his client . . .

> "Well, Mr. Ostrofsky, we've examined your entire operation from top to bottom. We've taken a very detailed look at your booking area. The people who work there are not able to give their best to the company. Staff turnover is high and the training budget is permanently overstretched."

"Now look here, Quantrill, I know all this. I certainly don't pay you these substantial consultant fees so that you can discover problems I am already well aware of. I want answers, man!"

"No problem, sir. The answers are all in our report—the fifty-five page document that we sent you last week. Didn't you get it? It was bound in red with a blue border. I had hoped that you would be able to find the time to read it before this meeting."

"Yes, I got it. But I certainly haven't read it. The last time I read a fifty-five page report was twenty years ago—and the next time will be the day after my funeral! I want answers, Quantrill, and I don't want to spend two days reading your report to get them. So, come on man, out with it, what are we going to do about the booking area?"

Quantrill, a man of great charm and external calm is beginning to get flustered—a very unusual occurrence. Ostrofsky observes the effect of his outburst and allows a small smile to cross features that seem to be set in a permanent frown of concentration.

"Now, now, Mr. Ostrofsky—no need to get cross. Would you like me to arrange a presentation of the report? It would not take long. I could phone the office and get Chris started on it right now."

Ostrofsky bangs the top of his desk with his fist. "I don't think you are listening to me, Quantrill. I want answers and solutions and I want them now. I do not, I repeat, do not want reports or presentations. Do you understand?"

Quantrill has recovered himself somewhat. He turns to the first pages of the red and blue report and studies them for a few moments while Ostrofsky glowers at him. "In a nutshell, Mr. Ostrofsky, we recommend that you link your computer system directly to your telephone system. That is the one recommendation that will really produce some savings in the ordering department. We further recommend that the time saved by the agents be directed towards active selling of your products. This will give the agents a more interesting and challenging set of tasks and should help to retain staff and provide more income to the company. We recommend that some of that increase be used to provide an improved, group-based incentive plan for the agents. If all of our assumptions are correct, that could save you between $100,000 and $150,000 per year in wages and retraining costs."

"Is that it?"

"Well, no. There are a lot of other recommendations, but those are the major ones. If you would just like to cast your eye over Section 3 of the report . . ."

"No, I would not like to cast my eye over Section 3, not at all. But this sounds interesting. How do we do it and what will it cost?"

"Fortunately, the telephone system you bought—on our recommendation, eighteen months ago—will link to the computer system that runs order processing. The only cost is the upgrade software from the telephone supplier." Quantrill fumbles through the thick report. "That will cost $30,000."

"Sounds good. We'll do it. Get Schmitt to prepare the order for you and . . ."

"Ah, there is something else, Mr. Ostrofsky."

"I should have known. Well, go on. What is it, how much will it cost, and how long will it take?"

"Yes, well it isn't that simple, I'm afraid. The order-processing software has to be modified to deal with the link to the telephone system. Oh, and the sales database will have to be modified, as well. I'm afraid I haven't got costs and time scales for those items. New territory for us this is, Mr. Ostrofsky. We could extend the study a little so that we can get some estimates for you . . ."

Ostrofsky groans.

5.3 How Do You Create Integrated Applications?

Who do you feel sorry for, Ostrofsky or Quantrill? Perhaps both. Someone once said that opera is made up of beautiful moments and boring half hours! Ostrofsky experienced his beautiful moment when he heard that his problems could be fixed for $30,000. Then reality hit, and he could see the boring—and expensive—half hours rising before him.

This is one of the dangers of CTI. It is possible to get carried away with the concept, which is simple enough—simple enough to be a consultant's dream—but to forget the practicalities of implementing the dream.

This chapter is concerned with those practicalities, not with the basics of implementation—how big, how many, is air conditioning required, and so on. Rather, it concerns the practicalities of implementing applications that take into account and can control the telephone world.

Applications are the key to CTI. Application software is often the most expensive element of a system, and it is also often the area in which a business has the greatest investment—in terms of money and in terms of staff training, support, and office procedures.

It is, of course, the availability of a wide range of applications that accounts for the popularity of the PC. It is the need to reuse existing applications and to purchase off-the-shelf new ones that fuels the continued popularity of the Windows operating system. Meanwhile, there is as yet no standard environment for CTI applications. Standardization of the link itself is covered in Chapter 8. Suffice to say that there is little commonality of service sets across the wide range of available switches, let alone a common protocol, at present. A glance back at Table 3.5, which lists a selection of the available protocols, provides a reminder of the actual situation. Assuming that a common protocol is adopted—and implemented in exactly the same way by all switch suppliers (a user's dream?)—even this does not guarantee a standard programming environment. The protocol is a definition of the messages to be transferred across the link; it is still necessary to define methods of sending and receiving those messages at an application programming level before they can be used to create an application. Standardization at this level would focus upon standardizing the functionality of CTI. Unfortunately, initiatives to standardize the application programming environment are generally supplier specific. Though this encourages reuse of applications within a supplier's computer range, it does not allow the transport of applications between ranges.

The generation of CTI applications presents a number of problems and opportunities—problems for Mr. Ostrofsky and opportunities for Mr. Quantrill, perhaps!

5.3.1 CTI in the Real World

In the real world there are relatively few green field sites. In the majority of cases the switch is already in place, as is the computer and its application software. A European survey asked potential users about this; 65% of those people surveyed expected to use their existing switch in CTI applications [1]. However, significantly more people expected to use their existing computers—80% of those surveyed. This was primarily because the computers are already in use for business applications—applications that would benefit from integration with telephony. Of course, this does not provide a complete picture of the proportion of green field sites. Existing businesses do change their computer systems for various reasons: moving from mainframe to departmental computers, moving from a centralized architecture to client/server distributed solutions, migrating to or moving from a single supplier to an open supply strategy. Businesses also change their switches to gain more capacity, to provide extra functionality, to save space, or even simply because the existing switch is becoming unreliable. The point about such changes is that they do represent an opportunity to introduce CTI. However, the probability of a simultaneous change of switch and computer system is very small. If it was large, this book would be concentrating on fully integrated systems, that is, a physically merged combination of the switch and the computer.

Summarizing the above is best done by listing the various possibilities that can occur when CTI is installed. Table 5.1 provides an example—and includes other CTI components.

There are plenty of options here, all of which present distinct compatibility demands, and all of which may require different approaches to the generation of applications.

5.3.2 Who Creates the Integrated Application?

The choices here are the following:

- The customer;
- The switch supplier;
- The computer supplier;
- The software house;
- The software package supplier;
- The system integrator;
- The telephone company.

The preferences identified by the respondents to the survey mentioned earlier are listed in Figure 5.3.

Table 5.1
CTI Components

	Old	New	Upgrade
Telephone system		X	
Computer system			X
Application software		X	
Voice messaging system	X		
Fax server	X		
IVR			X
Call recorder		X	
E-mail system			X

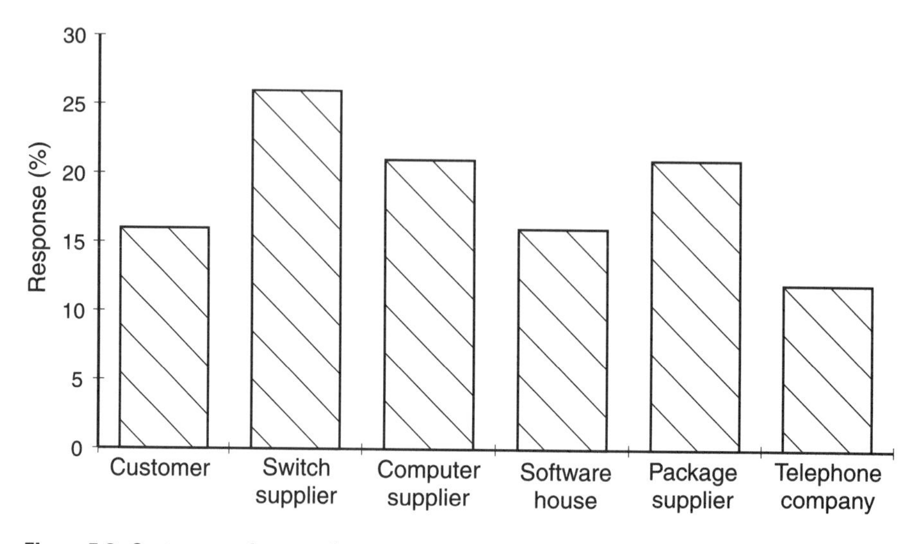

Figure 5.3 Customer preferences for provision of CTI applications.

It is interesting and rather surprising that Figure 5.3 shows a slight preference for the switch supplier over the computer supplier. It is also interesting that the software house is rated relatively low and that the telco, though the lowest of the five, is at least in there as a contender. I think the fact that the switch supplier is on top and the telco is actually rated as a viable supplier is indicative of a certain state of mind in the potential customer. Perhaps they think that the integration of their computer applications with telephony needs significant telephone-based skills. Perhaps they also think that the switch supplier, the telco, or both might have these skills. In considering the significance of this finding it is useful to compare the results of Figure 5.3 with the results that would be produced if the question had concerned the provision of, say, an order processing or an airline reservation application. It is not necessary to produce results for such a survey; needless to say, it would be a great surprise to find a switch supplier or a telco among the contenders. During the 1980s, the period of convergence, people often posed the seemingly rhetorical question, "Would you buy a computer from your local telco?" In our context the question is, "Would you buy a CTI software application from your local telco?" And, according to the survey, over one in ten customers would say yes! What is more, if you broadened the question to include the switch supplier, nearly four out of ten, by far the biggest single group, would say yes.

Is this confidence deserved? Looking at the responses to a rather different question—"Where would you expect to get your system integration skills?"—the picture does change. The majority vote then went to the computer supplier, but only just, and the percentage that expects to get these services from the switch supplier *or* telco does actually increase. In fact, two people in ten thought the telco was the best source for these services.

Obviously the telephony side of the information technology supply industry is not the best place to source either business application software or system integration skills. What do telcos and switch suppliers know about the world of business computing, the world of order processing, accounting, selling, and the like? The answer is: very little. However, they can learn and, more importantly, if they do not know they might "know someone who does"—and there is nothing to stop the telco or switch manufacturer from buying those skills. There is now evidence to suggest that both the telco and the switch suppliers have taken that route—with varying degrees of success. The battle lines of the convergent 1980s were redrawn in the early 1990s. The battle was about who supplies the equipment—CTI solved that by providing linkage—the battle was then over who supplies the application and who ties it all together. However, as the millenium approached, interest once again began to focus on systems—as the major switch suppliers began to roll out fully integrated systems.

Meanwhile, the environment is becoming more open—there is an inexorable shift to standard operating systems mentioned at the beginning of this section. There is also a conscious trend by customers to choose solutions that free them from dependence on a particular supplier for their computing needs. Computer suppliers had a pivotal position in the supply of CTI programming environments, but they were not able to exploit that position without encouraging a drift away from solutions based on their platforms. They therefore tended towards supplying solutions that encouraged the development of CTI applications on their machines. Of course, at the same time, they are keen to supply total solutions to business problems through CTI. The question then is: Do they have the skills to provide the necessary telephonic functions and to integrate the applications successfully with the switch. Once again, they probably do not. Some of the computer suppliers have dabbled in the switch world (see in particular the activities of IBM in the 70s, described in Chapter 1), but have later withdrawn. It is interesting to observe that, as the 1990s progressed, only IBM retained a significant foothold in the CTI application market.

Software package suppliers, particularly in the telemarketing area, soon offered CTI capabilities in their products—often by linking to the environment provided by the computer supplier. Generally, system integration skills are still required to coax the packages into supplying the customer's real needs. As the CTI market matured, a small number of specialist suppliers emerged—focusing on the call center. At the same time, a large number of small companies developed desktop packages.

This leaves the in-house solution and the software house. There is no reason at all why in-house application developers should not develop CTI applications. Given training and the right tools, in-house developers, with the built-in advantage that they are closer to the customer's actual requirements, are in a strong position to provide CTI applications. On the other hand, the world is moving in a very different direction. Business, generally, is interested in minimizing the amount of in-house information technology (IT) support expenditure. It is not looking for ways of spending more money on training its IT staff in new areas. It is more likely to be looking at methods of outsourcing the existing development capabilities, either through a computer supplier or a software house. However, there are exceptions to the rule. The British Airways Rapport call centers form a good example. The applications here are implemented by internal IT staff, with some support from the switch vendor [1].

Software houses have not been famous for their telephony-based skills. CTI is a new area to them, just as it is to our beleaguered consultant, Quantrill. Enlightened software houses have quickly become aware of the potential of this new opportunity. They have established pools of CTI expertise—usually targeting the call center market. Some of the CTI specialist software houses have been absorbed by switch suppliers. Others have continued to expand, and there are many new entrants.

5.3.3 Overview of Approaches

There are supposedly many different ways to kill a cat. There are also many different ways to generate CTI applications. You may recall that Chapter 1 referred to the magical way in which a call, plus its data, is transferred to another agent. Here, we stand behind the magician and all is revealed, although, of course, there is no magic really—just clever deceit and showmanship and, as you, the reader, will have guessed—clever software from the CTI applications programmer. Figure 5.4 reveals all in the CTI application creation field.

Five approaches are shown in Figure 5.4. They all rest on the *application programming interface* (API) that provides the programmer access to the facilities supported by the CTI protocol or, in first-party CTI, the line signaling system. This interface is the key to the implementation of all CTI applications. The figure shows five clouds surrounding the operating system and the solutions contained in these clouds introduce the various approaches to application generation. These will now be considered in more detail. The solutions range from a completely new design or rewrite to the toolkits that are available to create and modify applications.

As the CTI industry develops, so too do the tools that allow straightforward implementation. Software development generally follows this route. In

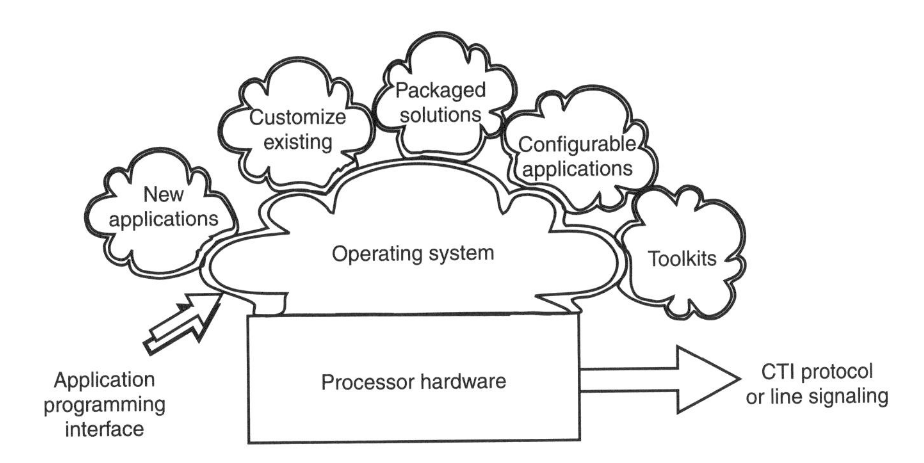

Figure 5.4 Approaches to application creation.

the beginning, programmers wrote operating systems before commencing the application. However, operating systems quickly became standard items. In the beginning, programmers used machine code or the assembly language specific to the machine being used. Later, machine-independent high-level languages like COBOL, C, and FORTRAN became the main tools for producing programs. Later still, languages began to move closer to the application (4GLs), and still more recently, the concept of reusable application objects has developed.

The availability of CTI APIs, which protect programmers from the hardware platforms on which applications will run, is a crude beginning in the cycle of CTI programming development—comparable to the standardization of operating systems. Already the toolkits are beginning to move out towards the application; that is, the functions of CTI are becoming abstracted so that they are quickly usable at application level.

5.4 Application Programming Interfaces

Because APIs are so key to this chapter and to the whole world of CTI, this section will begin by examining them and by giving a number of practical examples.

Given that most switches have different CTI interfaces and different protocols, how is a fixed environment to be provided that gives program-level access to the CTI functions for the application programmer? In this fixed environment, the programmer will hopefully not be aware of the nature of the switch that provides the CTI functions. Indeed, the application program may need to interface to a number of switches. Thus, the basic definition of the purpose of a CTI API is:

> To provide program access to all CTI functions in a hardware-independent forum.

Naturally, the API provides a view of the switch at layer 7 in the OSI model; the application program does not need to concern itself with the way in which messages are transferred between the computer and the switch and vice versa. In most circumstances, but by no means all, the API provides a buffer between the application software and the proprietary CTI protocols that exist on the various switches a computer may interface with. Ideally, the API should remove all switch variability, but this is a difficult objective to achieve in practice. Figure 5.5 demonstrates the means by which a multiswitch API can be achieved. For the purposes of this explanation, a single computer system is shown connected to three different switches. In practice, the computer system is usually connected to a single switch in any one installation. The multiswitch arrangement is used to demonstrate the need for normalization through the drivers. Each switch may have a different physical interface to the computer; that is not of relevance here.

There is a software driver within the computer for each supported switch. These drivers deal with communications on the link and are conceptually very similar to the print and disk drivers, which are also shown in Figure 5.5. The print and disk drivers transfer data to the printer and transfer files to and from the disk.

For received messages, the switch drivers terminate the link, remove all the lower-layer protocol elements from messages, and present the messages to the API. These messages are defined by the API, and each driver must translate the messages of each protocol (three in this case) into the presentation interface defined. Transmit messages are sent in standard form from the API to the driver. Once again, the driver must translate these into a form defined by the protocol of whichever switch is being addressed, and prepare for transmission to line by addition of all the necessary lower-layer protocol elements.

On the application side, the API offers a single set of procedure calls or messages to the application program, giving access to all the CTI functions it supports. Thus, the application developer exists in an environment bounded by the CTI API, the operating system calls, and the other drivers in the system. The developer has a defined library of calls that can be made to the telephony world—as shown in Figure 5.5—and a defined set of messages that can be received from it.

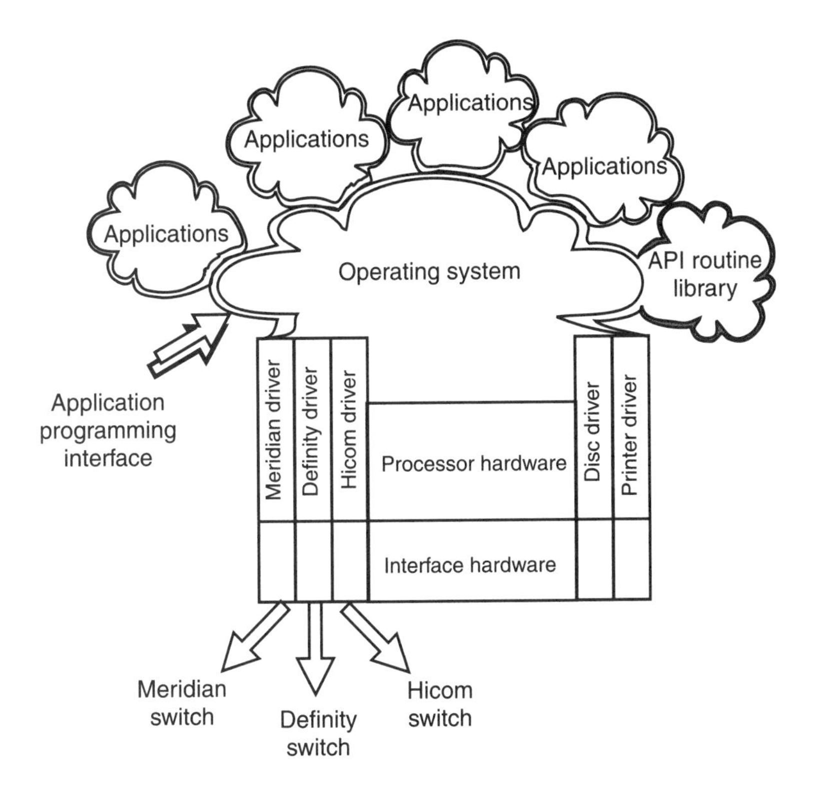

Figure 5.5 A multiswitch API.

5.4.1 Functions

The functions available through the API are already known, at least in general terms. They consist of the normal request and status functions found in most CTI protocols. Any of the protocols introduced in Chapter 3 could be used as an example. The difference here is that the set of request and status functions implemented in any specific API are those that the API definer decides to include. The other difference is that they are presented in a usable form for the application developer, as procedure calls with parameters and returned values, for example.

It is, of course, true that the application program cannot access a function that does not exist within a switch. It is also true that functions may exist within a particular switch protocol that are inaccessible via the API. The former can be dealt with by rejecting an invalid request; the latter tends to constrain the breadth of functionality available from the API. For this reason it is often said that multiswitch APIs tend to offer the lowest common denominator in functionality. It is important to understand how different computer suppliers deal with this issue of implementing multiswitch APIs. The examples that follow should be enlightening in this respect.

Naturally, single-switch APIs also exist. These may well be implemented by the switch supplier or are at least defined with a full knowledge of the switch protocol. They therefore tend to be richer in function (always assuming that the switch has a rich protocol), but they are switch-specific.

5.4.2 Examples of Application Programming Interfaces

Computer suppliers, software suppliers, and some switch suppliers provide APIs. Some of these have been selected for detailed consideration. The examples chosen are to some degree arbitrary—you should check with the supplier for up-to-date information. However, since the list includes one of the world's largest suppliers of computer systems, the leading supplier of PC software, one of the earliest switch suppliers to enter the CTI market, and a leading dumb-switch supplier, the coverage given represents a good overview of available APIs. The suppliers and the solutions are the following:

- IBM with CallPath;
- Novell with TSAPI
- Microsoft with TAPI;
- Mitel with MITAI;
- Summa Four with Assist;
- Sun with JTAPI.

It would be inappropriate to make any value judgment on the relative quality of any of the APIs examined. Their capabilities do differ at present. However, if a particular approach or function that exists in one supplier's API does prove to be essential in implementing CTI applications, it is evident that the other suppliers will adopt it. The first two examples have been influenced to some extent by standards activities in the CTI area. Others contain explicit switch-specific features.

Some of the examples focus entirely on call control, whereas some include automation and messaging functions. There are a number of APIs that focus on the latter; these are summarized in Section 5.4.3.

Other APIs not included in this book due to a shortage of space are the following:

- Call applications manager (CAM), which is Tandem's CTI enabler;
- CT-Connect, an enabler that Dialogic supplies to its developers and that is based on computer integrated telephony (CIT) the DEC initiative;
- CAPI, the common API for use with ISDN call control and a standard from the European Telecommunications Standards Institute.

5.4.2.1 IBM's CallPath Services Architecture

Overview. CallPath Services Architecture (CSA) is IBM's approach to the world of CTI. The general architecture applies to the entire range of IBM machines, from PC to mainframe. Most CallPath implementations are, in fact, client-server-based. These utilize the SwitchServer software, which runs on many different forms of server hardware and software. It also services clients that run a wide variety of operating systems and supporting software. However, this description will focus on CallPath/400. This applies to the midrange AS/400 machines and is a representative example. Note that the following is a mere glimpse of IBM's CallPath; the CallPath/400 programmer's reference manual alone contains nearly 300 pages.

Figure 5.6 depicts the architecture as portrayed by IBM. This is a very similar representation to that provided by Figure 5.5, in the general introduction to APIs. However, it is useful in that it does indicate some of the switches that worked to CallPath/400 and demonstrates a computer system supplier's view of CTI architecture (the computer is always on top!)

The CallPath/400 subsystem communicates with the switch and provides switch-specific protocol mapping, whereas the CallPath API presents a single interface to business applications regardless of the specific switch or switches attached.

In physical terms, the basic requirements of a CallPath/400 system were specified by IBM as the following:

- AS/400 system (any model);
- Operating system/400 and CallPath/400 program;
- Disk storage of 4 Mb for the program;
- A ½-inch magnetic tape reader;
- A communications adapter port or IBM token-ring adapter port for the AS/400 system attachment to the switch.

What the programmer sees. This description addresses early versions of CallPath and is included as an example only. Later versions were Java-based.

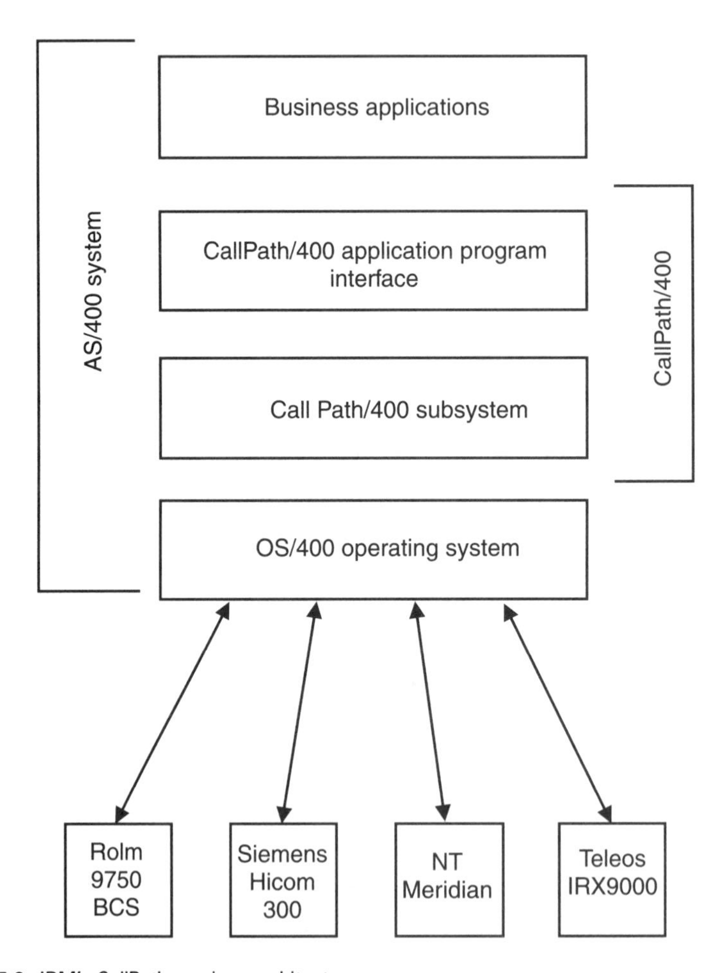

Figure 5.6 IBM's CallPath services architecture.

The application programmer is presented with a set of procedure calls that allow interaction with the telephony world. A programmer wishing to set up a call between two parties would cause a procedure call to be made by writing, in COBOL, for example:

CALL "STLMAKE" USING parameter0, parameter1,
parameter2,.......,parameterN

The parameter list and the means by which the programmer tracks the progress of the call will be covered later. However, before calls of this type can be made, the programmer has to cause the application program to sign on.

Signing on. In CallPath there are two sign on–related calls: initialize call profiles and register ownership. Call profiles can be regarded as templates that condition each of the API services. Call profiles are switch specific and there can be a number of them for each switch. They must be initialized, but individual settings can be subsequently modified through use of the set function. Some examples of the 20 or so characteristics contained in a call profile are given in Table 5.2. Each call profile has a unique identifier that is used in all calls that use the characteristics.

The register ownership procedure call registers parties (normally, via the directory number) that the application program wishes to act on. For example, the application program needs to own at least one of the parties if it wishes to execute a make call. Ownership can be sole or shared.

Not all switches support ownership and this spotlights one of the dilemmas of API implementations. Should the operation be globally barred because some switches cannot support it? CallPath deals with this particular function by supporting ownership itself if the switch does not. In other cases, CallPath simply passes the "not supported' response to the application and allows it to resolve the problem.

The API functions. There are two categories of functions within CallPath: commands, which are issued to the switch by making a procedure call (e.g., make call), and messages, which can be responses to commands, call progress event messages, and system messages. Messages are always prompted by the application program in some way. Command and message functions respectively define the transmit and receive vocabulary of the application program. The command set is given, with a few exceptions, in Table 5.3.

Table 5.2
Examples of Call Profile Characteristics

Characteristic	Initial Value
Call type	Voice
Make call forward override	Party one override
Party specification	Directory number
Return response	Response not returned
Automatic answer	Answer
Billed party	Null

Table 5.3
CallPath API Commands

Command	**Comment**
Add party	Add party to existing call
Alternate call	Toggle between calls
Answer call	
Conference call	
Disconnect	Remove part from call
Extend call	Attempt connection to another party
Hold call	
Invoke feature	Activate particular features at a phone
Make call	
Monitor	Request call progress messages
Query party status	
Receive	Receive messages
Redirect call	
Reject call	
Retrieve call	
Return control	Return control of call to the switch
Set . . .	Set the characteristics of a call profile
Transfer call	
Trigger	Trigger an action in the switch

Note that CallPath had no media processing commands; it did not extend to automation, although IBM had a range of voice systems in its portfolio. Later versions of CallPath migrated to Java and then encompassed media control. The function of most of the commands listed in Table 5.3 is self-explanatory, since conventional telephony functions are requested, though in this case they are initiated by the AS/400 application. Trigger does not belong to that category though. Trigger can, for example, be used to request that the switch send a request instruction message whenever a call is routed to a specified telephone. The application might then redirect the call, returning control to the switch after doing so. The receive command requests the return of asynchronous messages—this will be described later.

In addition to requesting actions from the switch, the API needs to monitor the progress of calls and the response to commands. CallPath messages are listed in Table 5.4; many of them will only be received if monitor has been requested on an owned device.

Note that most of the messages will arrive asynchronously, especially those that occur because a phone is being monitored. The question is, how does the API deliver these messages to the application program? This is the function of the receive command, introduced in Table 5.3. When the application program executes the receive command, it will be given the first message on the queue. Of course, there may be no messages at all. In this case, a parameter is attached to the receive command to specify how long the function should wait for messages before being returned to the application empty-handed. So, as inferred earlier, it is the application program's task to call in for messages.

Command structure. A more detailed examination of one particular function will shed more light on the manner in which application programs communicate across the CallPath API. Again, make call is chosen for this examination.

Table 5.4
Messages the API Provides to the Application

Message	**Comment**
Call alerting	Ringing
Call alternated	Party has toggled
Call conferenced	
Call connected	Party has become active in a call
Call held	
Call parked	
Call picked	
Call rejected	
Call routed	
Call transferred	
Disconnected	Party passed into idle state
Feature invoked	
Network reached	Outbound call has seized trunk
Party status	Reply (see commands, Table 5.8)
Request instruction	Switch needs advice on this call
Response	Acceptance or rejection of a command
Setup	New call from phone
Switch status	System message

The procedure call used for this function has already been given. What is examined here is the parameter string and the meaning of parameters within it. Remember that the function of make call is to attempt to establish a call between two parties; at least one of these parties must be "known" to the switch. Normally, this means that one party will be a phone connected directly to that switch. In CallPath, virtual parties can be used. This is useful in that a single-party call can be set up (i.e., connected to silence) for later association with a real party. The parameter string of make call is presented in Table 5.5.

Note that in Table 5.5 the make call command can reference the call by party directory number or party connection ID. Whichever is used is specified by the current call profile; the other value is ignored in the parameter string. The returned code provides details of the success or failure of the procedure call. Returned values can be OK, communications fail, not supported by switch, invalid parameter, and so on.

None of the returned values can indicate that the *telephone call* is successful. That can only be ascertained by monitoring the call and receiving call progress messages.

Supported functions. In introducing the API concept, the fact that switches do not all support the same request and status functions has been mentioned many times. Table 5.6 concludes this brief overview of IBM's approach to the provi-

Table 5.5
Parameter List for the CallPath Make Call Command

Parameter	Comments
Party one length	Length of directory number
Party one	Directory number
Party two length	Length of directory number
Party two	Directory number
Call profile ID	See Table 5.2 and related text
Account code length	
Account code	Billing point for this call
Party one ID	Connection ID
Party two ID	Connection ID
Request tag	Specifies variable for returned ID
Return code	Specifies variable for returned code

Table 5.6
Early Support for CallPath Commands across Four Switches

Command	9751 CBX	Hicom 300	Meridian 1	IRX9000
Add party				✓
Alternate call				
Answer call	✓	✓	✓	✓
Conference call	✓	✓	✓	
Disconnect	✓	✓	✓	✓
Extend call	✓	✓	✓	
Hold call	✓			✓
Invoke feature				
Make call	✓	✓	✓	✓
Monitor	✓	✓	✓	✓
Query party	✓	✓	✓	
Status				
Receive	✓	✓	✓	✓
Redirect call				✓
Reject call				✓
Retrieve call	✓	✓	✓	✓
Return control				✓
Transfer call	✓	✓	✓	✓
Trigger				✓

sion of a CTI API by indicating CTI command support for each of the switches initially covered by CallPath/400.

Note that Table 5.6 is simply a snapshot of the availability of functions from switches; it was established in 1992 and is included as an example only. The number of switches supported and the support given to commands by any one switch has increased with time. At the time of writing, CallPath supported over 20 switch models. Note also that a tick against any particular function does not imply both-way working; a number of the 9751 CBX-supported commands apply to outgoing calls only.

In 1997, CallPath had the largest installed base of call center integrations with more than 100,000 agents enabled by this software. Many third-party suppliers had written applications specifically for this platform, and IBM had established extensive test and support facilities. Also in 1997 the move to Java began together with the introduction of a JavaBean visual application generator.

5.4.2.2 Novell's TSAPI

Toward the end of 1992, Novell and AT&T (now Lucent) announced a strategic agreement concerning the distribution of CTI functions to PC clients via a telephony server and the LAN. Novell supplied the PC network software, and AT&T (now Lucent) provided the Definity PBX with its CTI interface. Though client server CTI was not new, this agreement between the then leader in LAN software and a giant of telecommunications seemed to be breaking new ground. The approach is similar to that adopted by DEC for client server CIT applications except that the DEC server supported a number of switches.

The agreement between the two companies outlined the development of a suite of APIs. These were specified to enable application developers to integrate telephony services into network applications or to develop telephony applications like PassageWay, an existing AT&T CTI application. The interface was called the *Telephony Services API* (TSAPI). An interface was also established to allow other switch suppliers to support the Novell approach on their products.

The platform created by this agreement was said to be open, and in 1993 Novell established a PBX manufacturers support program; eight manufacturers committed to integrating their systems into the NetWare environment. The initial partners were the following:

- Alcatel;
- AT&T;
- Comdial;
- Fujitsu;
- GPT;
- Interconnect;
- Mitel;
- SDX.

This represented a good spread of switch suppliers offering a variety of approaches to the PBX market. However, some of the bigger players were missing.

The Telephony Services Early Implementers Program, an essential and parallel program, had already recruited 24 providers of end-user telephony applications, so the show was very much on the road. Novell seemed to be becoming the market former in CTI. I recall telephoning one of the Novell evangelists and being told quite firmly that this initiative was quite different from the old CTI offered by IBM and others. I puzzled over that—but not for long.

Early in 1994, the Novell Open Telephony Association was formed. At the time, Novell was the focus of CTI activity, perhaps because the company then supplied 66% of the computer network software market. More and more switch suppliers saw the Novell approach as the means of distributing telephony across the LAN to PC users. One of the attractions of TSAPI was that it promised to present a common interface to the application developer even though most of the switch vendors provided proprietary links. Even better, TSAPI was itself based on the standard definitions established by ECMA in CSTA.

The secret of rolling out CTI to the masses had always lain in efficient distribution, and many people thought that Novell's TSAPI initiative held the answer. However, there was more to CTI supply than laying down driver and API specifications, providing a server, and recruiting switch suppliers to the cause. Novell clearly realized this and in 1995 began to establish Novell Telephony Solutions Centers. Datapoint in the United Kingdom was the first company to establish a center. Its job was to promote the benefits of CTI in order to build a mass market. It was also there to provide advice, training, and support to Novell's partners, which were implementing TSAPI solutions; it helped them to plan applications and provided general CTI knowledge.

That TSAPI gained significant mindshare in CTI is without doubt, but it did not lead the way to mass-market CTI. Basing TSAPI on the specification for a CTI protocol that has been standardized by ECMA, seemed a good idea at first. However, the ECMA standard had a weakness that might also be regarded as a strength—its flexibility. It had no clearly defined call model—in other words, nothing that provided a precise definition of the order in which status messages would be delivered from the switch. Hence, TSAPI was in the same state, and applications written for one switch did not always work when the application was moved to another switch.

Novell was forced into introducing switch normalization into its drivers at a late stage. Normalization is an attempt to ensure that the sequence of messages that the application developer receives from, for example, a Lucent Definity are the same as those received from, for instance, a Nortel Meridian. However, some damage had already been done: The Novell approach was perceived as switch-dependent, and this certainly set the software developers' teeth on edge.

Early in 1996, TSAPI was extended from the Novell LAN environment to the TCP/IP world—a move into a more open environment, but once again a little late. At around that time, Novell announced its foray into voice processing. It delivered an IVR platform based on Voysys technology and supported by an IVRDesigner Toolbox, a forms-oriented application generator.

Media processing is always alluring, but this seemed to be a distraction from the "CTI for the masses" campaign.

At the time, costs for TSAPI itself were $1,295 for five users and $26,995 for 250 users. This seemed reasonable, but application costs and switch upgrade costs added a lot to this price. Meanwhile, the support from switch suppliers had continued unabated; TSAPI then supported 16 different switches, and there were eight more in the pipeline. It was claimed that 43 different applications existed and were in use by customers worldwide. This was not quite the revolution expected when the 1992 announcements were made, but it was nonetheless significant.

That is a little of the history—but what exactly is TSAPI? Novell has an overall name for its approach to CTI: NetWare Telephony Services. This is a LAN-based CTI solution that enables communications between networked computers and telephones. It uses a CTI link between a NetWare server and the telephone switch, plus the services of several *NetWare Loadable Modules* (NLMs). It gives users the ability to intelligently place, receive, conference, and transfer calls using existing networked computers and information. It is a third-party approach to CTI in that it requires a special CTI link to the switch. The components of NetWare Telephony Services are shown in Figure 5.7.

According to Novell, NetWare Telephony Services does the following:

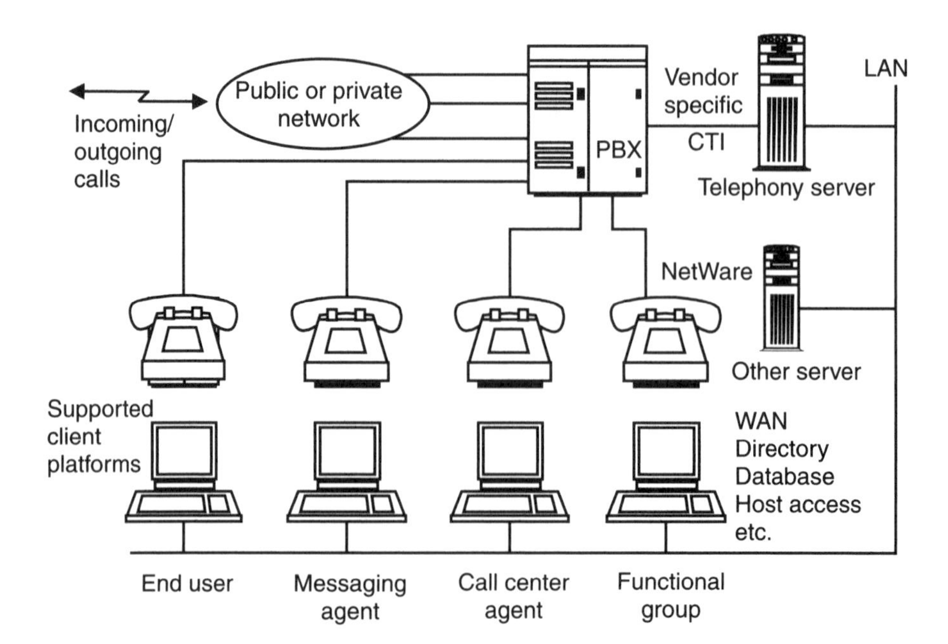

Figure 5.7 NetWare Telephony Services. (*Source* Novell)

- Offers a non proprietary, affordable CTI solution;
- Integrates with NetWare services;
- Provides a turnkey solution for non-NetWare environments;
- Uses an open architecture;
- Centralizes administration and management;
- Complies with CSTA standard;
- Enjoys broad industry support.

Most of these statements are true. However, it is not true to say that TSAPI complies with CSTA. The latter is a protocol, and TSAPI is based on its message set. However, the API does clearly use the CSTA definitions as the sample of calls shown in the following list of example calls and events indicates:

- cstaClearCall();
- CSTAClearCallConfEvent;
- cstaClearConnection();
- CSTAClearConnectionConfEvent;
- cstaConferenceCall();
- CSTAConferenceCallConfEvent;
- cstaHoldCall();
- CSTAHoldCallConfEvent;
- cstaMakeCall();
- CSTAMakeCallConfEvent;
- cstaMakePredictiveCall();
- CSTAMakePredictiveCallConfEvent;
- cstaTransferCall();
- CSTATransferCallConfEvent.

Note there are no media processing calls here. TSAPI was originally based on the first issue of CSTA, which dealt only with call control. As media control commands were added to CSTA, later versions of TSAPI were similarly extended.

TSAPI certainly did set the switch suppliers alight; following an initial pause, it seems that they could not wait to jump aboard this new CTI bandwagon. Table 5.7 provides a list of switches that supported TSAPI in 1996. The

Table 5.7
Some of the Switches Supporting TSAPI

Supplier	Switches
Alcatel	4400
AT&T (Lucent)	Definity G3S
BBS Telecom	416/308 CAT
Bosch Telenorma	Integral 33XE
NetPhone	PBX-618
Comdial	DXP & DXP Plus
Cortelco	DSP 1000
CSB-System GmbH	Siemens Hicom
Ericsson Business Networks	MD110
Fujitsu	F9600
Intertel	AXXESS
Iwatsu America	IX_LANLINK
Mitel	SX2000 Light
NEC	NEAX 2000
Nortel	Meridian 1
Panasonic	VB-43941
Philips	SOPHO iS3000 Series
Samsung	Prostar DBS
Siemens ROLM	9751 CBX
Sprint	Protégé
SRX	Vision
Tadiran	CoralLink

list has been constantly expanding, and others were in the pipeline as Table 5.7 was compiled. However, do note that Novell belatedly introduced accreditation in 1996 and that few of those listed had been through that process at the time the list was established. In those days, claiming TSAPI conformance was not the same as supplying it.

One of the tardiest of the larger switch suppliers to throw its weight behind TSAPI was Nortel. However, when it did so, it did so with great enthusiasm. Why did Nortel not throw in its lot with Novell from the outset? Maybe Nortel was not keen to back two horses—the Novell and Microsoft solution. There is evidence for this; when Nortel did announce TSAPI support, it also announced Tmap, a piece of software that provided a TAPI interface that maps into the Novell server. Nortel later donated this software to the ECTF. Tmap's purpose was to allow TAPI-compatible applications to be run in a TSAPI environment. Does this make sense? It would seem more logical to write applications for TSAPI in the first place. Nevertheless, Tmap is a bit like those adapters we all carry with us so that we can plug our laptops into the power supply of another country. Adapters are a bit of a nuisance—but essential where there are no accepted standards for sockets.

TSAPI provided good switch support, but this is only part of the CTI solution; the API is there to support applications. On the computer side, TSAPI provided client support for the following operating systems:

- Windows 3.1 and 95;
- Windows NT;
- Macintosh;
- OS/2;
- UnixWare.

Table 5.8 gives some impression of the applications supported in 1996. Note that some of these have since changed name and sometimes supplier.

TSAPI gained acceptance in two key industries: the switch industry and the application industry. Many users and commentators regarded it as a standard. It certainly helped to grease the CTI distribution routes but probably did not achieve threshold. Many mistakes were made in its evolution, but this should not minimize its achievements.

Clearly, TSAPI's success was directly tied to Novell's domination of the LAN market; that was what all the excitement was about when TSAPI was first announced. However, as that domination declined with the growing popularity of Windows NT, interest in TSAPI also declined. Users want to obtain their CTI enabler from the people who supply the server operating system—Microsoft in an NT world. Otherwise, they want their enabler to come from some operating system-dependent source. What about CallPath from IBM? It runs with most client operating systems and supports TCP/IP, so that much is okay. This leaves the network operating system. Initially, the CallPath Server ran on OS/2 or AIX—very IBM. Companies that do not use IBM software are loath to base their CTI enabler on what is, for them, a nonstandard operating system. However, IBM ported CallPath to NT so that selection of a CTI enabler depends upon the functionality, switch support, and application support for CallPath. In this role, it offers tough competition to TSAPI. Further com-

Table 5.8
Application Support for TSAPI

Product	Supplier	Novell Category	Function
MultiCall-CTI/ CD	Lansys Ltd.	Telemarketing	Flexible call distribution
MultiCall-CTI/ CR	Lansys Ltd.	Telemarketing	Intelligent call routing
MultiCall-CTI/ PD	Lansys Ltd.	Telemarketing	Intelligent predictive dialing
Page Prospect!	Page Telecomputing	Telemarketing	Predictive dialer-based telemarketing system
Rostrvm2	Intercom Data Systems Ltd	Telemarketing	Complete software system for call centers
Sales Call Software	Page Telecomputing	Telemarketing	Aggressive telemarketing software
Sales Call Software LIGHT	Page Telecomputing	Telemarketing	Telemarketing from the phone
Tele-Scope	Davis Software Engineering	Telemarketing	Integrated contact management and inbound telemarketing
Telescript	Digisoft, Inc.	Telemarketing	Call center and telemarketing software
VerSatility	VerSatility, Inc	Telemarketing	Telemarketing, sales, and customer service
Link	Page Telecomputing	Personal	Personal information manager
Day-Timer Organiser	Day-Timer Technologies	Personal	Personal information manager
Oasis	Oasis Technology Inc.	Help desk	Customer service and workflow management system
Support Express	Opus	Help desk	PC-based help desk system
CallXpress3 Automated Agent	Applied Voice Technology	IVR	Voice response development environment
CallManager	Q.Sys International	Power dialers	Telephone operator aid
Edition 1 Telephony	Dr. Materna GmbH	Power dialers	Telephone function control from the PC
Group PhoneWare	Q.Sys International	Power dialers	Telephone function control via groupware
Micrologica Communications Centre	Micrologica Computer Systems GmbH	Power dialers	Predictive dialing and incoming call management

Table 5.8 (continued)

Product	Supplier	Novell Category	Function
Personal PhoneWare	Q.Sys International	Power dialers	Drag and drop call management
PhoneLine	CCOM Information Systems	Power dialers	Telephone function control
Phonetastic	SoftTalk Communications (Now CallWare)	Power dialers	Intelligent call control and intelligent call routing
SoftPhone	AnswerSoft	Power dialers	Integrating the data with sophisticated call functions
Call Script	Q.Sys International	Tools/enablers	Inbound call center with automatic screen pops
FastCall	Aurora Systems	Tools/enablers	Call control and middleware
PhoneLink	TeleMagic, Inc	Tools/enablers	Complete telephone automation through PIM
PhoneWare DDE Phone 2.0 for TSAPI	Q.Sys International	Tools/enablers	Allows rapid development of scripts to telephony-enable your existing Windows applications
PhoneWare TSAPI VBX	Q.Sys International	Tools/enablers	Visual Basic TSAPI tool
Telephony Trekker	Fitzgerald Associates	Tools/enablers	Configuration and status information management
On-the-Go!	AnswerSoft	Utilities	Scheduling and notification
Sixth Sense	AnswerSoft	Utilities	Call control and middleware
SoundByte	Envision Telephony	Utilities	Selective record and playback conversations
CallWare	CallWare Technologies	Voice mail	Voice mail
TeLANophy	Active Voice	Voice mail	Incoming call handling, integrated messaging, database integration
Vpro NetVoice	Voice Processing Corporation	Voice recognition	Telephone function control using simple voice commands

petition that does not suffer from the anti-IBM syndrome is represented by Genesys T-Server, Royalblue's Rostrvm, Aspect's Telelink, Dialogic's CT-Connect, and so on (see later). CT Connect is particularly interesting in this context. It offers a native API together with TAPI and TSAPI options. Yet other solutions appeared as the years rolled on, the later versions of TAPI and Sun's JTAPI being the most significant.

5.4.2.3 Microsoft Telephony Application Programming Interface

Overview. In the summer of 1993, Microsoft and Intel announced the Microsoft Telephony Application Programming Interface (TAPI) and its support by 40 other companies. Fundamentally, these companies were establishing a programming interface that allows a PC program to manipulate the telephony world without being too aware of which switch it is working with. Interestingly, the interface was primarily developed by Intel; Microsoft only became involved in the final year.

You might ask why a hardware manufacturer was so involved with a software specification? The answer must be that it wished to develop a market that it can then supply. Microsoft's entry into the CTI marketplace was considered to be critical at the time, simply because Microsoft Windows offered the prime route to the PC software marketplace.

It was claimed that TAPI would allow Windows application developers to write to a single Windows telephony standard, and that telephone switch manufactures would be able to provide an interface which links to their various switches. This industry standard would ultimately allow the control of multimedia communications embracing audio, video, and computer data.

TAPI gives Windows PC programmers access to the telephony world through a telephony *dynamic-link library* (DLL). The DLL also provides the interface to what are described in the TAPI specification as service providers. A service provider in this context is simply a device driver; providing an interface to the line, the telephone, or a third-party CTI link. Suppliers who wish to interface their telephone or switch must conform to the *telephony service provider interface* (TSPI), whereas suppliers or users writing application programs must conform to the telephony application programming interface. Microsoft simply supplies the bit in the middle that links these two—and the specification to allow the application and device supplier to conform.

TAPI defines two device classes: the line device class and the phone device class. This represents a clear bias towards first-party CTI because the concept of splitting a phone from the line that connects it to the switch is not supported by most third-party implementations. TAPI is defined as providing a first-party call control model, and this is all that the initial release (TAPI1) does.

TAPI is interesting in that it provides various levels of functionality, which can be grouped into two categories: simple telephony and full telephony.

Simple telephony. A few limited but powerful commands provide a basic level of functionality. A simple telephone call is generated by a straightforward make call command—only the destination address needs to be specified. Programmers need to know very little of the telephony world to use this, and there is no feedback. The call is managed by a call control application; no progress messages are specified. A default call control application is included with TAPI, but a programmer can substitute an alternative.

The only other commands provided in this category allow programmers to make and clear media calls. This provides application programmers with the means to connect an application via a selected device to a specified media channel and to set up a call via that channel. In this case, feedback messages on the progress of the call are returned. They are written to the window specified by the application programmer. The call is managed by a call control application. One again, a default call control application is included with TAPI, but programmers can substitute an alternative.

Full telephony. This level of functionality includes multiple functions and multiple messages. Functions are requests that are issued by the application programmer; messages are responses or events that are sent to the application programmer by the service provider. Messages usually indicate some change of state in the devices that are being manipulated. Functions and messages are grouped into three service levels and are separated at each level into functions or messages that communicate with the phone and functions or messages that communicate with the line.

1. *Basic level.* This level contains functions and messages that are mandatory and allow the controlled setting up and clearing of telephone calls. This level corresponds to first-party basic CTI, as defined in this book.
2. *Supplementary level.* This optional level contains many additional functions and messages that allow control over more advanced devices and lines with access to a greater range of features. This level corresponds to first-party enhanced CTI, as defined in this book.
3. *Extended level.* This level allows for service providers to incorporate their own specific functions and messages. These are not part of TAPI and can be agreed upon by an application programmer and a device supplier. This level allows developers to extend beyond the definition of TAPI and yet conform to its basic structure. It corresponds to the

escape function included in the CSTA standard from ECMA (see Section 8.6.1.2).

As previously stated, there are a large number of TAPI functions and messages listed in the specification—both at the application programmer level and at the service provider level. Table 5.9 lists the number of functions and messages for each device class in the initial release of the specification and gives an example of each.

What the programmer sees. To make a simple telephony call the application programmer simply writes

tapiRequestMakeCall

This call is linked to the DLL, and the application programmer can supply the following parameters:

- Destination;
- Application name;
- Called-party name;
- Comment.

Only the first parameter is mandatory; the others are for the programmer's own convenience. The destination address can be specified as a full canonical description, that is, a universal address, or a dialable number.

To set up a media call, the programmer writes

tapiRequestMediaCall

Table 5.9
TAPI1 Functions and Messages

Device Class	Phone	Line
Number of functions	26	60
Example	Phone Set Display	Line Hold
Number of messages	4	5
Example	PHONE_STATE	LINE_REPLY

This call is linked to the DLL, and the application programmer supplies the following parameters:

- Application window for response messages;
- The identity of the requesting application;
- The device class that specifies the media type;
- Destination;
- Security flag;
- Application name;
- Called-party name;
- Comment.

The last three parameters are optional; they are there for the programmer's own convenience. The destination address can be specified as a full canonical description, that is, a universal address, or a dialable number.

To release the call, the programmer writes

tapiRequestDrop

This must include the identity of the application that originally requested the call as a parameter.

Tone and digit monitoring. TAPI provides functions for detection of specific tones. These can be reported by frequency and cadence. Detection of digits can be provided for pulse or DTMF.

Media monitoring. TAPI provides functions for monitoring the media mode of a channel. This can be enabled in the CONNECTED state via the line-MonitorMedia function. Media modes include:

- Interactive voice;
- Automated voice;
- Digital data
- G3/G4 fax;
- Data modem;
- Teletex/videotex;
- *Analog display signaling interface* (ADSI).

Dealing with specific switches and phones. TAPI leaves this to the device supplier. The interface specifies each of the functions and messages that are to be implemented at the service provider level. These correspond to the functions and messages that the application programmer sees, but the Telephony Service Provider Interface (TSPI) provides for the linkage of these, through the DLL, to the phone and line hardware.

Many early applications for TAPI were simple and used a modem as the first-party line intercept. Many of these applications used a basic driver that Microsoft includes with the operating system called Unimodem. This is described as the universal modem driver for Windows operating systems. Unimodem translates TAPI calls into modem AT commands to perform basic functions such as the following:

- Configuration;
- Dialing;
- Answering.

Unimodem enables basic CTI functionality and can be used to enable fax applications. An improved version of Unimodem, Unimodem V, added features for data/fax/voice modems including the following:

- WAVE file playback and recording to phone line and handset;
- Support for speakerphones;
- Caller identification;
- Distinctive ringing;
- Call forwarding.

Once again the parallels between first-party basic and first-party enhanced can be clearly seen here.

Telephony objects. TAPI recognizes the line, the phone, and the call as objects. All of these objects have an identity and a state. The call represents a connection between two or more addresses. TAPI does not define a call model; the state of calls is maintained by the DLL. Applications are informed of state changes via a LINE_CALLSTATE message. Examples of call states are

- Offering;
- Accepted;

- Dial tone;
- Dialing;
- Connected.

Incoming call information can be requested via the lineGetCallInfo function. The information includes: bearer mode and rate, media mode, caller ID, called ID.

The Evolution of TAPI

TAPI seemed to be all talk and no action for some years. Developer kits were available for early versions of Windows, but it was not until TAPI's inclusion in Windows 95 that things really began to happen. Even then many people were skeptical about Microsoft's entry into the telephony world. There were many cries of "Where are the applications and where are the drivers?"

In mid 1995, I began a search for TAPI applications and found 42 of them, though some were really application development aids. Here are some examples:

- ACT! 2.0 for Windows from Symantec Corporation, California: "Manage your business relationships more productively with ACT!, the top-rated, best-selling contact manager."
- Multimedia Connect from Synchro Development, Pennsylvania: "Multimedia Connect provides the look, feel, and function of a desktop telephone/fax right on your computer screen."
- WinPhone from MegaSoft GmbH of Austria: "WinPhone is a very comfortable dialer that acts like an assistant."
- ECCO Professional from NetManage, Washington: "ECCO integrates personal and group information, including, calendar, phonebook, project management, and group scheduling."

Generally, the applications were basic productivity tools that allowed user access to screen-based telephony; this suited TAPI's first-party orientation at that time.

I also found just under 40 service provider interfaces, some for the POTS line and some for digital lines. Here are some examples:

- PC-DT from Fujitsu Business Communications Systems, California: "The PC-DT is a TAPI-compliant card for Fujitsu's digital telephone."

- Alcatel TAPI Interface Kit from Alcatel Business Systems of Germany: "Alcatel Proprietary PBX phone set adapter with link to PC serial port, TSPI, and TAPI-based clipboard dialer."
- Axxessory Connect from Inter-Tel, Arizona: "TAPI service provider and PC phone for the Inter-Tel Axxess hybrid/PBX phone system."
- PowerPhone/N from Lexical, California: "PowerPhone/N provides easily accessible desktop telephony services for connection to a Northern Telecom Norstar or Meridian PBX."

The applications and drivers were becoming available, but there remained a big question over the whole TAPI initiative: Would the applications work with the drivers? Microsoft merely supplied the enabling software to link the two; it did no testing outside of this. This gave rise to the TAPI "Bakeoffs"—events that brought software and hardware companies together so that they could carry out some compatibility testing. At the first Bakeoff, over forty companies took their software and hardware to Dallas. In an event sponsored by Microsoft and Nortel, they began to test their products in a realistic environment.

Pressure to move TAPI into the third-party world became irresistible after the release of Windows 95. Microsoft gave presentations that explained that TAPI was indeed capable of providing third- and first-party CTI. The third-party diagrams that were displayed included layers in the client PC that converted first-party calls to third-party equivalents. However, the software for these layers did not exist.

Some suppliers began to provide this conversion layer, linking TAPI calls to their own servers. Genesys provided early solutions in this way. One large call center was based on the Genesys T-Server software. The SoftPhone API, which Genesys supplied for establishing the client interface, was interfaced to a TAPI service provider. The application works to the TAPI interface in the usual way. However, TAPI1 did not contain the necessary functions for call center operation. Genesys had to extend the TAPI command set using the "extended level" section of TAPI (see previous description of full telephony).

However, the job should be done at source, and Microsoft had become very keen to extend its market into the corporate world. This it had done with the introduction of Windows NT, and TAPI2 was released with this operating system in 1996. The TAPI2 architecture is shown in Figure 5.8. Initially, there were problems with the remote service provider, but these were resolved in later issues.

Major changes took place between TAPI1 and TAPI2. These included support for 16- and 32-bit applications, management functions for the client

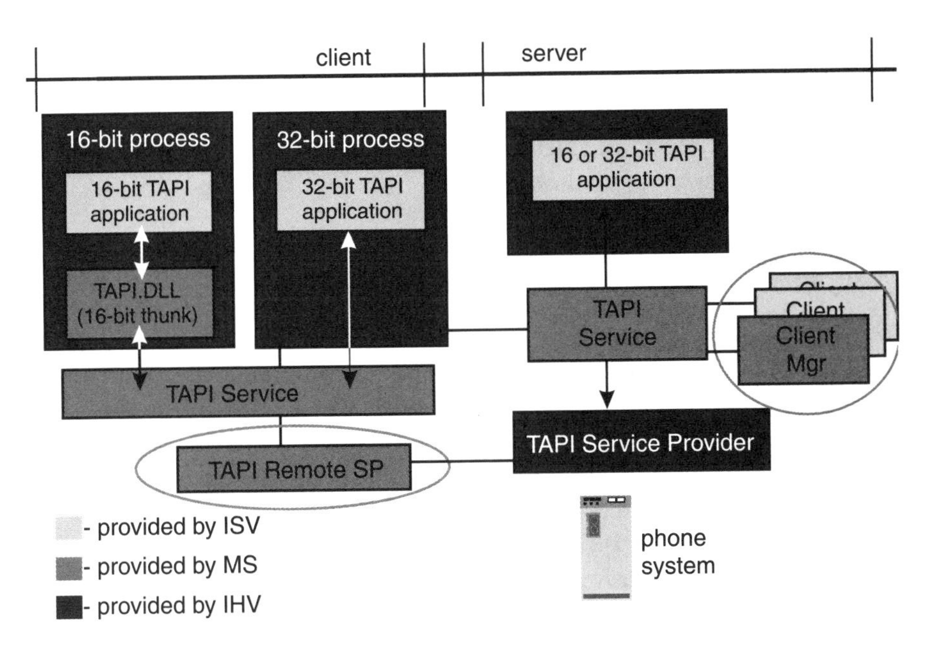

Figure 5.8 The TAPI2 architecture. (*Source:* Microsoft.)

server environment, and the move to third-party CTI support. The additions that were specific to the call center included the following:

- New device types, including predictive dialer, route point, and queue;
- Filtering of call state, call information events;
- Association of call and application data (for VDCA);
- Agent login (ID, group, password) and state monitoring and control;
- Message waiting and display control without using phone functions;
- Centralized timing of call state duration.

It was Microsoft's intention to begin packaging third-party drivers (service providers) with its operating system. This has tremendous implications for CTI. If Microsoft ships the drivers, it must also test and support them. Inclusion of driver software could make the enabling of a third-party CTI application almost as straightforward as adding a new printer—and perhaps as cheap. Unfortunately, Microsoft has not yet delivered on the intention, partly because the switch suppliers have not provided their drivers for testing. This development is of great interest in the CTI world.

In 1997, Microsoft announced yet another version of TAPI, TAPI3. This version is object-based and should therefore be language-independent. It also includes Internet protocol telephony functions and has extended the media control capability of TAPI. TAPI has certainly increased in functionality since its initial appearance in Windows 95.

5.4.2.4 Mitel's MITAI

Overview. Although many CTI implementations incorporating the Mitel SX2000 PBX utilize one of the mainstream APIs described in the chapter, it is interesting to recall that Mitel did implement its own solution. Mitel was one of the earliest PBX suppliers to provide a CTI protocol on a PBX and to produce an API. Called MITAI, the API resides within a PC and therefore allows system integrators to produce applications within a PC environment that is buffered from the raw HCI protocol. The MITAI software toolkit runs under UNIX and uses standard PC AT hardware. It is therefore an *open environment* for software developers who wish to develop software for the closed environment offered by Mitel switches that support the company's HCI protocol. It represented a move into the applications area by a PBX manufacturer. The company developed its own applications on this platform.

The physical arrangement used is shown in Figure 5.9 [2]. The HCI link is connected to a PC using a Dataset running at 19.2 Kbps. The Dataset is a Mitel device; it acts as a safety barrier between the PC and the switch. The Dataset connects to a PC intelligent communications card. Applications can

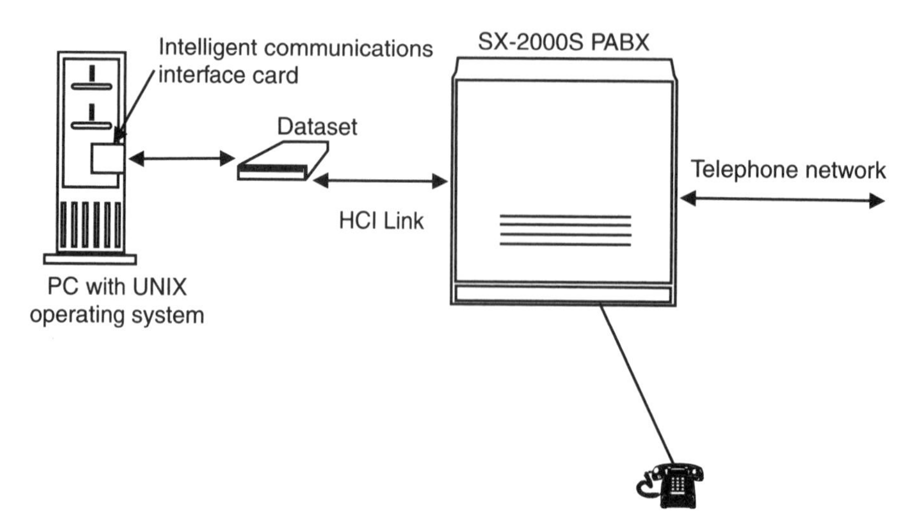

Figure 5.9 MITAI functional configuration.

exist within the PC shown in the figure or at any station on the LAN to which the PC may be connected.

Signing on. The first thing that has to happen in the MITAI environment is that an initialization routine has to be called. No other functions can be called before this. This call sets up default values within the MITAI internal structures and will be followed by the creation of a communication channel, which selects the switch to be used in the session and returns a reference number for the switch.

Telephony objects. MITAI supplies support for the following devices:

- Simple analog telephones;
- Mitel feature phones;
- Trunks;
- Call queues (including ACD groups);
- ACD agents.

Note the inclusion of the Mitel feature phone. These are called Supersets by Mitel, and the range includes a variety of specialized telephones. Each line or button on a Superset is considered to be a separate device.

MITAI also recognizes calls. Devices connected to a call have unique identities.

Monitoring. Devices can be monitored by executing a routine that specifies the device to be monitored. Devices are specified by the switch on which they reside plus a DN. The DN may refer to any of the following.

- A telephone;
- An ACD group;
- A line button on a Superset;
- An ACD agent.

Events can be polled by execution of a get event routine—if nothing is happening to the device, this will return no event. Alternatively, the events can be collected asynchronously through the execution of a set callback routine that establishes a call back procedure to which events will be returned as they occur. This routine also allows the specification of the events that will be returned—a filter facility. It is activated by an activate callback routine.

The API functions. Mitel divides the services supplied by MITAI into two groups. The first group is described in Table 5.10. These are services called *call manipulation services.*

The second set of MITAI functions is limited in number. These services are called *noncall-oriented services* and are listed in Table 5.11.

Command structure. Once again, make call is chosen for this examination. The function of make call is to attempt to establish a call between two parties; at least one of these parties must be "monitored object" and must therefore be local to the switch. In MITAI, in contrast to CallPath, the destination DN must be specified. MITAI is based on the C programming language—the call used to

Table 5.10
MITAI Call Manipulation Services

Service	Notes
Alternate call	Alternate between an active call and a call on consultation hold.
Answer call	Answer an incoming call that has been offered to a device.
Assign call identity	Once the identity of a caller on an incoming trunk has been established (name and number, via an attendant or IVR), this identity can be assigned to a call and will then be returned whenever the trunk information is reported.
Cancel consultation call	Return to original call that was on consultation hold.
Clear call	Release the call from a specified device.
Conference call	Merge two or more calls into a single conference call.
Consultation call	Place active call on hold and place a new call to another party. Can be used to simply place a call on consultation hold.
Hold call	Place an active call on hard hold—the call must be monitored.
Make call	Make a call by first calling a monitored device.
Outpulse digits	Send STMF digits from the monitored device.
Pickup call	Pick up a call that is ringing at another device and connect it to the monitored object.
Redirect call	Redirect a ringing call to another device.
Retrieve call	Retrieve a call from hard hold and connect it to the monitored object.
Send callback message	Send a message to a called party that could not be contacted.
Split conference call	Split a conference call into two calls.
Transfer call	Complete call between a consultation held party and the other party of the active call.

Table 5.11
MITAI Noncall-Oriented Services

Service	Notes
Alter device feature	Set or clear attributes, such as Do not disturb ACD agent login ACD agent busy Auto answer Cancel ACD work time
Query device feature	Determine the status of the attributes listed in alter device feature
Feature monitor	Return a feature event if a feature of the specified device changes

execute the make call routine is #include ai.h

Error = SXMakeCall (SX_SOURCE_DN, source_dn, SX_DN, dest_dn, NULL)

Naturally, the API has to be initialized and the switch opened prior to execution of this routine.

Table 5.12 presents the parameter string for the make call command. Note that in Table 5.12 the make call command can reference the originating device by monitor ID or source DN, but not both.

The returned error codes are the following:

- OK;
- Switch not open;
- Nonterminal Comms Error;
- Missing Required Attribute;
- Attribute not Supported;
- Unknown Attribute;
- Cannot Get Shared Memory;
- Invalid Monitor ID;
- Invalid Switch ID;
- Invalid DN;
- Congestion;
- Invalid Device;

Table 5.12
Parameter List for the MITAI Make Call Command

Parameter	Notes
Switch identifier	Not required for single switch applications
Monitor identifier	Obtained if the source has been placed in monitor mode
Source DN	The line from which the call will originate
Target DN	The DN to be called
Account code	The code to be used in charging for this call

- Privilege Violation;
- Unable to Originate Call;
- Invalid Called Device;
- Feature not Allowed.

None of the returned values can indicate that the *telephone call* is successful. That can only be ascertained by examining the call event messages.

Test tool. Mitel provided an Experimenter's Interactive Test Tool with MITAI. It allows up to three monitors to be set up on specified DNs, ACD groups, and so on. The details of all events occurring at the monitored objects are then displayed in a window of the management terminal. The user can do a number of things with the three monitors, ranging from starting and stopping them to issuing call manipulation commands.

The display of monitored events is characterized in Table 5.13. The first line contains the name of the event, the state, and the cause. The next gives call

Table 5.13
Example of Output from the MITAI Test Tool

CallReceivedEvent	ReceivedState	NewCall
Callid=93	hmonitor_obj=1212072 number=153	Mon Mar 18 02:45:45 1992
Calling:/2000 dialed_digits:/3000	Margaret Debenham	

identity, monitor reference numbers, and a time stamp. Subsequent lines provide the values of any event-specific data fields.

The MITAI test tool allowed a newcomer to the CTI application generation world to watch the establishment and progress of telephone calls from an applications viewpoint. It further allowed a budding application engineer to try out some ideas before designing and coding new programs.

5.4.2.5 Summa Four's ASIST/API

Application Software Integration Support Tools (ASIST)/API was defined as a C language representation of the Summa Four Host Control Link protocol. The API is itself written in the C language and claims to be independent of any specific host operating system. It is a set of subroutines that communicate with a communications driver via a link manager. It supplies these functions:

- Command generation;
- Response confirmation;
- Message interpretation;
- Network layer exchange;
- Application event logging.

The higher-level functions of command buildup and report parsing are also included, and these allow all the command and report functions to be organized under two function calls. The API uses a single data structure for all the command and report functions.

The API lists some 36 command/report functions. All of these are a direct rendition of a protocol function, and as such, a description of the API becomes simply a description of how to exercise these functions at the application level in the C language.

5.4.2.6 JavaSoft's JavaTel

The success of the Java language is undeniable. A recent visit to my favorite book shop brought three things home to me. First, the book shops still do not know where to place books on CTI; second, the computing section is orders of magnitude larger than the telecommunications section; and third, the computing section overflows with books on Java. What makes Java so popular? I am sure that there are many answers to this—not all of them logical. However, a major attraction is the platform independence that Java applications possess. Java programs are not compiled into machine code but into an intermediate code that is interpreted by a Java Virtual Machine. Java Virtual Machines are software packages that are platform-specific. They interpret the same intermediate code for a Wintel environment or for a Macintosh environment. This portability of applications is a useful characteristic generally, but even more so in a network computing environment. Furthermore, application portability is supremely important in building a stock of CTI applications that can be used in all environments.

To make the Java approach usable for CTI, however, a Java-based API is required; this is what JavaTel, the Java CTI enabler solution, is all about. Its API is called the *Java Telephony API* (JTAPI). The specification was defined by a number of leading software, telephony, and computer companies in addition to JavaSoft itself, and the first issue was released at the end of 1996. The objectives from Sun and its collaborators for this approach are listed as follows:

- To support both third-party and first-party CTI;
- To complement existing call control standards, TAPI, TSAPI, CallPath, etc.;
- To be compatible with the Java Media framework;
- To support all environments capable of running a Java Virtual Machine;
- To allow telephones, computers, and other devices to be linked through the Internet/Intranet.

Java is an object-oriented language, and it should be no surprise to readers that the various API calls are themselves objects. The JTAPI objects are broken up into functional groups. The heart of the JTAPI packages is referred to as the "Core." The Core package defines basic telephony objects, the call model, and methods to perform very basic telephony functions.

Examples of telephony functionality found in the Core package are originating, answering, and disconnecting telephone calls. Advanced telephony functionality and other application-specific functionality are found in a set of class library packages referred to collectively as the Standard Extension. Figure 5.10 indicates all of the functional groups that are envisaged.

Table 5.14 provides a few examples of the core group of functions in JTAPI form.

Introducing a new API into the CTI scene late in the 1990s does not seem a particularly good idea. The scene is already strewn with competing solutions, and any newcomer is faced with the daunting task of interfacing to all the different switch protocols that exist. Novell and Microsoft expected the switch sup-

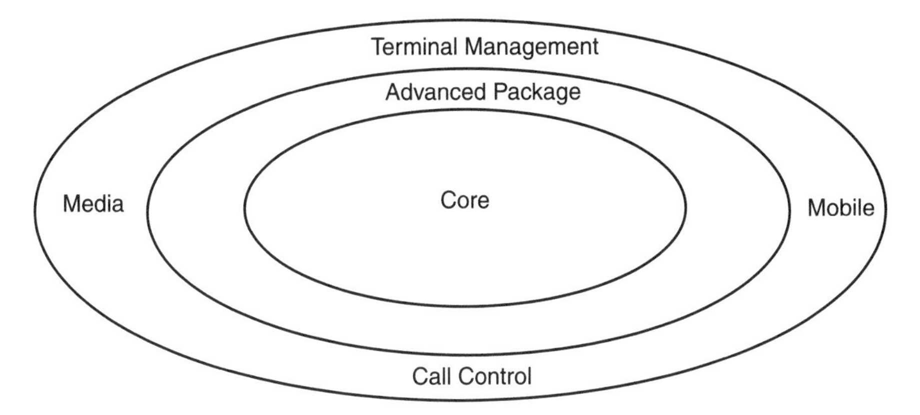

Figure 5.10 The JTAPI object groups.

pliers to write the drivers that enabled TSAPI and TAPI for their switches. On the whole, the suppliers delivered, but they were left exhausted by the effort.

Fortunately, the JTAPI approach does not require a new set of drivers. The specification proposed that JTAPI interfaced to the existing CTI APIs: TSAPI, TAPI, XTL, CallPath, etc. In this way, the need to provide switch drivers is removed. The drivers are supplied by the existing API software running on a central CTI server and communicating with JavaTel clients via the remote methods invocation protocol.

That was the theory. Unfortunately, however, the commercial world does present some barriers to this approach. The API suppliers do not normally make their software available at nil cost. Therefore, the JavaTel approach is likely to need a license from, say, Novell, for the use of TSAPI. Cost aside, this hybrid approach has support implications. If problems occur between the application

Table 5.14
Example Core Functions

Function	Description
Answer	Answers a telephone call
Drop	Drops a connection from an active telephone call
GetCall	Returns the call object associated with this connection
GetDevice	Returns the device object associated with this connection
GetState	Returns the current state of the connection

and the switch, these might lie in JTAPI implementation or the TSAPI implementation or the switch driver—all of which have different suppliers.

Possibly to avoid commercial and support issues, Ubiquity, one of the first suppliers to provide JTAPI enabled applets, wrote a driver to work with the switch rather than working through an existing API. However, IBM provided CallPath-based JTAPI and endeavored to resolve the support issues by providing the CallPath-to-Java interface software as an extension to its existing CallPath Enterprise Server product.

5.4.3 Media and Message Processing APIs

There are many APIs in this area, and many of those described above do provide some media control functionality. Though, there are no independent standards for call control APIs, there is at least one set for media processing. This is the S.100 standard from ECTF; it will be covered later in the standards section of this book.

Possibly, the most significant API in the messaging sphere is the *messaging API* (MAPI) from Microsoft. MAPI is said to provide an easy-to-use and consistent interface for working with a whole range of e-mail systems. In this world, the drivers translate between, for example, MAPI commands and ccMail or Lotus Notes commands.

Simple MAPI provides the most common functionality required in a messaging system, including e-mail and data routing facilities. It consists of 12 functions and can be used to create messaging-aware applications. Sample functions are the rather obvious ones, *send mail* and *send file.*

Extended MAPI has a richer function set and provides a full interface to messaging systems and subsystems through an advanced set of API calls. It includes the ability to create forms, to manipulate message store data items, and to customize address books.

Also in the Microsoft stable is the *speech API* (SAPI), which provides an interface to speech recognition and synthesis hardware. There are plenty more from Microsoft. One set of interfaces is generally directed to media control and is grouped under the general title DirectX. This is said to be a type of API that forms a hardware abstraction layer that acts for Windows 95 and various types of hardware. The DirectX interface includes Direct3D (which speeds up texture mapping and other 3D graphics processes), DirectSound (for audio cards), DirectDraw (for vector graphics), DirectVideo (for AVI files and other moving pictures), and DirectPlay and DirectInput (which simultaneously support sound, drawing, video, networked gameplay, and joystick standards).

The SAPI interface has some competition in SRAPI, a speech recognition API agreed by a group of vendors and interested parties. The same group has also published an API for speaker verification. Will the list of APIs never end? Fortunately, most of the specialized APIs mentioned here are covered by the ECTF S.100 specification.

5.5 Application Creation in CTI

It is expected that eventually the features of CTI outlined in the APIs examined in Section 5.4 and those introduced by the various protocols examined in Chapter 3 will become standard options in most packages and environments. CTI calls will be stripped of their arcane reputation and absorbed into mainstream computing at personal, departmental, and corporate levels. Microsoft's lead in packaging TAPI with the operating system represents a step in this direction. However, there is still a long road of change to be traversed before the ultimate goal is reached.

Writing new applications that have CTI capability requires an understanding of the operation of telephony and, of course, the needs of the business application itself. A necessary preliminary is the establishment of agreed-upon customer requirements—a statement that is true of all application developments. But CTI is special in respect to requirement specification in two rather opposite areas. At one extreme, users are unaware of the potential provided by linking their switches and computers; at the other, they may overestimate the ease with which this can be done and over-estimate the functionality that may be achieved.

With a clear requirement specification and a well-specified API, the developer is all set to go. If the application is simply concerned with call control, then the only question to answer at this stage is whether the application is to work with one switch or many. If it is to work with one switch only, the API specification for that switch can be used without constraint. However, if the application has to work to more than one switch or, as is more likely, has to work to one switch initially and others later; certain constraints should be imposed. The constraints will help limit the functions to a common subset of the switches to be covered. For example, if the application were being implemented in IBM's CallPath/400, as Table 5.6 indicates, the following functions were common to the Rolm CBX, Siemens Hicom, and NT Meridian switches:

- Answer call;
- Conference call;
- Disconnect;
- Extend call;

- Make call;
- Monitor;
- Query party status;
- Receive;
- Retrieve call;
- Transfer call.

This represents 10 of the full set of 18 functions that were then supplied by CallPath and a reasonable subset for the implementation of CTI applications, though it reinforces the point made earlier that multiswitch APIs do limit functionality to the lowest common denominator.

Surprisingly, some of the earliest application creation tools operated at a high level. The *service creation language* (SCL) was developed by the U.S. telephone company Nynex at the end of the 1980s and was used by Aurora Systems to generate CTI applications. Itself written in 'C,' the language provides English-like commands to create such applications as the following:

- Operator services;
- Order entry;
- Message delivery and voice mail;
- Telemarketing;
- Call center automation.

It was claimed that the language incorporated the concepts of object-oriented programming, 4GL, computer-aided software engineering, and state machine generators—which seems to cover most of the popular concepts in software development today! In SCL, a new service was prototyped and executed in a similar manner to that in which an interpreted program would be executed. SCL transformed single thread tasks into multitasking programs.

The application was created from a number of sessions, each of which consisted of a simple call script and was equivalent to an individual telephone call. Events were part of the SCL language. A caller pressing a touch tone key is an example of an SCL event. Events were attached to particular actions; actions could be created by the developer or selected from a built-in library. Examples of the latter are the following:

- Compare;
- Convert;

- Copy to;
- Send to;
- Time;
- Add to;
- Divide;
- Locate (database);
- Update (database);
- Wait for.

SCL supported its own local databases, which were simple collections of records. It also provided support for Informix, Oracle, and Sybase databases by defining them as objects. The objects recognized by SCL included hardware devices and software programs (a database manager, for example). The language itself was supported on a wide range of computers.

SCL, which may have been one of the first languages of its kind, formed the backdrop for the development of a plethora of service creation languages for use within the intelligent network. Many of these are now GUI based—the service Design Center from Alcatel is a good example.

Application production outside of the intelligent network took a rather different course, although graphical application generators are also used here. Vicorp, for example, produced a graphical application generator that covers most of the CTI basic functions. The technology, which is called QuikScript, is described as follows.

- Combines telephony integration, voice, agent, and transaction functions;
- Creates and maintains scripts and applications;
- Allows development of new application objects;
- Downloads scripts to voice response and agent environments;
- Based on OSF/Motif GUI running on UNIX workstations;
- Operates in a 'C' development environment;
- Has integrated voice recording, mixing, and editing functions.

Visual Age for Java provides another example. This allows Java beans to be placed upon the screen and connected to produce Java-based CTI applications.

Graphical application generators aside, the major thrust has been around the establishment of layers that provide stable interfaces. The interfaces then form the foundations for approaches that CTI enable existing applications rather than creating applications from scratch. Section 5.5.1 describes this evolution and is a useful summary of this chapter.

5.5.1 Application Enabler Layers

Much of this chapter has been devoted to describing the CTI APIs. However, the API is the most basic of enablers. In Figure 5.11, the API is shown as application-enabling layer one (A1). It may well be the layer that the CTI application sits upon; however, other application layers may exist, as shown in Figure 5.11.

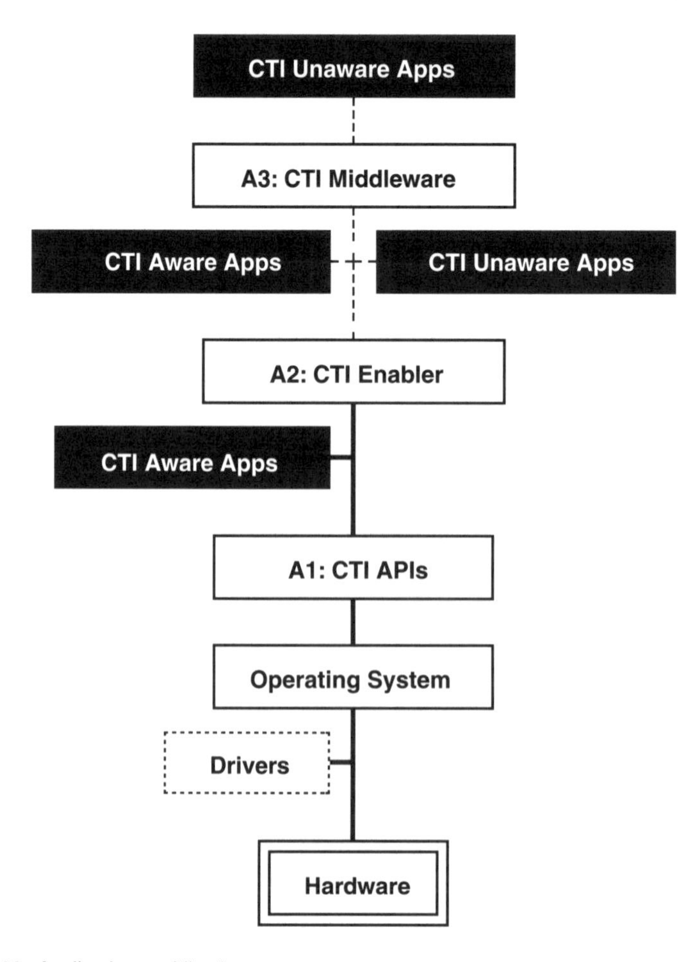

Figure 5.11 Application-enabling layers.

Applications that sit upon the enabler layer itself (A2) may have been modified to insert CTI API calls, in which case they are termed CTI-aware. Alternatively, the enabler may link to an unmodified application which is CTI unaware. CTI middleware generally links to unaware applications.

In Figure 5.11, the driver is the software that deals with a switch-specific protocol or media hardware, providing a common internal interface to the basic API.

The API level can consist of a basic API—software that presents basic telephony functions such as make, transfer, or conference a call—available in a precise form to applications. The basic application interface will usually be in the form of a library of routines, each function being defined as a procedure call. However, it may also exist at client or server level. If the former is the case, then another layer may exist to transport it across the LAN—over TCP/IP perhaps. The diagrams describing TAPI and TSAPI will clarify this.

A number of API variants have been considered in some detail in this chapter. However, there are more. One of the most significant omissions may be CT-Connect from Dialogic. This software has historic significance in that it was derived from the CIT software developed by DEC. In its day the DEC initiative played a leading part in the development of the CTI market. I believe that Digital was one of the first companies to provide a book listing all of the applications that operate with its API. I have one of those books. I also have the most recent applications listing for CT-Connect [3]. It contains well over 100 applications from well-known CTI application software companies such as VerSatility to specialized banking software from companies like ICR. There are similar listings available for CallPath, TAPI, and TSAPI.

CT-Connect is an interesting API for another reason. Besides supporting its own CSTA-based API, it also supports the TAPI and TSAPI interfaces. It is therefore a hybrid in the layer A1 world.

All of the applications in the CT-Connect handbook are CTI-aware. This means that they have specific CT-Connect calls embedded within the program and code to support messages that may be sent to them from CT-Connect. The next layer up in Figure 5.11 can support applications that are not CTI aware.

The enabler layer (A2) contains one or two optional software functions placed between the API layer and the application software, usually at the client. One of these is sometimes called the funnel. Its function is to simplify the API down to a few key commands and to convert to a first-party interface from a third-party implementation. The other is called the linker. It provides a method of linking the application programming functions to an existing application program without modifying the latter. This is a highly desirable method of incorporating telephonic functions into application software. If the application already has its own CTI calls, the linker translates these into the API equivalent.

If it has none then the linker may, for example, translate a call arrival event into an application initiation message and provide call-based data to write the first screen. As the CTI market has progressed, enabler layer software has increasingly been incorporated into the next layer above it—the middleware.

5.5.2 CTI Middleware

The middleware layer (A3) is usually a combined linker and funnel together with a CTI user interface. Middleware is usually an application in its own right, providing screen-based telephony and call-based data selection along with simple screen popping and voice and data call association. It is called middleware because it sits between the API and applications that are not CTI aware. It is often configured by the activation and modification of rules; generally, the rules are manipulated by Windows-based menus and buttons.

Middleware is a much abused term—it is so general that it is almost too useful. But there are examples of individual packages that exactly fit the description provided for layer A3. Among these, Aurora's FastCall was probably the first to address both the call center and desktop markets. It was soon followed by Phonetastic from SoftTalk and many others. One of the major innovations that FastCall introduced to the CTI market went beyond the control of calls by rules; it also allowed a degree of user training. With FastCall, it is possible for the user to do what he or she would do in processing an incoming call—and FastCall learns and remembers. The user training is at a very crude level. FastCall simply records the actions that the user makes and then replays them for each call from a *keystroke macro.* Crude though this may be, it is fast and effective. It has the usual flaw inherent in any keystroke macro. If anything on the user's desktop changes, the macro may not work. Still, it does not take long to rerecord so practicality might be preferred to perfection in this case!

Middleware can be much more sophisticated. Some solutions provide a complete CTI server solution that embraces the middleware concept but supports it with a selection of application layer modules. These modules are optional; they perform such things as the following:

- Screen-based telephony interfaces;
- Outbound dialing of various sorts;
- Call and data synchronization;
- Inbound call routing, which is often application-based;
- Coordinated collection and processing of statistics;
- Control of the IVR function;

- Voice or fax server control;
- Legacy interfacing.

Suppliers that provide much or all of this functionality sometimes supply all three layers of the CTI stack, together with switch drivers. Others concentrate on the A3, middleware layers, obtaining layer A1 support from established CTI API suppliers.

The correspondence between layer A3 functions and the basic functions of CTI that were elaborated at the beginning of this chapter should be apparent. Some suppliers that have developed solutions of this type are listed, with their solutions in Table 5.15. Some of these are system integrators in the CTI field; others are purely software suppliers, and one is a switch supplier that bought a system integrator.

The Genesys product is an extensive suite for introducing CTI into the call center. The main components are described as follows, based on the definitions contained in the company's web site:

- *T-Server:* Integrates public and private switches, ACD units, IVR units, voice mail systems, and predictive dialers with call center software applications to provide customer applications with CTI capabilities.
- *InterActive-T:* A set of standards-based integration toolkits that enables the integration of desktop applications; its interfaces include support for industry-standard tools and protocols, including Java, ActiveX, OLE, OCX, DDE, COM, CORBA, and TAPI.

Table 5.15
Some System Integrators and Their Enablers

Supplier	Middleware
Aspect	Telelink
Co-Cam	Euphonic
ETSI	Avid
Genesys	T-Server
IBM	CallPath Enterprise Server
Nabnasset	VESP
Royalblue	Rostrvm

- *SoftPhone:* A user interface and a client API that usually sits within the agent's PC, providing screen-based telephony or CTI buttons and feedback into an existing application.
- *Genesys Intelligent Call Delivery (ICD):* An intelligent call routing product that performs both skills- and data-directed routing. ICD decides the best agent or agents to handle specific customer requests by checking caller information against agent records. Routing can also be based on service-level objectives and a number of other parameters.
- *Genesys Campaign Manager:* A predictive dialer, designed to perform true call blending; it allows outbound campaigns to run on top of inbound services.
- *Genesys Call Center Pulse:* Provides real-time statistics and reporting of call center operations and agent activity.
- *Genesys Data Analysis and Reporting Tool (DART):* Supports intelligent, long-term reporting and information analysis for call centers. It includes built-in reports, custom statistics, and SNMP support.
- *Genesys Net Vector:* Allows a company to leverage its Web sites to deploy innovative Internet-based customer service solutions. Customers navigating a Web site can click a button and request an outbound call from the company. The agent can also view the same Web pages as the customer.
- *Genesys Video ICD:* Enhances customer interactions by adding video capabilities to the voice call. Calls are routed to the appropriate video-enabled agent. The agent and caller can also view the same data on their screens.
- *Genesys Agent Pulse:* Provides a mechanism to display individual agent performance in comparison to a group. This may be used by both the agent and supervisor to gauge performance and set performance objectives.
- *Genesys IVR Server:* Provides a standard IVR device driver interface to T-Server. It should provide a fixed interface regardless of changes in the IVR or client applications.
- *Genesys OCX Toolkit:* A graphical package for developing CTI applications for the call center.

Another supplier, Royalblue, offers a similar set of components under the general title Rostrvm. Once again, the following descriptions are based on the contents of the company's Web site.

- *Rostrvm CallManager:* The core element of Rostrvm responsible for interfacing to all telephony hardware components (PBX, IVR, etc.) and for the routing of information within a Rostrvm-enabled call center.
- *Rostrvm AdVisor:* Rostrvm provides full on-screen call control for agents. The AdVisor provides a call control window, call history window, real-time directory services window, activity codes window, and a Rostrvm configurable CLIPboard viewer. The Rostrvm MiniVisor provides the same facilities but is presented as a floating tool bar to minimize use of screen real estate.
- *Rostrvm ScreenPhone:* For switches that are either PBXs or ACD/PBX hybrids, Rostrvm provides a software telephone that allows individual user positions to be CTI-enabled. Calls can be freely transferred between agents and management of back office staff.
- *Rostrvm SuperVisor:* A full suite of supervisor functions is provided allowing call center supervisors, managers, and team leaders to view call center activity in either real time or recorded time.
- *Rostrvm ControlCenter:* Rostrvm provides a comprehensive suite of GUI drive configuration and management functions for the integration of the telephony and computing environments. For example, separate components are provided to handle agent profile configuration, call class selection, skills-based routing, network configuration, security, and so on.
- *Rostrvm CallDirector:* CallDirector provides sophisticated off-switch call routing, based on call information from the PSTN or from information collected by a VRU. CallDirector supports different routing algorithms that ensure the call is delivered to an agent best placed to handle that call.
- *Rostrvm CallBroker:* Rostrvm supports integrated voice and data transfer between agents on the same PBX, between agents on different PBXs, and between agents on different manufacturers' PBXs. CallBroker mediates between multiple Rostrvm enabled call centers to ensure that the voice and the data element of every call is delivered reliably.
- *Rostrvm OutBound:* A comprehensive outbound dialer supporting preview, progressive, and predictive dialing. Rostrvm OutBound is a software-only component that handles all of the management of various outbound campaigns and provides the software-driven dialing engine used to make calls. Rostrvm OutBound is an integrated component of the Rostrvm product suite and can therefore provide outbound dialing services to any agent within the call center environment.

- *Rostrvm InterFace:* Rostrvm provides a comprehensive suite of APIs, available in a variety of formats for use by application developers to integrate the telephony and computing environments. The APIs are switch-independent, are available on a variety of hardware platforms, and present a CSTA-based telephony model to applications.
- *Rostrvm SwitchSimulator:* The Rostrvm switch simulator allows application developers to test the functionality and the performance of their applications without the need to access a real switch or to interrupt activity in the live call center.
- *Rostrvm AutoAgent:* AutoAgent acts as an automated agent so that applications can be CTI-enabled without having to be modified.
- *Rostrvm CallGuide:* CallGuide helps agents navigate through complex conversations. It interacts with the dialogue between the agent and the business application to ensure that the agent is kept on track and understands what is needed throughout that conversation.
- *Rostrvm InterAct:* Visitors to the company's Web site can request call backs. Rostrvm will automatically schedule the call back, make the call, connect an agent, and allow the agent to see the same Web pages as the caller.
- *Rostrvm ReportWriter and AuditLog:* These components provide comprehensive recording and analysis of call center activity and ensure that companies are able to understand what is happening in their call centers.

Table 5.16 attempts to map these functions against the generic functions of CTI as defined in Chapter 1 and at the commencement of this chapter. The mapping is my own and is based upon the definitions listed above.

Table 5.16 should not be taken as a feature or function comparison; it is far too crude for this. Many functions that seem not to be supplied at all are undoubtedly contained in one or other of the components rather than packaged as separate components. Interestingly, there were some Rostrvm components that could not be mapped, specifically Rostrvm ControlCenter, which is a management component, and Rostrvm CallGuide, which is an application component. Maybe this indicates a need to expand the scope of the CTI generic functions into this field.

It is also interesting to note that there are no entries for DT or ME. Once again, the functionality may exist, but there are no separate components to cover these items—perhaps there should be.

Table 5.16
Mapping Generic Functions Onto Commercial Middleware

Generic Function	Genesys Mapping	Rostrvm Mapping
Screen-based telephony (SCB)	SoftPhone	Rostrvm ScreenPhone, Rostrvm AdVisor
Call-based data selection (CBDS)		
Application-controlled routing for incoming calls (ACRI)	Genesys ICD	Rostrvm CallDirector
Application-controlled routing for outgoing calls (ACRO)	Genesys Campaign Manager, Genesys Net Vector, Genesys Video ICD	Rostrvm OutBound, Rostrvm InterAct
Voice and data call association (VDCA)		Rostrvm CallBroker
Data transport (DT)		
Coordinated call monitoring (CCM)	Genesys Call Centre Pulse, Genesys Agent Pulse, Genesys DART	Rostrvm ReportWriter and AuditLog, Rostrvm SuperVisor
Interactive voice response (IVR)	Genesys IVR Server	
Message exchange (ME)		
Basic API and drivers	T-Server	Rostrvm CallManager
General interfacing	InterActive-T	Rostrvm InterFace
Linker		Rostrvm AutoAgent
Development tools	Genesys OCX Toolkit	Rostrvm SwitchSimulator

In the first edition of this book, I wrote that many application programmers dream of the green field where their structures and flows can develop untrammeled by established programs. I went on to say that, sadly for them, CTI system integration is, in most cases, a matter of incorporating the CTI functions into a large and critical application suite resident upon a mainframe. After all, the key data that is central to a business has usually been computerized for many years so there is little doubt about who is Mohammed and what is the mountain. The evolving CTI support tools recognize this fact and attempt to interpose between the application program and the CTI API. In the best solutions, CTI functionality is shoehorned into the overall system without the mainframe applications being changed. This is difficult to achieve—and smacks of magic. How can you change something without altering it? The secret lies in that little word *interpose.* If you were transported to the world of Oz, the wizard would ensure that you were issued a pair of spectacles. The Emerald City then becomes green. You have not changed and your brain certainly has not; yet you perceive things in a very different way. Perhaps the wizard can design a pair of CTI spectacles that do not require the original application to change and yet introduce CTI functionality to the user in some way. Is this possible?

Of course it is possible. The original application exists in a bounded environment. Physically interpose a PC between the application and the human operator, and the operator then sees whatever the PC is programmed to present of the original application. Of course, the physical interposition of the PC requires network connection. Connect the PC, with others, to a LAN which is also connected to a CTI server and a mainframe gateway, and the necessary client server network is established.

In this environment the agent logs on, and the PC signs on to a CTI server *and* opens the application. A call arrives and the server allocates it to the agent's PC. The PC receives the calling line identity from the server and then translates this into a customer identity that is recognized by the original application, which is then requested by the PC to open the data records for that customer. The agent processes the call and then decides that a specialist agent should deal with it. The agent keys in the necessary transfer command. The PC instructs the CTI server to transfer the telephone call *and* transfers the session established with the original application to the PC associated with the telephone to which the call is transferred.

This entails very powerful "green glasses." The "glasses" reside on a PC, an environment that is change-oriented, while the original application resides on a mainframe, an environment that is change-resistant. However, a large snow-covered mountain can seem pleasantly green if one wears the correct spectacles!

It is at least *possible* to "interpose" a CTI sliver between an existing application and the real world. The questions posed by this solution are the following:

- What are the limitations imposed by this approach?
- What is the cost of providing the sliver of CTI control software?

In the early 1990s, there were not too many practical examples of this approach so the questions remained largely unanswered. Now there are plenty of solutions, and few limitations are observed. The cost of providing the "sliver of CTI control software" has dropped significantly, and the sliver has become an impressive array of components, as demonstrated by Table 5.16. The cost has still not reached the point at which third-party CTI can be effectively deployed outside of the call center, but the pressure continues to reduce prices.

In the early 1990s, complications were envisaged over the provision of VDCA. Problems do arise in transferring data sessions, but these have mostly been addressed by "sticky data." Each call has an existence within the computer, and relevant data is stored with the call. This data can then be used by a receiving application to build a new screen or whatever. The Rostrvm middleware abstracts this to a clipboard function that operates in a manner similar to that of the cut and paste mechanism with which we are all familiar.

The problems presented by transferring calls and data were seen to be more acute when the transfer involved the movement of sessions across a network. These difficulties were neutralized to some extent by transferring sticky data. This left the challenge of associating data and call across the network. Fortunately, ingenious minds have arrived at ingenious solutions that make such transfers seamless to the users.

Applications can be developed in any language that allows access to CTI functionality. Most new applications are written in some variant of the C language or perhaps Java. Many are prepared in some higher-level application generation language.

References

[1] *Computer-Supported Telephony in Europe,* London: Schema, 1992.

[2] *Mitel Telephony Application Interface (MITAI),* (9180-951-572-BA), 1991.

[3] *CT-Connect Solution Directory,* Dialogic, 1997.

6

Applications and Case Studies

6.1 Finding the Right Application

A young man was walking along a street in postwar Japan racking his brains to find the reason why no one bought his newly invented voice recording machine. Everyone who saw and heard the machine loved it. Demonstrations were a great success. People clamored to attend them and were visibly impressed by what they heard. Nevertheless, they didn't buy.

Admittedly the machines were large and expensive, but that was the best that could be done with the existing technology. Surely somebody wanted to own the ability to reproduce their own voice and those of their friends and families?

It was in such a reverie that the young man began to observe someone staring at a Ming vase in the window of an antique shop. The person, a man of medium height and good bearing, eyed the vase from various vantage points, and then finally entered the shop. After a discreet interval had passed, the young man followed. The man was in deep discussion with the owner of the shop. After some haggling and some discussion as to the vase's true origin, the man bought the vase! He bought the vase for an amount that made the cost of the young man's recording machine seem trivial.

This sequence set the young man thinking. The man in the shop obviously had a need for the vase—a need that potential customers for his recording machine did not have. Why was this?

All night long, he tossed and turned, unable to free his mind of this conundrum. After all, he had created something with a function—something that actually did something rather than sitting on a shelf looking beautiful. Why did the man buy the vase, and why didn't people buy the recording machine?

By the next day he had the solution. He made an appointment with the clerk of the local magistrate's court and demonstrated the recording machine to him. The clerk was delighted. He was impressed by the technology, but, more importantly, he had a real need for such a machine. Recording court proceedings in the Japanese script was labor-intensive and error prone. The young man's machine saved labor and provided better results. He had made his first sale and found his first niche.

Version two of the recording machine was cheaper and could be sold into the language departments of schools and colleges. Many versions later, the machine became the Sony Walkman, and Sony's president, Akio Morita, became one of the best known businessmen in the world [1].

The Ming vase incident convinced Akio Morita that he needed a simple and practical application for his new technology. The fact that a voice could be recorded and replayed was interesting in itself, but it was not sufficient to persuade people to part with significant sums of money. The application that Mr. Morita found was simple but important. His machine improved efficiency and, more than that, was more accurate than a human recorder.

It is not entirely clear from this tale whether CTI is comparable to the Ming vase or the early tape recorder. One thing is certain; even as I write the second edition of this book, CTI has still not reached the Walkman stage!

The other thing that is certain is that CTI needs applications. Without applications, it simply provides a link between a switch and computer—it is like a railway bridge across a river with no railway lines running to it or away from it!

You may recall from Chapter 1 that early applications of CTI date from the late 1960s—as do many of the major changes of this century. There is little doubt that these early applications were genuine applications, albeit that they were driven by the need to link the IBM PBX to data applications. The obvious question thus is: Why did this not take off like wildfire? This is a question that will be asked here and at other places in this book. One relevant answer is the lack of applications, although, when looking back to the 1960s and 1970s, the answer has to be coupled with the sheer cost of supplying solutions and the general lack of widespread computer automation. In the 1980s, the cry was often, where are the applications, the case studies, the real benefits? CTI demonstrations were generally greeted with genuine interest, but people were wary of committing themselves to something that they did not entirely understand and something in which the benefits could not be convincingly demonstrated.

During the late 1980s and early 1990s, that situation began to change. Early adopters began to install CTI pilot installations. Then, having measured the benefit, they moved on to fully fledged systems. In the mid 1990s, the concept of CTI middleware—software that was capable of CTI-enabling any applications—became better understood. Such enablers opened up an entirely new sector of the market. CTI middleware products are applications in their own right, but they also provide links into applications that are not CTI-aware.

This chapter begins with a general view of CTI applications and a more detailed look at the generic functions that underpin them. This leads to a categorization of applications into a general set and an examination of some of the applications that are available—off the shelf. Finally, the chapter presents a number of case studies that span a whole decade of CTI implementation in call centers around the world. The early case studies will be examined with a view of determining the original business application, the way in which CTI was implemented, and the benefits that accrued from it. A similar treatment has been given to installations that happened much later in the CTI cycle. Some of these do make use of the middleware enablers.

6.2 Application Integration Potential

Which applications are most suited to integration with telephony? Any application in which users are involved in a significant level of telephone communications and in which the application itself is computerized are candidates. This roughly covers every desktop and every call center agent position in the developed world! However, there are other criteria that can be applied. One consultant group assembled a list of characteristics that successfully integrated applications share. These characteristics are listed as follows.

- A close relationship existed between users and suppliers;
- The applications were central to business requirements;
- The user's main motivation was to reduce cost and increase the competitive edge;
- The systems were continually enhanced;
- Users achieved a quicker payback than predicted.

With the exception of the fourth characteristic, this list almost exactly describes the early implementation of predictive dialers—does it apply to CTI in its wider sense? It certainly did so once. It clearly applies to the call center world, where there is a real need for an ongoing relationship between user and supplier. In other areas, that relationship is more akin to that we have with the owner of the local supermarket chain. It is both distant and tenuous. On the desktop, CTI is unlikely to be close to central business requirements, unlikely to reduce cost, unlikely to be continually enhanced, and has a payback period that is difficult, if not impossible, to calculate. Thus, identifying desktop applications is very different. Successful desktop CTI applications are likely to feature the following:

- Low or bundled cost;
- Simple installation;
- Ease of configuration and use;
- Compatibility with existing desktop packages.

6.3 Application Categories

The concept of the underlying CTI generic integration functions is introduced in Chapters 1 and 5. The functions are repeated here as a reminder, as they form a key approach to the examination of applications and case studies. The generic functions are the following.

- Screen-based telephony (SCB);
- Call-based data selection (CBDS);
- Application-controlled routing for incoming calls (ACRI);
- Application-controlled routing for outgoing calls (ACRO);
- Voice and data call association (VDCA);
- Data transfer (DT);
- Coordinated call monitoring (CCM);
- Interactive voice response (IVR);
- Messaging exchange (ME).

The majority of third-party CTI applications currently installed are call center-based. Call center applications are very telephone-intensive and almost always require that the agents access and modify database information during the progress of each call. This is the equivalent of Akio Morita's law court application. The benefits of applying CTI to call centers are generally well understood, quantifiable, and accepted. Most, though by no means all, of the risk in implementation has been removed by lessons from the experience of early adopters and by the use of the development tools and interfaces that have evolved.

In Chapter 1, the call center was defined in the following terms. A call center is based on a group of people, usually called agents, whose sole or main task is to do business via the telephone. The business that is transacted is almost always automated via a computer system. That is to say, the computer contains the business application software and the business database. Each telephone call, whether it be incoming or outgoing, will require access to data from the computer system and will normally require that the data be modified. Agents are typically equipped with a headset and computer terminal. The whole process is generally heavily monitored.

The case studies presented by this chapter cover all aspects of this definition. However, what relation has the call center to the CTI-generic functions? It is important to note that the definition covers both inbound and outbound calls. Most call centers are concerned with customer service, and most customer service call centers do deal primarily with inbound calls only. However, the faster growing area is in outbound calling and in bidirectional or blended calling. It is in the latter that most of the functions of CTI are found to be valuable, as can be seen in Table 6.1. For each function, Table 6.1 indicates whether that function relates to inbound calls, outbound calls, or both types of calls, also referred to as *bothway.*

So, applications can be categorized by the direction of the calls they support: inbound, outbound, or bothway. They can also be examined with respect to the generic functions required to support them.

Beyond this, there are the various categories of CTI platforms that support the applications to be considered, first- and third-party CTI and the various subdivisions. Also, there is the matter of where the switch resides—in the customer's premises in the server or on the public network.

Table 6.1
Generic Function Requirements for Call Centers

Generic Function	Inbound	Outbound	Bothway
SCB		✓	✓
CBDS	✓		✓
ACRI	✓		✓
ACRO		✓	✓
VDCA	✓	✓	✓
DT			
CCM	✓	✓	✓
IVR	✓		✓
ME	✓	✓	✓

6.4 General Applications

Over and above all of this detailed categorization, there are some basic general CTI applications that are identifiable. These are horizontal applications, in that they are used in most businesses—though there may be major differences in the data that is being manipulated, in the call holding times, and so forth. They are described below with an abbreviated list of the generic functions that are needed to support them.

6.4.1 Desktop Communications

This is a broad category. It refers to a general set of applications that provide much the same functionality, although they may look quite different on the screen. The functions usually supported are the following.

- SCB;
- Telephone directory management;
- Call placement scheduling;
- Individual incoming call handling;
- Screen popping;
- Call detail recording;
- Screen-based voice messaging;
- Screen-based telephony feature control.

Example products will be covered later. The desktop communications application is primarily a personal productivity tool. As can be seen from the above list, it allows a user to do everything that he or she could previously do from the telephone but via the screen, keyboard, and mouse. However, it goes further: It provides a management environment for communications and a mechanism for controlling telephony-related services such as voice messaging.

6.4.2 Inbound Call Handling

Inbound transaction processing has always been one of the most popular application areas for CTI. This covers the area of selling goods and services over the telephone in which the customer makes the initial telephone call—usually in response to a conventional advertisement or direct mailing. It also covers help desk functions that, in turn, cover anything from fault reporting and remedy to product inquiries and advice. Inbound transaction processing efficiency is much improved when agents are provided with information prior to answering the calls and with the ability to transfer calls and data to specialists when necessary. It is also improved by *smart routing* where the data held on a caller is used to determine the best place to answer the call (CBDS, ACRI, VDCA, or CCM).

Related to this is the more mundane, but equally challenging, task of dealing with calls to the desktop. This overlaps with desktop communications above. Here the smart routing is likely to be configured by the desktop owner, usually in the form of *rules.* The owner specifies which calls are important and which are not, which calls should go to the voice messaging system and which should bring up an urgent indication, and when calls should be rerouted to a secretarial answering point. More than that, desktop control should provide a simple means of popping up other applications with relevant data—often without CTI awareness in the application that is used (CBDS, ACRI, VDCA).

6.4.3 Outbound Call Handling

In the call center, where selling is telephone-based and the call list is a database of existing customers, CTI can provide significant reductions in call times and, at the same time, increase the closure rate. The same approach can, and is, used in areas as widely separated as debt collection and customer care. Often, this category of applications is supported by some form of power dialing; these forms are described below (ACRD, CCM).

On the desktop, the needs are usually less oriented toward overall efficiency and more toward ease-of-use. However, there is still significant commonality. Desktop calls are unlikely to be generated automatically so SCB is probably the dominant function here, but the others can also be useful (SCB, ACRO, CCM).

6.4.4 Automated Inbound Call Handling: Interactive Voice Response

There are many IVR applications that link the telephone to the computer without using a CTI link. These are straightforward voice automation applications, such as banking, placing an order, or determining the current location of the parcel you sent yesterday. The applications are horizontal in that the basic functions of playing prompts, detecting tones, and so forth are the same regardless of the business conducted. IVR applications are increasingly used with integration, especially where a call is transferred to a human agent from an IVR system together with the data associated with that call (IVR, VDCA).

6.4.5 Automated Outbound Call Placement: Power Dialers

A power dialer is a system capable of automatically generating a number of outbound calls and connecting these to agents when they are answered. This is a general definition; there are many variations—some people claim that power dialing itself is a very specific form of dialing. People often refer to dialers as a general term. The most basic form of a dialer is a manual or screen dialer. The first is very basic; it simply uses the keyboard rather than the telephone keypad as the dialing mechanism. Screen dialing is a component of SCB. It involves dialing from an object on the screen (e.g., name or number) and receiving an indication of call progress on the screen. Neither manual nor screen dialing has power dialing options. The origins of power dialing are traced in Chapter 1, and the generally accepted definitions are provided in Table 6.2.

The effectiveness of power and predictive dialers is dependent upon their ability to detect that a person has answered the telephone. Good dialers attempt to detect calls that terminate on the following:

Table 6.2
Dialer Mode Definitions

Type of Dialing	Description
Preview dialing	A list of customers to be called is maintained by the computer system, and the call list or data pertaining to the next person to be called is presented to the agent. The agent then initiates the call, usually by a single keystroke or mouse click.
Power dialing	The power dialing controller has a list of numbers to be called, a number of outgoing telephone lines and a group of agents; it launches as many telephone calls as it possibly can and, as soon as an answer is detected on a particular line, attempts to connect the call to an agent; if no agent is free, it drops the answered call and launches another; in certain countries, a recording is played to the called party encouraging them to wait until an agent is free.
Progressive dialing	Like power dialing, but at least one agent must be free before a call is launched.
Predictive dialing	Also very similar to power dialing but far more subtle; the difference is that, rather than launching a mass of telephone calls regardless of agent availability, the controller uses a pacing algorithm such that the rate of launch is based on the probability of answer and the probability of an agent being free.

- No answer;
- Busy tone;
- Network messages;
- Modems and fax machines;
- Answering machines.

These calls can be returned to the list or retried.

Power dialers are loaded from a database containing, say, a list of debtors. This linkage is shown in Figure 6.1, which also displays the other system connections that are made to a closed power dialer. At the end of a campaign, a list of successful contacts is returned. In operation, the predictive dialers bring a free agent into an answered call within one second and presents details about the called party on the agent's screen. If no agent is available, it is possible to place the answered call on a hold queue. If this occurs, the called party is played a recording that states that all agents are busy and encourages the caller to wait. Playing recordings in this way is not allowed in many countries.

The power dialer is connected to telephone lines through a local switch or through a direct connection to public exchange lines. Figure 6.1 shows the former. Agent terminals are often dual-ported so that connection to the power dialer and the host machine can be achieved.

Early power dialer products were closed. They contained the controller, switch, and call progress monitor line units. They were, and still are, closed

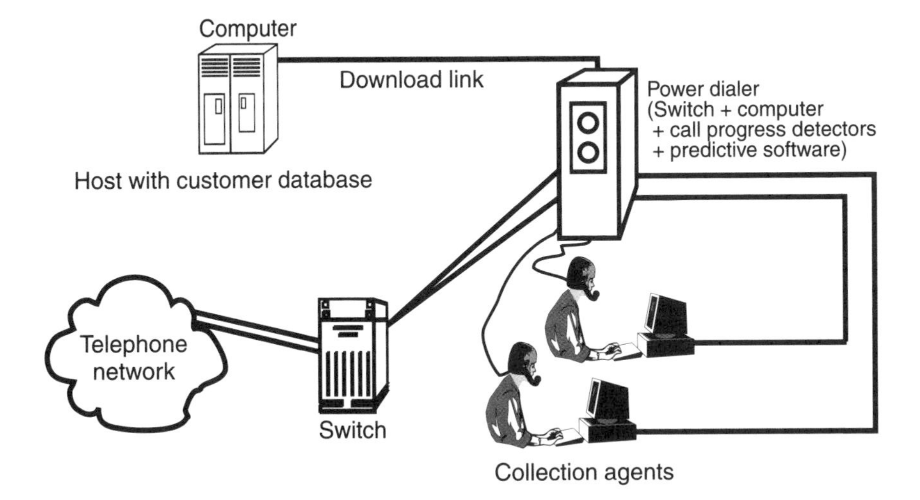

Figure 6.1 Power dialer system and its connections.

third-party-dependent CTI systems within a single box. They do not offer an interface to the switch (except through normal extension ports) and do not integrate with the business computing infrastructure (except through some form of terminal dual porting). During the 1990s, two things began to happen to dialers: first, they became more integrated with the call centers in which they were installed, and second, open designs were created. Both these trends encouraged the use of blending—allowing an agent to perform inbound and outbound work, with automatic switchover dependent on inbound traffic load. The open dialers are sometimes called software dialers. They use traditional components of the call center, including the CTI link, rather than specialized units.

6.4.6 Soft Automatic Call Distributors and Call Routing

In *soft ACDs*, the telephony software and the business application software are in the same computer environment and in complete control of a dumb switch. This is an implementation of third-party-dependent CTI.

One benefit of this arrangement is that, since call processing and data processing software are in the same machine, synchronizing the two is no different from synchronizing any other two software processes. Similarly, a common database is used so the configuration of the switch and application software can be unified—as can the logging of performance statistics. In short, the management of the entire business activity can be integrated. All changes can be made at a single management terminal.

How appropriate this approach is depends on the application in question. In telemarketing, the application may be a particular selling campaign. Coordination of the resources required to mount a new campaign will almost definitely involve making changes to the configuration of the switch and the computer. This is always a fertile place for misunderstandings and misinterpretations. In operation, the telephony and data applications will need to be synchronized in some way—either by the human operator for unintegrated solutions or via the link protocol for conventional CTI solutions. Both of these solutions are inefficient. The soft ACD is an agent-based approach to the provision of CTI applications.

The soft ACD is a natural step toward the creation of a single box call center. Here there is no external dumb switch—the switching function has been absorbed in to the computer.

Call routing is not as extreme as the soft ACD. In call routing, the computer enhances the routing capability of an existing telephone switch. The call routing application may supply advanced skills-based routing, affinity routing, or data-directed routing. The computer system-based software is working in parallel with the switch-based software. Some switch suppliers use this approach to provide ACD working via a PBX that has a CTI link but no ACD software. In other cases, it is used to supplement a switch's call routing by supplying skills-based routing or other more sophisticated routing techniques. Routing to the agent with whom a caller usually deals forms another example of supplementary call routing.

6.4.7 Multimedia Messaging

Sometimes called integrated or universal messaging, this application extends conventional text messaging. In the latter, incoming mail belonging to a PC user is usually viewed via an in-tray screen. The in-tray displays useful information about the messages such as: arrival time, name of sender, length of message, etc. A multimedia message in-tray simply extends the scope from text to cover voice and fax messages. The in-tray contents can be built up from voice and fax message waiting indications sent from the voice processing system, together with the more conventional text messages placed in the in-tray by the electronic message handling system.

Text messages are opened simply by pointing and clicking the mouse; this causes the relevant file to be opened and its contents delivered to the screen. When a voice message is opened, the file containing that message may actually reside within the voice messaging system or the PC. In the case of the latter, then the message may be played via the sound card. In the case of the former, the voice message may be delivered to the telephone. If it is delivered to a telephone, then a command must pass from the PC to the voice messaging system, which causes the voice message to be delivered to the telephone associated with that PC. Similar arrangements are made for recording and sending messages.

One of the advantages of an integrated messaging solution is the ability to view and manage multimedia messages. All of the voice-related functions that can be carried out from the desktop can also be achieved via the telephone. The advantage of this integration is the improved control interface and the advanced capability of associating voice, fax, and text messages. A simple example of the latter is provided by the transmission of compound messages—a fax might be forwarded to another user together with a covering voice message.

6.5. CTI Application Examples

The following examples are simply that—examples. There are many CTI applications on the market—in fact, a constantly expanding set. The one to choose is the one that suits a particular set of requirements best, that will work in a particular environment, and that does not cost too much. The items in the following list illustrate some particular aspect of the CTI application world; their inclusion is not to be taken as a recommendation to purchase. They concentrate on the desktop rather than the call center, although some of those featured can be used in either environment. There are, of course, many CTI-enabled applications specifically for the call center, though many of them can be used with or without integration. Early examples were the Brock applications, those from Early Cloud (later part of IBM), and IMA's Edge. As the call center market for CTI expanded, so did the range of applications. Niche applications became available for banking, insurance, telecommunications, travel, and so forth. There are too many of these to list, but examples are included in the case studies described in the next section. The applications covered here are:

- Call from Teleint;
- Phone from Microsoft;
- Phonetastic from CallWare;
- FastCall from Aurora;
- Group PhoneWare from Q.Sys;
- Unified Messenger from Octel/Lucent.

6.5.1 Call from Teleint

Almost too simple to include, this application is one of the most basic and is little more than an autodialer. The product consists of a small red module that connects to the printer port of a PC together with a disk containing the Windows application. Installation is very easy, and the user retains the existing telephone. The Teleint software needs to run all of the time in the background. It is brought into action when the user selects the "hot key." If there is a document on the screen that contains a telephone number and the user wishes to call this number, then it is highlighted and the hot key is pressed. At this point the screen would look something like Figure 6.2.

If all works well, then the number you want appears in the Teleint box. The user simply presses enter or clicks on the dial button and the number is called from the little red module. The line is connected to the PC speaker so that a distorted form of audible feedback on call progress is provided. Teleint Call will then detect the fact that you have lifted the telephone handset and cut off the speaker.

This is truly little more than an autodialer; the little red module is a simple line intercept for first-party CTI. However, the product also allows you to store 10 frequently used numbers and has DDE capability so that applications can

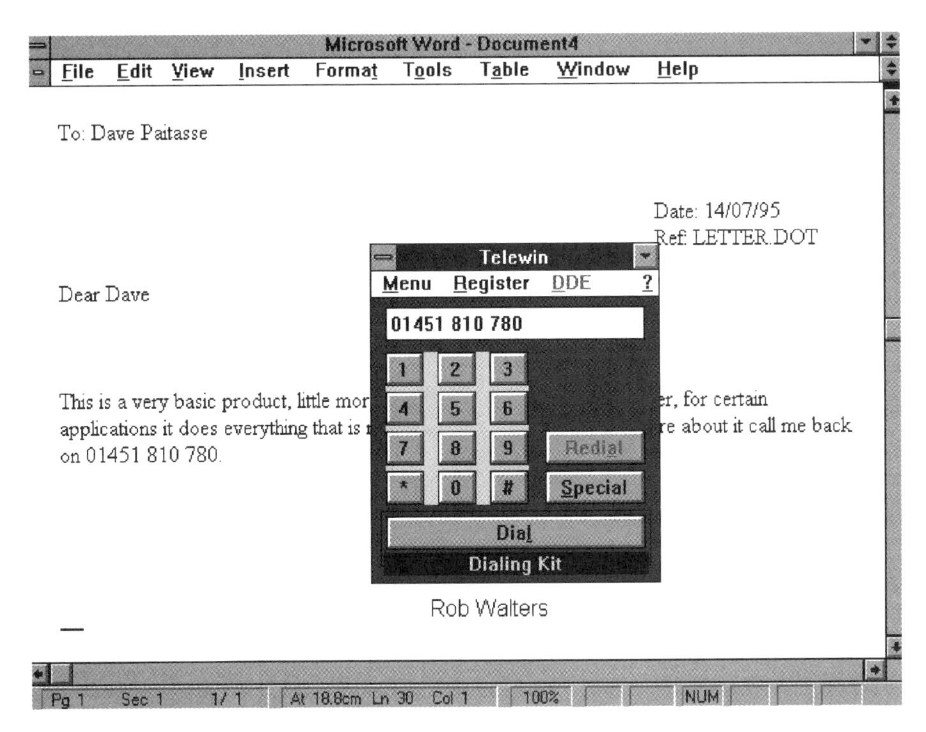

Figure 6.2 Making a call with Teleint Call.

supply numbers directly to the dialer. A similar application is included within Windows 95 as Phone Dialer.

6.5.2 Phone from Microsoft

Microsoft Phone is much more than a simple dialer. It works with any standard modem and merely requires a PC with sound card [2]. The on-screen display is shown in Figure 6.3. It has a keypad, tool bar, and a file index for flipping through different telephone directories.

Phone can be used to make and take calls. It also supports voice messaging and call management functions. It is voice-enabled through Microsoft Voice, the company's approach to providing speech technology access. It can therefore be used for voice dialing and for speaking documents over the phone. It provides multiple mailboxes for fax or voice—each with different announcements.

Phone can divert calls to a mailbox as required and can use calling line identity for simple screen popping. It keeps a list of recently dialed numbers, and it provides speed dial buttons. The user can extract numbers from any MAPI-compliant address book.

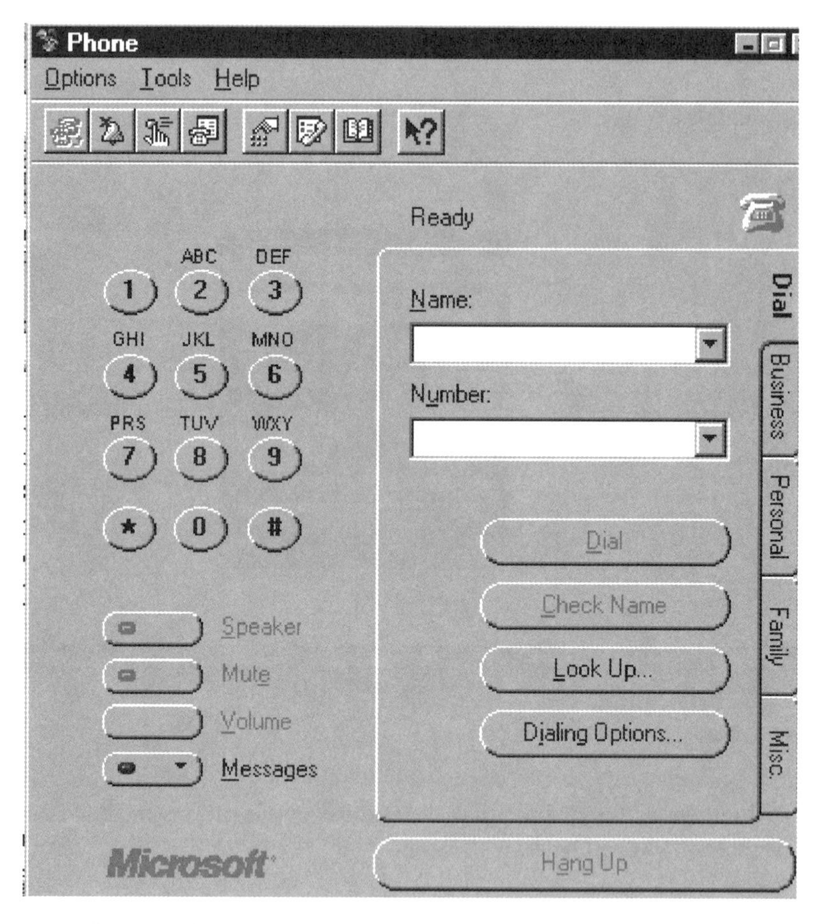

Figure 6.3 The Microsoft Phone Screen.

Microsoft is always keen to show synergy between its various packages. Phone works with the following:

- Unimodem/V, an extension to the Universal modem driver, this is a TAPI service provider that makes supporting voice modems easier.
- TAPI itself.
- MAPI, for storage and retrieval of messages and accessing the Windows 95 MAPI address book.
- Microsoft Exchange Inbox, which allows all incoming messages, whether voice or data, to be located in a single place.

- Microsoft Fax features, included in Windows 95, which allow users to create a fax-back system and to forward fax messages.
- Speech API, a new API for developing speech-recognition applications. Microsoft Phone was the first Windows 95-based application to use the Speech API.

6.5.3 Phonetastic from CallWare

The Phonetastic product was one of the earliest CTI products to use TSAPI, the Novell client server approach to integration. Whereas earlier products tended to provide very basic SCB, Phonetastic was one of the first products to provide a really intuitive approach to call handling. The screens are relatively uncluttered and meaningful, and the user controls the whole thing via the mouse.

Phonetastic provides desktop integration. It allows the user to do all of the CTI generic functions directly from the PC screen. It can make call manipulation so easy that users do begin to use all of those PBX facilities that otherwise need training in order for users to master them—conferencing, for example. Figure 6.4 shows the general screen layout. Rob is currently making a call to Margaret; he initiated this by clicking on "Margaret" in the dialer window.

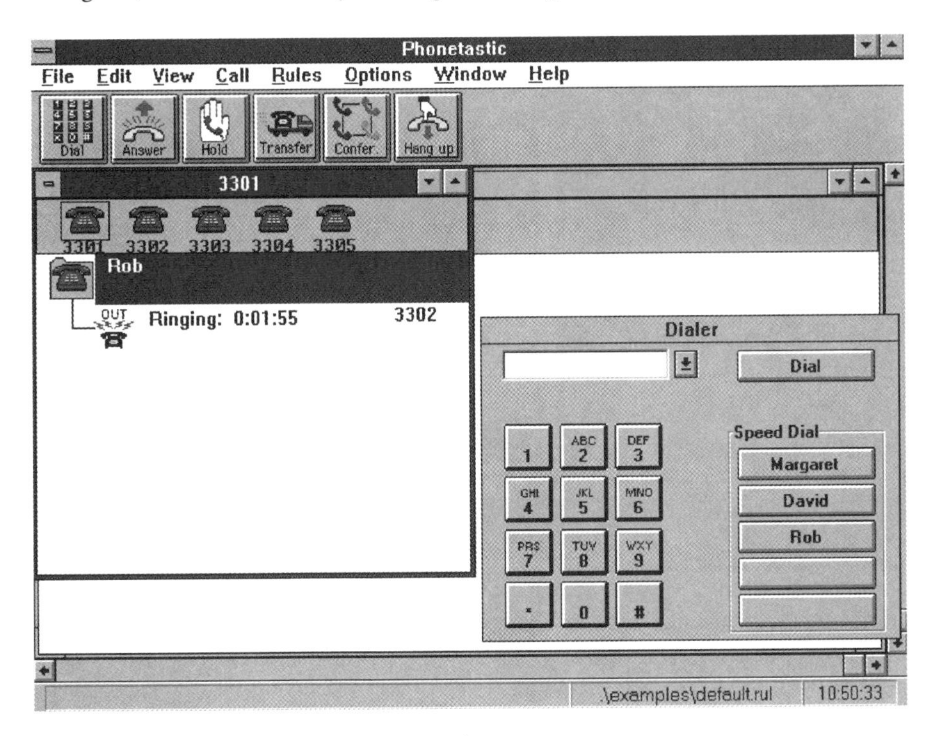

Figure 6.4 Screen-based calling from Phonetastic.

As mentioned in Chapter 5, Phonetastic is an example of middleware—it is an application and an enabler. An important part of the product in this context is its use of an expert system to create the rules that govern call handling and the activation of other applications. Using Phonetastic, the user does not program the system to change the way that it deals with calls. Instead, a selection of rules from a simple forms-based editor is built up to achieve individual requirements.

In Figure 6.5, we can see Rob's original window together with Margaret's. Margaret has invoked a rule whereby incoming calls kick off her card file application and begin a search for the number of the calling party. You can see that Rob's number was found and his card displayed accordingly.

Figure 6.6 shows how Margaret set this option by clicking the top box in the rules manager. This "modify rule" portion of the figure gives good insight into the way that rules are created. Notice that they read almost as an English sentence. In the rule shown, calls are forwarded to the night attendant.

Most telephony applications require the creation of a unique telephone directory. This contains telephone number and general contact details. This is a major disadvantage since the information often already exists in other databases.

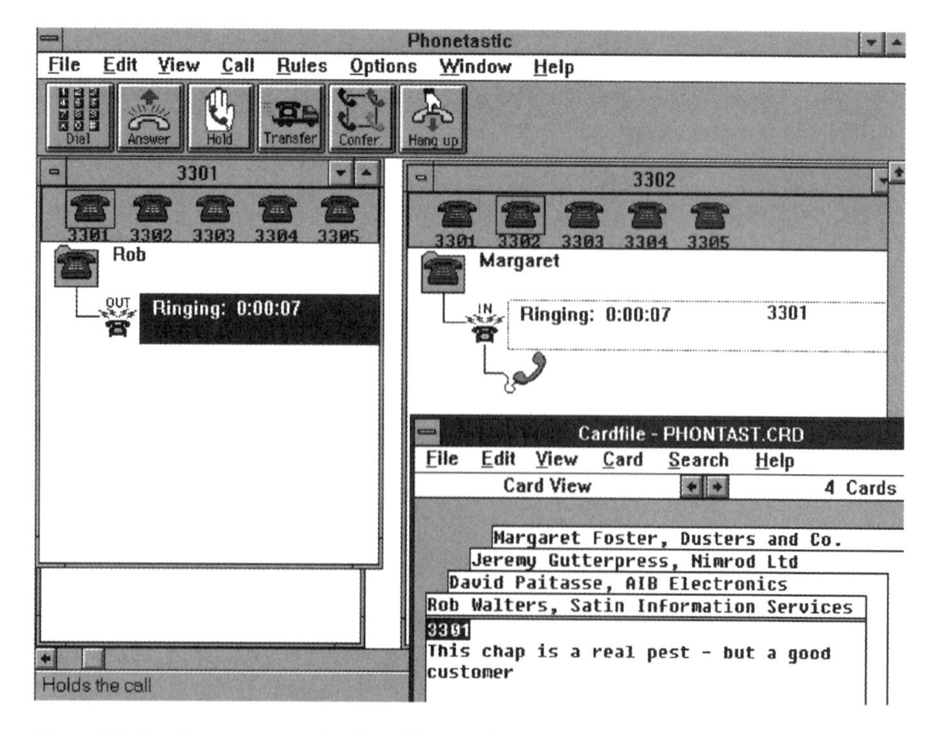

Figure 6.5 Desktop screen popping from Phonetastic.

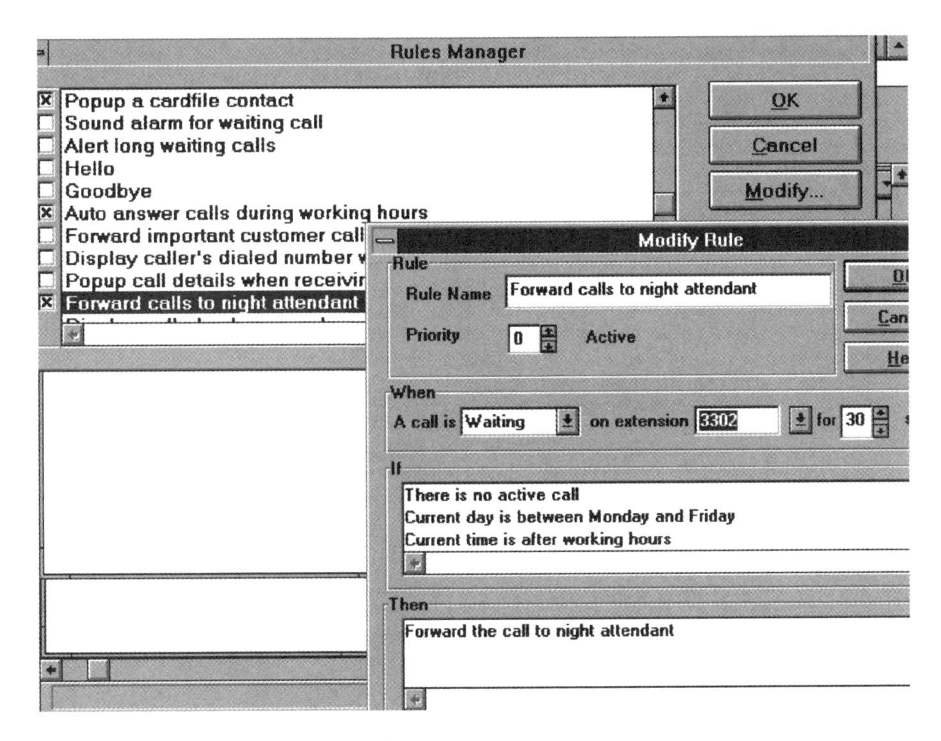

Figure 6.6 Desktop rules in Phonetastic.

Phonetastic does allow users to access the existing database, though it also provides a local, short-code database. Once again, Phonetastic accesses other databases through the use of rules. The rules will normally be written by the supplier or a system integrator.

6.5.4 FastCall from Aurora

At one time the FastCall product was distributed by so many suppliers that it began to be regarded as a standard. Originating from Aurora, it addresses the same market as Phonetastic but takes a different approach. The FastCall interface is more key-based than that of Phonetastic. The main view that a user has of the product is shown in Figure 6.7. Note the top line that carries the detail of the current call. Currently, Rob is calling Margaret, and the phone is in the ringing state.

Other screens are equally simple. Clicking on the dial key or pressing the F8 key will present the user with the screen shown in Figure 6.8. Calls can be made from a private list of names or from the keypad.

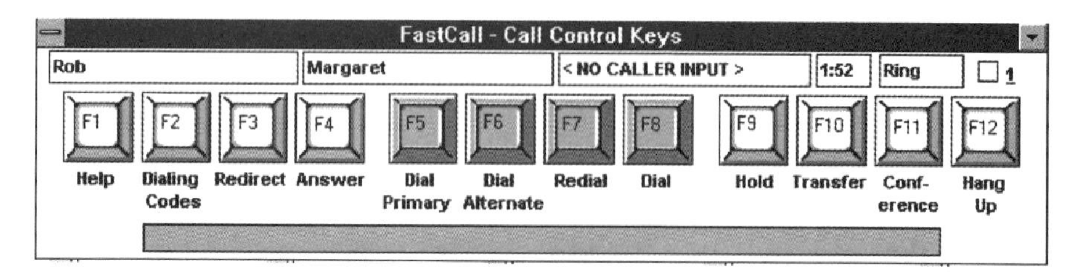

Figure 6.7 SCB from FastCall.

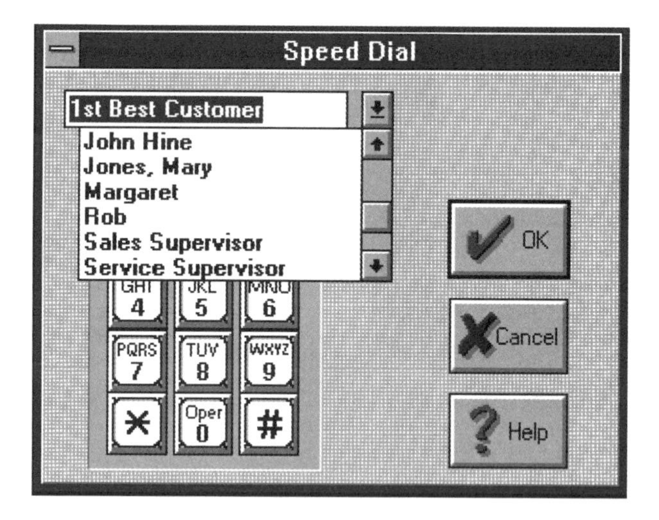

Figure 6.8 Dialing from FastCall.

Like Phonetastic, FastCall is an example of middleware—an application and an enabler. It too has a rules-based system that governs call handling and the activation of other applications. Here, a selection of rules from a rule administrator is built up to achieve individual requirements. There are rules for dealing with incoming or outgoing calls.

A particular rule set-up is shown in Figure 6.9. This screen is a little packed with detail because the figure is trying to show the alert window at the same time. However, it is well worth some study since it offers another good demonstration of the power that rule-based call handling provides.

The particular rule shown in Figure 6.9 can be translated into normal English. Basically the rule requires that a call from Rob to Margaret between the hours of 8 a.m. and 5 p.m. should alert Margaret by playing an urgent tone and triggering a Windows application. The trigger selected is "Pop up sales information card." Triggers are created by the user, and FastCall can be "taught" what

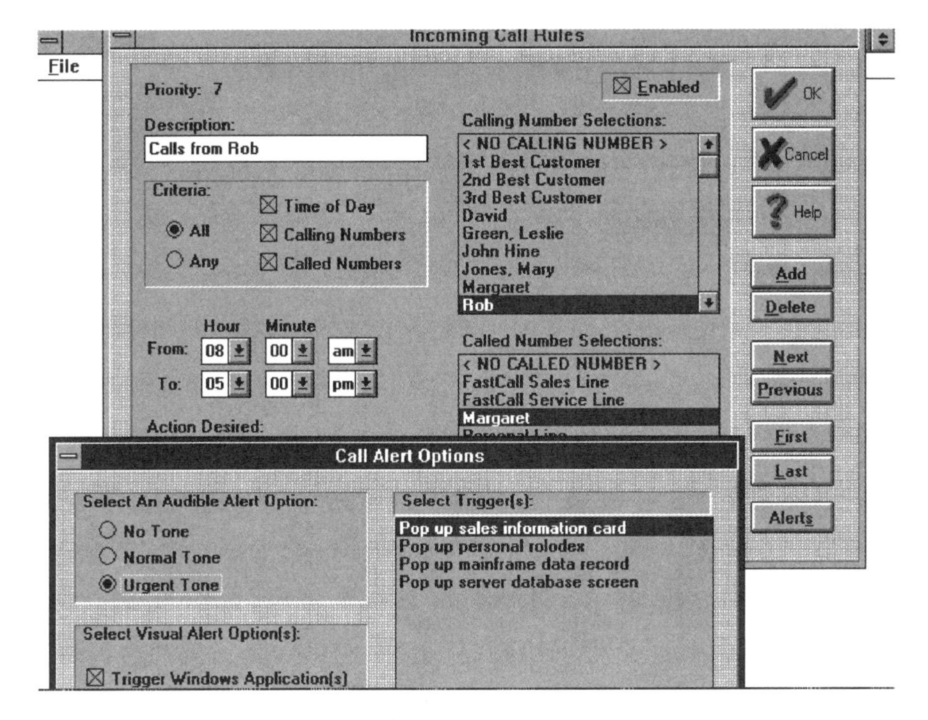

Figure 6.9 Call processing rules in FastCall.

to do by recording a keystroke macro of the actions to be taken when the call arrives. For example, part of that action may involve the conversion of a calling line identity to an account number and the transfer of this information to the triggered application.

Note the importance of priority in FastCall rule handling. The rule described applies between 8 a.m. and 5 p.m. What if Margaret wanted to do something different with the calls from Rob during the lunch time? She could establish a rule that activates between 12:30 and 1:30 p.m. to steer Rob's calls to an answering service. The current rule has priority 7—it is the seventh rule that is evaluated. If the additional rule were allocated priority 8, it would never be reached. It must have a priority that ensures that it is evaluated before the broader rule.

The rule demonstrated by Figure 6.9 involves a screen pop. The effect of the rule can be seen in Figure 6.10. When the call arrived, Margaret was using the word processor. FastCall popped up its usual interface screen and then brought up the card file index with Rob's details selected via his calling line identity.

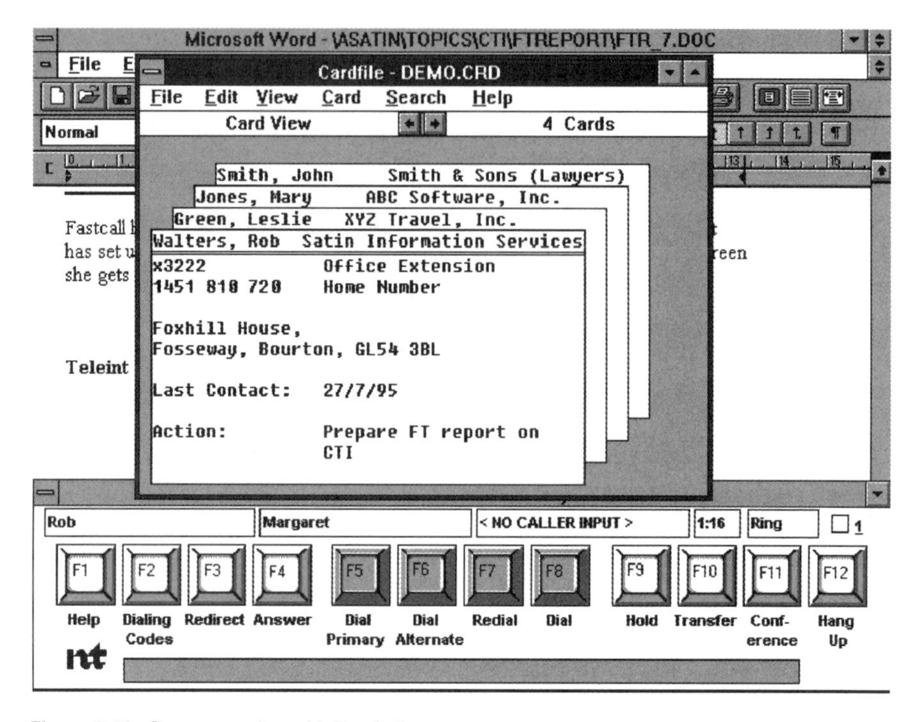

Figure 6.10 Screen popping with FastCall.

FastCall is supplied as a first-party or third-party CTI application. It usually needs to work in third-party mode to use the called number in its rules.

6.5.5 Group PhoneWare from Q.Sys

Group PhoneWare was one of the first CTI applications to introduce drag-and-drop onto the CTI desktop. It is one of a range of applications from Q.Sys that provide CTI solutions for various requirements.

It can use calling line identity from incoming calls to display any relevant customer attributes from an existing database. It allows you to pop up other applications when a call is either ringing or answered. Group PhoneWare uses DDE to link to other applications. DDE-aware applications can be driven from Group PhoneWare via DDE links.

What makes this product particularly "groupy" is that it has user-configurable group features such as the following:

- Icon-based "busy lamp field" for other group members;
- Support for messaging;
- Support for what Q.Sys calls "itinerary management."

Itinerary management is all about entering and displaying the daily diary of group members. The itinerary of other group members can be viewed in a number of ways. For instance, the user can configure Group PhoneWare to display the group member's itinerary when the cursor is on that group member icon. This is useful if calls are being managed for other group members.

Group PhoneWare was designed for users who need immediate visual access to members of their department or call coverage group. This is a fairly common need. The status of the phones of team members is displayed on-screen. This can be seen as the bottom row of telephone icons in Figure 6.11. The color of the icon indicates the status of Susan or Conrad. An incoming call can be dragged to an available team member who can then take over the call. Also, an itinerary balloon pops when a call arrives at your team member's phone.

This application is not rule-based—but it can be configured to do many things. Screen popping, for example, can be configured in various ways [3]. First of all, the user can choose to pop one application for external calls and

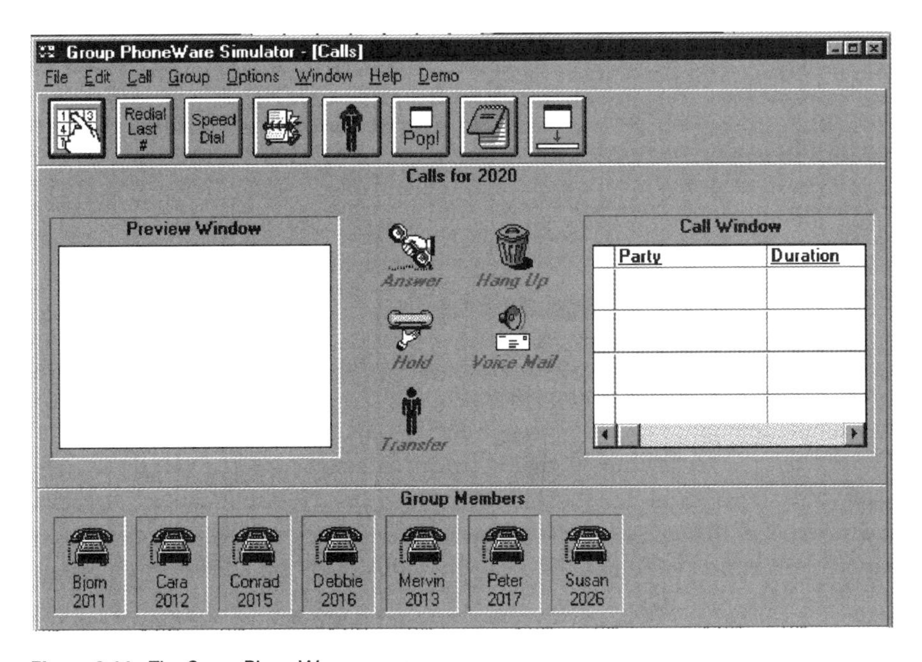

Figure 6.11 The Group PhoneWare screen.

configure a different application to be popped for internal calls. Second, there are various screen pop activation options, listed as follows:

- *Simple application selection:* This just opens the application on an incoming call. If the application is already open, Group PhoneWare will bring it in focus as the front window. If the application is reduced as an icon, Group PhoneWare will restore it as an open window.
- *DDE script:* Here Group PhoneWare executes a DDE specified script on an incoming call. The DDE scripts are configured separately.
- *Macro script:* In this case, Group PhoneWare executes a specified macro script on an incoming call. The macro scripts are prepared separately.

This product also provides a simpler solution still. This is called Database Preview, and the screen for viewing this information is part of the screen shown in Figure 6.11 (preview window). The preview window allows a user to view any information in the local database about an incoming caller before answering the phone.

Basic information on incoming calls is shown in the call window, on the right of Figure 6.11. It is from here that a call is picked up—using the mouse in drag-and-drop mode. Once a call has been picked up, the cursor becomes a telephone handset icon. This can be dropped on the answer icon. Alternatively, if you see from the name displayed in the call window or the preview window that this is someone with whom you do not wish to speak immediately, the call can be dropped on the voice mail icon.

Group PhoneWare also supports e-mail integration, itinerary management, and a messaging module, so it is very much a CTI application.

6.5.6 Unified Messenger from Octel/Lucent

Messaging has gradually extended itself by bringing together voice, fax, and e-mail at the desktop and at the telephone. Octel's Unified Messenger provides an example that emphasizes a scaleable and reliable architecture.

Integrated versus unified messaging causes many arguments, and there are many definitions for both. The difference for the user is often unclear. The idea is that you can manage all of your messages from one or another of two interfaces—the phone or the PC. This can be achieved by middleware that brings together voice, fax, and e-mail at the same user interfaces. The preferred solution should store all the different mail messages in the same store, use a single directory, and have a common administration point.

Unified Messenger is Windows-based and hence uses Microsoft Exchange Server as the message store and Exchange Client as the user interface. All that is added to the latter is a telephone icon to indicate a voice message. Clicking on the voice message brings up Octel's player interface, which can be configured to play messages through the sound card or the telephone. Users can also use it to record a voice annotation, which can then be dropped into a Word document, if required. The user interface and the recording interface are shown in Figure 6.12. Note the modified tool bar and the mixed media messages in the inbox.

The architecture relies on a unified messaging server, which connects to the PBX through PC voice cards. It also connects to the LAN to allow the compressed messages to be sent, or recovered, from Exchange Server. This means that messages can be stored anywhere in the enterprise—and recovered from any PC. The basic architecture is shown in Figure 6.13. Multiple messaging servers provide resilience and scaleability—up to 24 ports can be connected per server. The voice servers can continue to take messages in a reduced functionality mode even if the LAN fails.

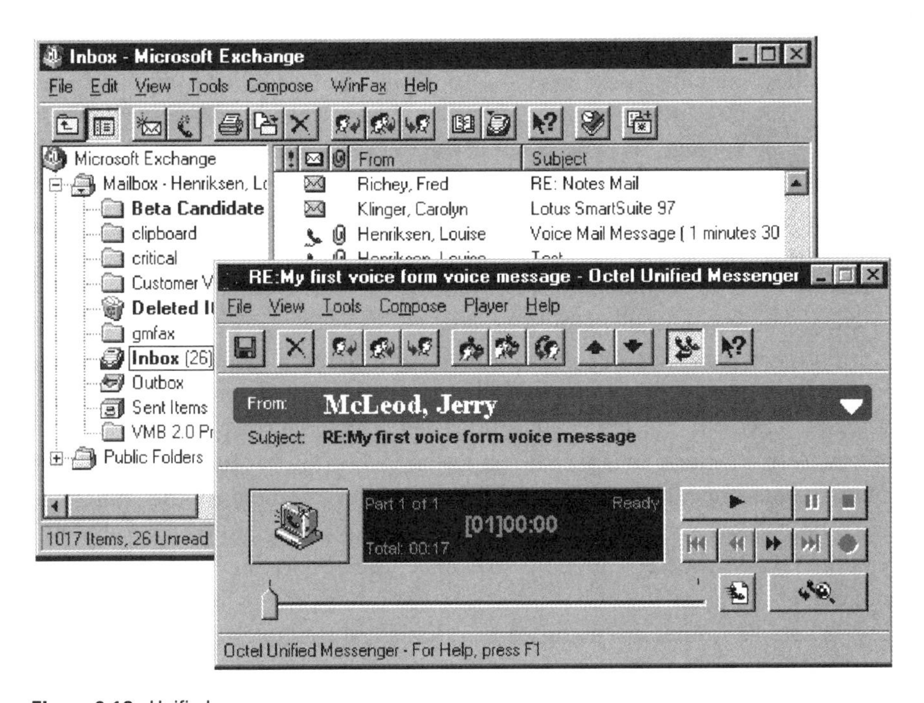

Figure 6.12 Unified messenger screen.

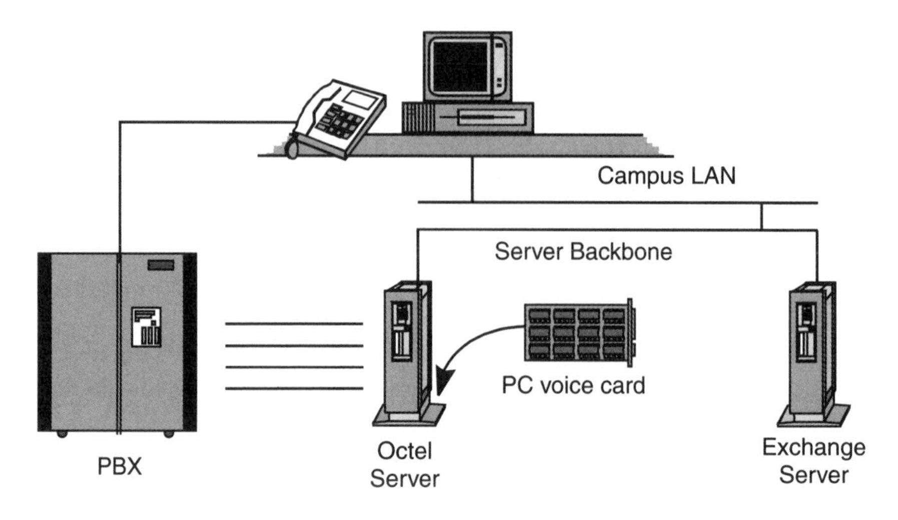

Figure 6.13 Unified messenger architecture. (*Source:* Octel.)

6.6. Sample Case Studies

This section contains some case studies from the first edition of this book. These are the classics, and there is still much to be learned from them. However, a number of contemporary case studies are also included to cover the use of up-to-date CTI technology. Most of the case studies selected are third-party compeer implementations. This is simply because most of the existing CTI installations are of this type. However, there is a perceptible swing toward the use of first-party CTI to implement PC-based turrets in the call center and PC-based keyphones in smaller establishments.

For each case study, Table 6.3 indicates whether calls are inbound, outbound, or bothway; the type of CTI platform used; and whether the implementation is based on the public telephone network or on a privately operated switch. Also included in Table 6.3 is an indication of the prime CTI-generic functions used in each application. This list is not exhaustive; other functions than those listed may be used at a particular location. However, Table 6.3 concentrates on the key functions that make the application one of particular interest.

The case studies are included to exhibit real examples of CTI in use, to examine the practical implementation of installations, and to extract lessons from each one by examining the benefits obtained.

Table 6.3
Summary Table of Case Studies

Application	Call Direction	CPE/PSTN	Functions
Automating banking services	Inbound	CPE	CBDS, VDCA, CCM
Campus emergency	Inbound	PSTN	CBDS
Car company collection	Inbound	CPE	CBDS, ACRI, VDCA
Efficient charity	Inbound	CPE	CBDS, ACRI
Quicker help desk for fast cars	Inbound	CPE	CBDS
Selling books	Outbound	CPE	ACRO
Automated holiday bookings	Bothway	CPE	ACRO, IVR
Campus emergency and integrated message center	Bothway	CPE	SCB, CBDS, ME
Operator assistance in the public telephone network	Bothway	PSTN	SCB, CBDS
Tour operator	Bothway	CPE	SCB, CBDS, ACRI, CCM, FAX
Telemarketing bureau	Inbound	CPE	CBDS, ACRO, IVR
Happy travel agents	Inbound	CPE	CBDS, ACRI, VDCA, CCM
Customer service from a utility	Bothway	CPE	CBDS, ACRO
Investment management	Bothway	CPE	CBDS, ACRO, CCM
Personal service in collections	Bothway	CPE	SCB, CBDS, ACRI, ACRO, CCM
Preventing car thefts	Inbound	CPE	CBDS, CCM
Fully integrated finances	Bothway	CPE	CBDS, VDCA, ACRO, IVR

Note: See Chapter 1 for a translation of abbreviations used.

6.6.1 Automating Banking Services

In the early 1990s, this case study was the classic—quoted by all and sundry as the road map for modernizing banking services. The then named Security Pacific National Bank was one of the first California banks to establish a 24-hour phone line, giving customers access to numerous bank services around the clock [4, 5]. The centralized customer service center assisted retail customers and more than 500 banking offices. The bank was absorbed by others during the 1990s, but the case study lives on as a seminal example of the benefits of CTI in action.

In 1988, Security Pacific turned to call center technology because it wanted to offer competitive service yet did not have the budget or staff to keep branch offices open at all hours. Because information requests and other branch functions had been centralized into a consolidated service center, service representatives could answer queries, solve problems, verify and trace information, provide tax reporting information, open new accounts, stop payments, and order checks for retail customers at any time of the day or night.

At the centralized service call center, 450 representatives and several hundred interactive voice response ports handled approximately five million calls per month. The agents sold a variety of financial products through inbound and outbound sales programs. During its first year of operation, the service center handled 43 million calls.

Prior to installing the system, Security Pacific service representatives had to log on separately to three different systems and access multiple applications and databases on different computers to obtain all the data needed to serve a customer. When agents completed a transaction on one host system, such as a stop payment request, they would then have to rekey all the data into the PC-based database from which forms were routinely printed. The data would then have to be entered into the host application, which actually stopped payment. Such complicated processes were completely eliminated with the new solution.

6.6.1.1 The Implementation

The automated call center ran on a Tandem NonStop computer system and *telephone delivery system* (TDS) software from Early, Cloud & Company. The fault-tolerant Tandem Cyclone system provided the continuous availability necessary to guarantee 24-hour service. The TDS contact management software provided scripting, voice/data integration, strong management reporting capabilities, and flexible screen presentation so that screens could be easily changed without programmer intervention. The Tandem system managed front-end presentation services so that a representative needed to log on only once to access the various systems and databases in order to complete multiple transactions. The TDS software accessed multiple applications on the back-end hosts, searched through the dissimilar data platforms, extracted the desired information, and presented it to the agent in a common screen format. The Tandem system interfaced to third parties that supply services, such as printing checks, enabling Security Pacific to route orders electronically to the supplier and turn orders around more quickly.

The NonStop system communicated with multiple host mainframes, dissimilar operating systems, various terminals, IVRs, and a Galaxy ACD from

Rockwell. Functioning as the information hub for the call center, the Tandem system supported more than 500 IBM 3270-type terminals. It also connected to multiple back-end IBM databases to access all the account data needed to answer queries, process transactions, and support the agents. This system enabled Security Pacific to simplify screens and scripts for service representatives and provided enhanced service to customers.

The system was introduced in phases, listed as follows:

- *Phase 1:* The call center used the Galaxy ACD.
- *Phase 2:* IVRs were introduced.
- *Phase 3:* The CTI link was introduced.
- *Phase 4:* IVR capability was enhanced.

The total cost of the system was around $7 million, 57% of which was spent on application software. It was not possible to pinpoint the actual cost of introducing CTI within this overall figure.

6.6.1.2 The Benefits

Security Pacific was able to reduce its staff by 33% and standardize procedures while performing the same functions with a higher degree of quality. Centralizing services reduced branch traffic and freed branch tellers to handle interactive transactions, such as check cashing. The key benefits of the overall automation project are that it accomplished the following:

- Extended business hours to 24 hours a day, seven days a week;
- Decreased staff costs by 33%;
- Reduced traffic in the branches; Simplified access to multiple host systems and databases;
- Improved customer service with more efficient operations.

The bank's centralization project vastly simplified the agents' jobs, increased efficiency, and cut training time for new representatives. The productivity improvements were achieved progressively with each of the phases, described above. The changes are presented in Table 6.4.

Because the Tandem system could interface to multiple databases on different platforms and TDS presented this varied data to the user in a common screen format, Security Pacific staff simply created interim screens overnight, loaded the conversion information onto the distributed platform, and

Table 6.4
Productivity Gains at Security Pacific National Bank

Measure	Phase 1	Phase 2	Phase 3	Phase 4
Number of centralized agents	335	580	563	556
Number of agents required without automation	395	714	759	829
Agents saved	60	134	196	273
Productivity improvement	15	19	26	33

immediately had online access to the new customer and account information. This enabled Security Pacific to provide very rapid access to accounts, allowing a higher level of service to the customer while saving considerable expense.

6.6.2 Campus Emergency

This is another classic—thought to be the first Centrex application of CTI. Syracuse University in the United States, with a total of 16,000 students and staff, wanted a system that could deal effectively with emergency calls [6]. If an emergency call was received from somewhere within this very extensive campus, the attendant needed to know immediately the location of the caller on the campus and indicate where the nearest exits were.

6.6.2.1 The Implementation

The system suppliers in this case were IBM, Northern Telecom (now Nortel), and the local telco—then Nynex. The computer system used was an IBM AS/400, and it was this machine that was able to relate the calling line identity to the location of the caller and other relevant information. The physical linkage between the AS/400 and the Nynex-operated DMS 100 Supernode was a basic rate ISDN channel. The protocol used was CompuCALL—one of the few public network-to-host protocols known to be available. The application, which used a limited subset of the CallPath Services Architecture capability, was prepared jointly by NT and IBM personnel.

6.6.2.2 The Benefits

It would not be possible to quantify the benefits of the campus emergency system in financial terms. However, the advantage of high-speed response, together with an indication of location of the caller, and the availability of exit paths added significantly to the more effective processing of emergency calls.

Previous systems were paper-based, and it could take up to three minutes to locate emergency information—if it existed at all. With the new linkage, the information was displayed immediately.

6.6.3 Car Company Collections

In the early 1990s, Nissan USA had one of the problems of success—people owed it money. On any one day, it was estimated that the company received as many as 5,000 calls and made 10,000 calls purely concerning overdue accounts. Serving this large number of debtors was time-consuming and expensive. Until the introduction of a CTI system, Nissan had a debt collection system that was primarily manual. The cycle of collection began with the issuing of an overdue notice to the debtor. Naturally, not everyone paid on this notice so the notices were followed up within 20 days by a telephone call to remind outstanding debtors of the owed amount. One of the problems that immediately concerned Nissan was the amount of time wasted by these calls. A significant amount of time was lost in calls to busy numbers, unobtainable numbers, and so on. Nissan needed a means of weeding out these calls before the agents were involved—a classic power dialing opportunity.

Of course, even weeding out the unsuccessful calls did not guarantee that the agent contacted the debtor; someone else could answer the phone, in which case a message was left to ring Nissan back. In this way a significant number of the outgoing calls become incoming calls. These were added to the calls that are made to Nissan in response to the overdue notice. Together, this produced a significant inbound load. Nissan wanted to shorten the time that agents spent on each call. The company wished to do this for two reasons: First, it wanted to reduce the wait incoming callers experienced—once having encouraged the debtors to ring in, the last thing the company wanted to do was lose them. Second, Nissan wanted to reduce the number of inbound operators. One method of achieving these two goals was to obtain the customer's account number before involving an agent.

6.6.3.1 The Implementation

Not surprisingly, automating this process focused on the computer that contains the debtor database. Here it was an IBM 3090. To automate placement of outgoing calls, a Davox power dialer was employed. This did not require any special connection to the switch. The power dialer was an independent system; its only connection with the other elements in the Nissan system was a data link to the IBM 3090. This was used to download the debtor list. There was, however, a CTI connection between the mainframe and the Rolm 9751 switch. This was a CallBridge link that allowed calls to be transferred from the IVR ports to live agents. The IVR system was connected to the mainframe and to the switch. It received incoming calls and offered a number of services. These included the following:

- Collecting an account number or social security number;
- Accepting a loan payoff;
- Requesting that a debtor promise to pay loan by a certain date;
- Gaining information about the debt;
- Transferring the call to a human agent.

The IVR systems were IBM 9270s. They were Syntellect devices; IBM has subsequently designed its own IVR system, DirectTalk. Development and testing of the Nissan call center took, in all, about twelve months. There were 100 agent positions in the system, split between the outgoing and the incoming task. The application itself was developed by Nissan.

6.6.3.2 The Benefits

At first the IVR systems were only used to collect debts from people who were less than 20 days overdue. Longer term debtors were referred to the human agents. However, experience showed that IVR was just as good at collecting debts as the human agents, so the period was extended, first to include 30-day debtors, and then, up to 35-day debtors. Interestingly, it was claimed that debtors who promised the machine that they were going to pay within a certain time proved more likely to do so than those who promised a human being the same thing!

One startling result of the Nissan installation is that the planned recovery time on an investment was 18 months, yet this was achieved in only six months! In real terms the benefit derived from the swifter and more successful collection of debt. In addition, however, a significant improvement in customer relations was observed, accounted for by the fact that the call center dealt with debtors in a more efficient manner.

6.6.4 Efficient Charity

The World Vision Telecommunications Center had the task of acquiring new donors, a task that it carried out via televised appeals that asked people to call in on toll-free numbers [6]. This organization needed to raise a lot of money—it funded 6,000 projects in 94 different countries. The call rates received were staggering. The call center over weekends dealt with 1,800 calls per hour and received a total 500,000 calls per year.

To cope with this level of telephone traffic, the center needed to do the following.

- Route calls according to the language of the caller;
- Give the agents some indication of what the call is about;
- Identify known donors prior to answer;
- Block unwanted calls;
- Keep track of the success of individual campaigns.

6.6.4.1 The Implementation

The installation consisted of a 60-position Aspect ACD linked by the Application Bridge to a Unisys 6000/70, which ran the Brock Activity Manager and an Informix database. The system was updated by tape from a remote AS/400, which supported the donor database.

Callers wishing to use a language other than English dialed different numbers and were routed to agents that could speak one of the languages supported. Callers who, for example, want to sponsor a child in Somalia called one number, while callers wishing to receive a tape describing the charity's activities called another. The called numbers were used to select the correct greeting data for the agent's screen. Calls could be routed to any agent.

Caller identification and call blocking both utilized calling line identity. The received number was applied to the database prior to answer, and the outcome was used to pop a personal greeting screen or to reject the call.

Campaign tracking was achieved by storing the time the call arrived, the number dialed, and the zip code.

6.6.4.2 The Benefits

One of the major benefits was the agent efficiency achieved, due, in particular, to the fact that agents could be used across multiple campaigns. Costs were reduced by call blocking the calls that arrived on toll-free numbers based on the use of more accurate donor and campaign data. Increased customer satisfaction was recorded as a benefit, and, of course, customer satisfaction is just as important in the charity world where people are giving as it is in telemarketing where they are buying.

6.6.5 Quicker Help Desk for Fast Cars

In 1991, Porsche Cars, in North America, had about 270 dealers. The dealers used an extensive network of personal computers to send parts orders, car orders, warranty claims, and other requests to a central host computer. Dealers generated over 300 help desk calls each week directly concerning the operation of this vital network.

Prior to the adoption of CTI, software, hardware, and communications problems were reported to a central help desk. The first thing the help desk agent had to do was to ask for the dealer code number. The agent then keyed this number into the computer system to bring up the dealer profile. This took some time to appear on the screen, and, while waiting for the profile, the agent asked about the dealer's problems, taking notes by hand. When the dealer profile finally appeared on the screen, the agent could enter the notes.

The arrangement was slow and obviously error-prone. The CTI system used the caller's telephone number to select the profile before the agent answered the call and then presented both the call and the relevant profile at the same time. The agent was then ready to enter details of the problem directly into the computer.

Porsche wished to extend the system to provide owner support. Porsche owners with a problem called the center directly, and their profile needed to be displayed on the screen prior to the agent answering the call. Many Porsche owners have more than one of these cars so the profile was designed to help the agent determine which car the owner was calling about. Since more than 60% of Porsche sales are to existing owners, there was a clear need for Porsche to maintain a very high degree of customer satisfaction.

6.6.5.1 The Implementation

The computer at the help desk center was an IBM AS/400, and it was connected to a Northern Telecom (now Nortel) Meridian 1 switch via Meridian Link. Customer information was held on a Telemagic database. The system proved very easy to program; the application was written in one week, and this, combined with very fast installation, enabled the implementation to be up and running in a very short time.

6.6.5.2 The Benefits

This solution helped Porsche reduce the time needed to process dealer calls by 20%. At the same time, it reduced the number of keystrokes required to carry out the tracking function, speeded up system response, and improved service to Porsche dealers.

The time saved was recycled. The help desk load was expanded to provide technical support to roughly 200 headquarter employees on the company's IBM token-ring LAN.

6.6.6 Selling Books

In the early 1990s, the U.S. publishing company, TimeLife Libraries needed to automate its telemarketing operations, which were spread across the United States [7]. As part of the automation process, each of these operations became a telemarketing center, and these supported a total of 225 agent positions. The company dealt with a quarter of a million leads each week, and the details of these were held at the central mainframe. The company needed an application to perform the following functions:

- Distribute leads to the correct centers;
- Automatically place calls to potential purchasers;
- Maintain an association between the calls and the relevant data so both can be transferred to a supervisor or conference.

At the same time that it supported these tasks, the application had to meet the challenge of fulfilling new orders within 72 hours.

6.6.6.1 The Implementation

The solution was a DEC VAX-based application that utilized DEC's CIT enabler. A central VAX was connected to an IBM host that contains the publisher's customer database. One quarter of a million leads were downloaded to the VAX each week via a SNA link.

The system used NPRI's *TeleTech Marketing System* (TTMS), which distributed the leads among the seven telemarketing centers. The centers have Meridian switches. NPRI subsequently changed its name to Versatility—it still supplies solutions to Time Life.

At the centers, the same TTMS software was used to place calls automatically from the list of leads via the CIT software and, where necessary, to transfer these calls with the relevant data records to a supervisor. Predictive dialing was provided by the TTMS software, call progress detection being carried out by NPRI-provided voice detection units. This was one of the earliest examples of a soft dialer. The TTMS software also provided facilities for creating telemarketing scripts, providing agent performance data, and updating the database.

6.6.6.2 The Benefits

TimeLife Libraries experienced significantly improved revenues through increased call volumes in this application, together with productivity increases. An improvement of 20–25% was observed in the proportion of closures of business, in addition to an increase in the number of leads contacted. A further benefit was the ease with which inexperienced staff could use the system and become effective quickly.

6.6.7 Automated Holiday Bookings

Butlins, the United Kingdom-based holiday company, used to process over 5,000 calls per day during the busy winter months. To cope with this level of traffic it established a call center. The call center employed approximately 120 people and had nearly 100 positions, dealing with reservations, general inquiries, and brochure requests. This workload was quite sufficient to keep the agents occupied during the busy periods. Unfortunately, the problem with incoming calls of this sort is that they do come together, so there were periods of intense activity, followed by periods when relatively few calls arrive—the usual call center story.

The company needed to do some proactive telephone selling to sell holidays to customers on its database. This indicated a requirement to provide an outbound telemarketing capability that call center agents could use during the less busy periods.

The company also needed to allow callers to request brochures at any time and, where possible, to minimize the involvement of agents in this repetitive task.

6.6.7.1 The Implementation

The switch was supplied by Aspect Telecommunications and the Application Bridge CTI link was used to implement this bidirectional telemarketing application. The Application Bridge linked to a telemarketing package called Telemation supplied by Database Systems. Telemation automatically located the next customer to be contacted from the current calling queue.

Agents used a hot key to switch over to the Butlins' reservation system to check availability of an outbound call capability. The system then determined whether conditions were correct for making an outbound call—for example, that there were no inbound calls waiting, that it was no later than 8:45 p.m., and that there was an outbound line available. If any one of these checks failed, an appropriate voice announcement was played to the agent by the switch. If all the checks were successful, then a call was automatically made by pressing a single function key. This represents a very simple example of blending.

Once the call was placed, the agent was guided through it by scripting provided by Telemation. Information about the call could be displayed via certain function keys, as in most implementations of screen-based dialing.

The integral IVR capability of the Aspect switch was also used in this installation—to entirely automate brochure requests for callers with touch tone telephones.

6.6.7.2 The Benefits

The implementation provided a means by which Butlins could obtain an outbound telemarketing capability without increasing the number of agents at the call center. Preview dialing minimized the time spent in setting up outbound calls. Also, the system dynamically balanced the load between inbound and outbound since it gave priority to any waiting inbound call before placing an outbound one.

The automation of brochure ordering provided a 24-hour service and made the agents available to deal with the telemarketing activities.

6.6.8 Campus Emergency and Integrated Message Center

When emergency calls were made at the University of Chicago, the authorities needed to know as much as possible about the call, to speed the contacting and dispatching of the correct emergency services [8]. Rather than interrogate the caller or refer to telephone directories, they needed an automatic indication of exactly where on campus the call was coming from. They also wished to know as much as possible about the caller so that special action could be taken, such as contacting others who could help in an emergency or who needed to know that a problem had occurred.

In addition to providing information on callers making emergency calls, the university authorities wished to improve the handling of incoming calls. A campus message center was needed. When an incoming call arrived for someone who had forwarded his or her calls to the message center, the attendant needed to know as much as possible about the called party. The attendant needed to know why the called party had forwarded calls to the message center, as well as any relevant remarks the called party wanted passed on to incoming callers. Also, attendants wanted to be offered a message screen automatically. This enabled them to input a message without inputting call information already known to the switch (calling/called party information, time of day, and so on). Finally, the called party required an indication that messages had arrived and, if they had a telephone with a screen, the ability to recover the message and/or dial the original caller at a single keystroke.

6.6.8.1 The Implementation

These services were mounted on IBM RS/6000 processors connected to an Intecom IBX switch via the *open application interface* (OAI) protocol. The campus emergency system relied on the availability of calling line identity. This was used to pop the correct screen to the attendant.

When a call had been diverted to the message center, a special screen was automatically presented to the attendant. In this implementation the popped screen provided prompting information for the attendant and provided direct access to the message-taking application.

6.6.8.2 The Benefits

The benefits of the campus emergency system derive from the higher speed response and more appropriate response achieved.

The message center application increased efficiency by encouraging callers to leave messages. This was achieved because the attendant was able to relay up-to-date information about the reasons for the called-party absence. It avoided the problems of *slam down,* which often occurs with voice messaging systems, and allowed fast and automatic recovery of messages by the user. It also provided a very effective mechanism for returning calls because the called party was presented with a *dial caller* button on the telephone itself.

6.6.9 Operator Assistance in the Public Telephone Network

Many telcos are keen to reduce the wage bill incurred by employing the large numbers of operators required for the following tasks.

- Directory inquiries;
- Operator assistance calls;
- Emergency calls;
- Changed-number interception;
- Cashless calling services.

Some telcos offer the more basic services fully automatically by utilizing applications based on speech recognition. Others are taking advantage of CTI to speed up the interactions between customers and operators. This case study concerns a system that provides operators with immediate access to the name and address of incoming callers, the ability to route incoming callers to the retail outlet when they call a freephone number, and speedy connection to emergency services.

In this system operators are presented with an incoming call together with a screen of data that contains the following:

- The caller's name;
- The caller's address;
- An indication of the status of the caller's telephone line;
- Information about the distant inquiry operator.

The operator therefore starts the call with all the relevant information at hand.

Operator-based freephone calls are directly related to a particular service. For example, a caller might call in to ask for "freephone Pizza Hut." The application allows the operator to connect the caller to the nearest Pizza Hut directly.

When emergency calls arrive the operator has only to press two buttons to connect the incoming caller to the correct emergency authority: one to select the authority from the screen and the other to connect the caller.

6.6.9.1 The Implementation

This is a U.K. development that began in 1988, commissioned by the lead telco, British Telecom. After a number of trials, the first installation was brought into use late in 1990, and the whole of the United Kingdom was converted to the operator assistance service in March 1992.

The screen popping and geographic routing functions were both based on calling line identity which, though sometimes unavailable outside of the public network, is always available within it. Routing to emergency services was a good example of SCB in action.

The public switches used were GPT System X switches, which provided a CTI operator support services link to a VAX computer running Invex software from IDS (now Royalblue). Invex provided the interface to the remote databases that contain customer data.

6.6.9.2 The Benefits

The major quantifiable benefit lay in reduced call handling times. Savings of £3M per operator second per year were quoted. Obviously, there were other side benefits, including these:

- Service generally improved;
- Incoming callers were dealt with more successfully since the operators knew more about them;
- Emergency calls were connected more quickly;
- People got their pizzas from the nearest shop rather than one located at the other end of the country.

6.6.10 The Tour Operator

In the early 1990s, there were 25,000 travel agencies in the United States, so competition was high [9]. Using CTI, Classic Hawaii personalized service, improved the efficiency of reservation agents, and reduced average call length, therefore reducing costs. The call center had 70 agents and received over 40,000 calls per month from travel agents calling freephone numbers.

6.6.10.1 The Implementation

The switch in this case was a Meridian 1 from Northern Telecom (now Nortel) using ACD Max software to distribute calls and provide the usual ACD functions. The computer involved was a fault-tolerant Tandem Non Stop machine, which connected to the switch's Meridian Link and used the Tandem *call applications manager* (CAM) software.

Automatic number identification (ANI) was used to identify the telephone number of the travel agency that called. This information was passed to the computer, which used it to retrieve profile information on the travel agency, including the following.

- The agency identification number;
- The names of agents;
- Active reservations.

This information was popped onto the screen, ready for call answer. The travel agents liked the personal touch and appreciated not having to answer a number of questions before stating their business. The dialed number was used to route the calls through to the relevant agent group that dealt with that particular travel agent. The system maintained statistics on average call lengths for the different travel agencies, and these were used to manage the individual relationships, improve customer service, and decrease call length.

6.6.10.2 The Benefits

The company estimated that by reducing the average call length by 30–36 seconds, it has gained the equivalent of four more reservation agents without hiring extra people.

The introduction of CTI was justified in this case solely by improved service levels. However, other benefits include reduced data entry errors, more productive and happier reservation agents, the ability to handle increased call volumes and peak loads, and improved statistics and reporting.

6.6.11 Telemarketing Bureau

British Telecom (BT), the United Kingdom's primary telco, established a call center that managed a complete range of telephone-based services, ranging from direct response inquiries stimulated by advertising to specialist help desks and sales campaigns [4].

The BT Customer Communications Unit's bureau was a large call center with 600 lines spread over three different centers located in the Bristol area of the United Kingdom. It had the capability of dealing with 1,000 different DDI numbers and managed over 300 different campaigns per year. As will be seen, the level of integration was extremely high. This call center was capable of great precision in routing incoming calls to the agent with the correct skill. It dealt with multiple campaigns from many different countries and companies.

Typical of the applications that it supplied for its own business was the help desk for customers planning to buy, or who had bought, an ISDN connection. Callers dialed the specified ISDN number that routed them to a dedicated ISDN call center agent. The agent recorded their details, then gave some first line advice. If that advice was not sufficient, the call could be transferred to a specialist within BT or an external advisor.

In another application, the call center supported customer and retailer inquiries concerning Mondex, a cash-on-a-card payment system. The center ran an information line and a help desk. Any of the Mondex 2,000 trialists could ring a freephone number and be automatically routed to an agent trained in the Mondex field. The agents were automatically presented with "call guides" on their screens to enable them to respond correctly and provide the necessary information.

6.6.11.1 The Implementation

The call center was specifically designed by BT and was first piloted at the end of 1992. The pilot phase identified the need for improved digital line interfacing hardware. The improved design underwent heavy testing and has shown itself to be capable of carrying more than 83,000 calls per day. It entered service at the end of March 1995.

In CTI terms this implementation is categorized as third-party-dependent and forms an early example of a soft ACD. It is based on a Summa Four "dumb" switch and Periphonics' voice processing system, the whole being controlled by DEC VAX computers.

The Periphonics voice processing system is used for overflow handling and related services. It can take the details of callers when the call center is exceptionally busy.

6.6.11.2 Benefits

The flexibility of this design allows full attribute routing. This, in turn, provides complete routing flexibility. Each agent and call can have a unique set of attributes. John, an agent in an insurance call center could have the attributes *experienced agent, French speaker, health insurance, car insurance.* A call, by virtue of its entry point and DDI digits might pick up the attributes *experienced agent required, car insurance.* This call would be queued onto John and any other agents that have the *experienced agent, car insurance* attributes. However, a call with the attributes *experienced agent required, Spanish speaker, house insurance* would be routed to a different set of agents.

This is multidimensional routing, and it is dynamic—as agents log in and out then the skill mix varies; as the call arrival pattern varies the skill requirements vary. Planning the matching of one to the other seems to add a new dimension to workforce planning. However, in BT's application where many campaigns are being handled by many agents, the flexibility provided by this distributed call center is highly desirable.

The voice processing system can also be controlled by the VAX computer and can therefore have attributes that will allow calls to be routed to it where voice automated call handling is appropriate. The system supports a mid-call diversion facility that enables callers to speak to a person if the automated system does not satisfy their needs.

The claim made for this call center is, "If you can write it down then the call center can be programmed to do it." With such a varied workload BT needed complete flexibility, and this solution certainly represented one of the most flexible in the integrated call center world. The degree of integration was high and the benefits were delivered through a completely soft approach to call handling that could be managed from one point. The system can respond to the dynamic and varied requirements of a large telemarketing agency environment.

6.6.12 Happy Travel Agents

British Airways' business is air transportation. However, to support that core business, it is also the operator of a large number of call centers. This case study describes an application used in some of the call centers. It is called "Rapport" and supports a form of help desk. Travel agents call British Airways agents with a complex array of queries—some simple, some complex [4].

From a business point of view, British Airways is keen to give a very good service to its travel agents since they are, to a large extent, its shop window. The call center agents are therefore well-trained and well-motivated.

The U.K. Rapport call center network deals with 10,000 offices and processes about 10,000 calls per day. It is sectioned so that a particular group of call center agents deals with a small number of travel agent companies. It is arranged in this way so that the call center agents become accustomed to particular companies and are therefore knowledgeable about the manner in which they do business. Calls are routed accordingly—based on the CLI of the incoming call where this is available. Those calls that do not have a CLI are routed to a VRU that allows callers to key in their reference number from a touch tone telephone.

The incoming CLI is translated into a unique travel agency number that identifies a particular travel agent office. This can then be used to prepare the initial screen that the agent sees. If necessary, calls can be transferred to another agent or supervisor and the screen of information transferred with it.

6.6.12.1 The Implementation

The switches used in this implementation are Aspect CallCenters. They are linked to the computer system via the Application Bridge. The main computer system is remotely situated and supports the corporate database. The CTI link is not connected directly to the computer system but terminates on a gateway. There are a number of computers involved here, tied together by the Activate software written by BA. The Activate software is layered and provides a switch-independent API. Though it originally worked only with the Aspect Application Bridge, it has now been extended to work with the Nortel Meridian Link and Rockwell Transaction Link to cover switches that BA already has in countries outside of the United Kingdom.

6.6.12.2 Benefits

The reduction in call lengths due to screen popping is the main benefit here, but there are others. Call center agents gain an immediate view of the travel agent's booking capabilities. This is important since some have very basic terminals, whereas others have much the same interface as the BA agents—the nature of guidance and aid that is necessary varies enormously with the travel agent's capability.

The ability to transfer a call with the relevant screen is a benefit for the travel agents. Sometimes they wish to speak to a particular Rapport agent; sometimes a problem needs to be escalated. In either event, the travel agent does not have to start all over again when the call is transferred. The call center agent has immediate access to the relevant travel agent details and the booking that is being discussed.

Finally, the integrated system provides BA the capability of managing call flow. Much more is known about the needs and skills of the travel agents that the call center agents work with.

6.6.13 Customer Service From a Utility

Welsh Water, part of Hyder plc, supplies water and processes sewerage in the country of Wales. An efficient call center was key to its vision of "one-stop shop" for billing and operational inquiries where one call should be sufficient to deal with most problems. The call center was also designed for outbound customer contact: to encourage the payment of overdue accounts, to provide customer follow up, to generate pollution warnings, etc.

The call center is based at Nelson with a satellite unit at Bangor. This case study focuses on the former which had capacity for 96 agents and handled over one million calls during 1995.

Service level ambitions are to answer 80% of all calls within 20 seconds and to complete 80% of queries without passing the caller on to another operative.

Wales has two official languages, Welsh and English. The call center can detect that a caller is likely to need a Welsh speaking agent from the number that is dialed and routes the calls to an appropriate agent. The area of a call's origin is displayed on the agent's turret.

There are, in effect, three applications running at the Welsh Water call center covering account queries, operational queries, and credit management. The first is a fairly conventional mainframe-based application. Callers provide their account codes, and agents input this to the mainframe and then receive all contact details about the caller. The caller might, for example, wish to query the amount of a recent bill, change address information, or whatever.

Operational calls may report a pipe burst or be made by someone who has no water at all. Once again a mainframe contact database is used in dealing with these calls, but this is supplemented by client server tools that allow access to a bulletin board and other information-based applications. The bulletin board has screen button access to all of the company's areas. Clicking one of these will provide the agent with details of any work in that area—color-coded according to the number of customers that the work affects.

Collection calls are generated by a predictive dialer. When an overdue account holder answers one of the calls generated by the dialer, the agents are presented with a screen of information. The screen contains all essential details, including the customer's details, the amount owed, and installment arrangements. An area of the screen is reserved for the outcome of the call, and overdue accounts are temporarily removed from the active list when a pledge is made.

The company has implemented its predictive dialing application in an interesting way. There is an arrangement that ensures each call is validated with the mainframe before it is launched—checking that the account is still outstanding and thus reducing the possibility of calling people who have settled their bills within the last few days.

6.6.13.1 The Implementation

The center is fully enabled for CTI. Welsh Water chose an integration based on the Rockwell Spectrum. An ICL mainframe is connected to a Rockwell VarCTI to provide predictive dialing. VarCTI is a Tandem Guardian computer containing predictive dialing software and high-level CTI interfacing software that has been anglicized by software company Twinsoft. It is connected to the Spectrum via Rockwell's Transaction Link. When contact is made with an overdue account holder the relevant screen is popped onto the agents terminal and a session is established with the mainframe. The agent terminals are PCs running Windows.

6.6.13.2 The Benefits

Many of the benefits to Welsh Water result from the use of a centralized call center per se—rather than an integrated call center—although there are very definite plans to move along the integration path.

The planned use of CLI to pop customer details onto the screen is expected to remove 15–30 seconds off those calls that are made from the customer's own telephone.

The use of predictive dialing has already demonstrated a 500% improvement in productivity—that is, the agents are now able to contact four times as many debtors. More generally, the implementation of presummons contact by telephone and by visits in order to help customers plan the repayment of debt has led to a 77% reduction in disconnections compared with the previous year.

6.6.14 Investment Management

Bristol and West provide loans to people who wish to purchase houses. To finance those loans it must attract investors, and it is here that CTI is employed [4]. Asset, the investment division of Bristol and West, decided to reduce its rate of turnover. It estimated that the cost of obtaining a new customer is of the order of £50, and the cost of retaining one about £8 per annum. A business case based upon these figures and desired retention level of 80% led to the introduction of an integrated call center together with other changes, including the issue of a quarterly investment performance report.

The Asset call center introduces telephone-based service into many aspects of customer management and in doing so has caused a redesign of business processes:

- For any type of call, it is possible to directly access images of recent correspondence through a record attached to the customer's details. Most transactions are now paperless within the center.
- Potential new investors can call in on a freephone number. Agents are automatically presented with the relevant screen and are provided with call guides in order to successfully process the application.
- Established callers can call in on a local tariff line where the call guide will lead them through strict security measures before access can be made to details of their account.
- New account holders are called soon after making an investment to ensure that they are happy.
- Investors who specify closure of accounts at specified times in the future—when a bond matures for example—are contacted a little before the date with the intention of persuading them to re-invest.
- Investors are contacted with suggestions that they might move funds to more effective accounts.

The motivation here is in building good customer relationships and hence good customer retention. The degree of integration is not high but is effective for this application. On the incoming side, callers are discriminated by the type of call (freephone or local rate) and the correct screens presented.

On the outgoing side, the agents need to make calls for all of the reasons listed above. Additionally, calls need to be made in response to received correspondence. All of the different outbound calls are allocated a priority through a sophisticated workflow scheduler. Calls are then presented to agents in priority order and according to the ability level of the individual agent. For example, when an agent has no incoming traffic, he or she can opt to accept outbound calls. Call details will then be presented to the screen. If the agent is experienced then that call may be to process an account closure. However, if the agent is a novice, such calls will be retained, and others, which may, for example, concern an address change, will be presented. The agent then clicks the "call" icon, and the call is placed automatically. This is a good example of skill-based routing for outgoing calls and of preview dialing.

6.6.14.1 The Implementation

The Asset call center is based upon an SDX 420N switch linked via a PC-based CTI server connected to a Novell network of agent PC's and two database servers. Other servers provide a gateway function to the Bristol and West Mainframe and run the various central applications. The agent software is a system called Teledirect Service System from AIT. The switch is connected to the network via 60 ISDN lines that are used for incoming freephone and local rate calls and for outgoing calls. All correspondence is scanned, allocated a priority for outbound calling, and stored on optical disks. There is provision for 30 agents, and the center has handled 1,500 calls per day.

6.6.14.2 The Benefits

The major objective of the Asset call center project is customer retention. It has been in operation since January 1994. A market survey covering 1,500 of the company's customers demonstrated that 86% of customers thought that the Asset service was good, very good, or excellent.

6.6.15 Personal Service in Collections

Pitney Bowes Credit Corporation is a subsidiary of Pitney Bowes and acts as a credit collection organization. To fulfill its business objectives, Pitney Bowes installed a call center targeted at increasing collection efficiency, with the primary activity being outbound dialing. By increasing the number of calls made by agents, collections can be increased. In addition, by reducing or streamlining repetitive tasks, the agent can become more efficient.

The corporation wanted to implement a solution that would automate the agents' efforts—outbound dialing, receiving inbound calls, tracking call histories, updating individual directories, and more. As usual, data stored throughout the enterprise needed to be easily accessed, and databases needed to be shared with other areas of the company.

The company also recognized the need to improve efficiency in its knowledge worker environments. This translated to telephony functionality on the PC and automated access to existing applications.

6.6.15.1 The Implementation

The call center can be characterized as follows.

1. Three hundred and fifty call center agents;
2. One hundred and fifty seats as knowledge worker management;
3. Both inbound and outbound credit collections;
4. Lucent Definity G3 PBX;
5. DMS Unisys host running proprietary database;
6. DCA InfoConnect;
7. Novell Network;
8. Windows/DOS on desktop;
9. Applications include cc: Mail, Lotus Notes, and 3270 legacy applications.

The CTI solution is based on AnswerSoft's *business process automation* (BPA) software and TSAPI. This provides automation of existing applications, databases, and communications tools. The solution includes the use of SoftPhone as a telephony application for each agent's PC and Sixth Sense as the automation technology. Sixth Sense can be customized so the call center is able to configure solutions specific to its collection needs, and scripting can be used for other automation implementations—such as paging, faxing, and sending e-mail.

6.6.15.2 The Benefits

Most importantly, more outbound collection calls are made than ever before, and a boost in the number of calls means a boost in the amount of collections. Moreover, the call center agents are now able to provide more personal service, even in the business of credit collections. As calls come into the center, customers can be directed back to the agent they have been working with, rather than spending time briefing a new agent each time they call the collection department.

SoftPhone has helped enormously in providing a means to track previous conversations. The application's call notes, call logs, and easy access to multiple directories allows agents and knowledge workers to save time on both inbound and outbound calls.

Information is automatically retrieved from multiple databases so that all details are prefilled on a service call report before it is delivered to the agent's desktop. Changes to on-screen information are automatically reflected in the main database, so all areas of the enterprise have access to the most current information on a customer. As a result, the company has increased the response time to customer inquires; alleviated the tedious, redundant tasks of accessing and updating data; and lessened the amount of time it takes to complete a call.

6.6.16 Preventing Car Theft

Car crime is a problem the world over; it is therefore good to hear that CTI can help in the fight against car theft! In 1997, the Skynet company introduced its Skynet 2000 system [3].This technology uses a whole range of technologies to identify the current position of a vehicle, to communicate with the car itself, and to efficiently run a call center.

Car owners that subscribe to Skynet have minor modifications made to their vehicle. If the vehicle is broken into at any time, its location is automatically transmitted to the call center. Similarly, if the car is started by someone who does not know the simple security procedure, the call center is contacted automatically.

Once contacted, the call center agent can talk directly to the driver of the vehicle in order to ask for a security code. If this is not given then it is assumed that the vehicle has been stolen. The call center agent can then cause a number of things to happen, such as the following:

- Sounding the horn;
- Causing the hazard warning lights to flash;
- Activating the engine immobilizer.

Meanwhile, the owner may be contacted, and the police can be informed of the car's location. Naturally, the use of the Skynet service is sufficient to keep most thieves away from the owner's car.

6.6.16.1 The Implementation

Skynet uses a GPS receiver within the vehicle to establish its location. It also uses GSM cellular technology to contact and extend control to the call center. A small microcontroller unit is installed within the car to link the GPS receiver and the GSM transceiver to the car's engine management system. The GSM car telephone can be used in the usual way, but is commandeered by the microcontroller if it believes the car is being stolen.

The call center itself is extremely secure. Sixteen agent positions are in use at the time of writing, but this is expanding rapidly. The call center is based on equipment and software supplied by Siemens, IBM, Brock, and Action Information. IBM, the system integrator, used CallPath/2 and CallCoordinator/2 as the CTI enabler.

Calls from the vehicle are routed to the call center, and the system automatically identifies the number of the GSM phone. This is used to retrieve the vehicle description and details of the owner from the Brock system. This information is popped onto the screen of the first available agent. At the same time, a map appears on an adjacent touch screen display giving the exact location of the vehicle. The agent can then communicate with the driver of the car and control the car's operation.

6.6.16.2 The Benefits

Clearly, it would be difficult, if not impossible, to supply this service at all without CTI so any discussion of benefit cannot be comparative. Additional benefits, over and above those of basic operation, have been identified. These include the storage of information relating to the number of calls received and the result of each call. These are all logged by the system and form a valuable source of management information—this is classical coordinated call monitoring.

6.6.17 Voice Response in Banking

The most common use of IVR has traditionally been in the banking world. It would be quite difficult to find a retail bank that does not offer an automated service in some form or another in the developed markets. This case study is included mainly for completeness. The United Kingdom's National Westminster Bank has been operating its ActionLine service for over four years. The functions provided are similar to those offered by other banks, although there are always differences in detail:

- Balance inquiry;
- Payment inquiry;
- Bill payment;
- Statement request;
- Fund transfer.

The service is popular. At the time of writing, it had seen 40% growth in usage over the previous 12 months.

6.6.17.1 The Implementation

This IVR implementation is based on the Periphonics voice processing system. It is accessed via free phone or local charge numbers.

6.6.17.2 The Benefits

The benefits of an IVR system to the bank are fairly obvious: It does not need to employ people to deal with these calls. This installation was dealing with 1.4 million calls per month at the end of 1997. It is estimated that the equivalent load would require two 200-seat call centers to provide a similar service.

General figures [10] in relation to banks with more than $1 billion in assets indicate the following.

- That a transaction dealt with at a branch costs $2.93;
- That a similar transaction dealt with by a call center agent costs $1.82;
- That the same transaction handled by IVR will cost 0.24 cents.

Also, as indicated by the growth rate experienced by National Westminster Bank, telephone transactions are growing rapidly—in one survey by 39% as against 6% growth in branches.

The benefit to the customer is that the service is available 24 hours per day; that, for simple transactions, it is quicker than dealing with an agent; and that some people actually prefer dealing with a machine for certain transactions. Added to this, service is consistent. Naturally, there are many customers who would prefer human contact, but also a growing number prefer automation.

6.6.18 Fully Integrated Finances

Lloyds TSB PhoneBank is a large call center with installations at Newport in Wales and Glasgow in Scotland. This installation was given the ACTIUS award for the best integrated call center in 1997. The call center is a networked arrangement with negotiated overflow from Newport to Glasgow and vice versa. The sites provide telebanking and other services to Lloyds TSB's customers, over 600,000 of whom are registered with PhoneBank.

6.6.18.1 The Implementation

The telephone system used by TSB is the Lucent Definity. Its CTI link is connected to a UNIX Sequent server that integrates telephony into Edge agent software from IMA and provides host and interactive voice response access. The latter provides a telephone-based automated banking service called PhoneBank Express. Customers that meet difficulties with the automatic system are transferred to agents—together with the data collected on the call.

The PhoneBank application requires access to a number of different mainframes but the implementation ensures that a consistent interface is presented to agents. Calls are recorded centrally and can be recovered from any permitted desktop—together with the data that was input during that call. This is used for training and audit purposes. In an extra touch of integration in this call center, the agents log on by swiping a card through a keyboard-based reader.

6.6.18.2 The Benefits

From a business perspective, Lloyds TSB has reduced the average duration of calls very significantly through the use of integration. The time that agents spend in wrap up—tidying up after a call—has been reduced to almost zero. This has been achieved without upsetting the customers. In fact, Lloyds TSB claims a customer satisfaction level of 97%.

References

[1] Morita, Akio, *Made in Japan.*

[2] Q.Sys Inc. Group PhoneWare Help. GPSIM 2.4., June 1996.

[3] "BM. Skynet: Calling a Halt to Car Crime," *CQ Magazine,* July 1997.

[4] "Computer Telephony Integration," Financial Times Management Report, ISBN 1 85334 400 1, 1995.

[5] "Security Pacific Cuts Costs and Provides 24-Hour-a-Day Service with Centralized Call Center," Call Center Application Sheet 600132, Tandem Computers, Inc.

[6] *Computer-Supported Telephony in Europe,* London: Schema, 1992.

[7] "Computer Supported Telephony," Eurodata Foundation Conference, 1991.

[8] Salter, Linda, "Quantifying and Evaluating the True Benefits of a Telephone-Based Strategy," *Cost-Effective Debt Collection and Recovery Conference,* London, England, 1992.

[9] *Voice IN Europe Newsletter,* ISSN 09669922, November 1992.

[10] ABA's Retail Banking Survey, 1995.

7

The Market for CTI

7.1. How Many Fish in the Sea?

This story came originally from the "Perishers," a strip cartoon in one of the English daily newspapers. It is very old, but unlike market estimates, it does not seem to age much!

Picture a little boy holding hands with a slightly larger girl. Their backs are toward you.

They are staring out over a blue sea. Waves lap gently at their feet.

"How many fish are there in the sea, Maisie?" pipes the little boy.

"Don't know," replies Maisie.

"'Course you do, Maisie. You know everything like that."

"I don't know, and I don't think anybody does, really."

"Well, about how many?"

"I don't know. Can't you believe me? I don't know," says Maisie, shaking her head.

"Well, have a guess."

"No, I couldn't even guess."

" 'Course you can, Maisie. Everybody can guess things."

"Well, about . . . uhm, about a thousand," says Maisie at last.

"Huh, that's a silly guess," the little boy says in obvious disgust as he walks off to look for suitable stones to throw into the fish-filled sea.

7.2. Market Definition

Maybe Maisie's answer is not very good. I am sure that many of you have been in this situation at some time or another—forced to answer the unanswerable and then castigated for providing the wrong answer. However, in fairness to the poor, browbeaten Maisie, was the question reasonable? In the first edition of this book, I pointed out that Maisie could have informed the little chap that the worldwide catch in 1989 was 99,534,600 metric tons—but this is probably of little interest to him. Though the catch does bear some relation to the number of fish in the sea, it is a dangerous base from which to extrapolate. In retrospect, he may have been more interested in the number of different types of fish. I understand that the most often quoted estimate is 20,000. However, there may be as many as 20,000 more according to Fish FAQ on the Web!

And, what does he mean by fish? Are whales included? At his age, I expect that he thinks they are fish. Are shellfish included? Is plankton? Plankton contains krill, and this is regarded as fish.

How many CTI systems are there in the world? Furthermore, what is the potential number of CTI systems that will be installed in forthcoming years? To begin to answer that question, it is essential that we define the market we are examining and, therefore, redefine CTI. Chapter 1 introduced Lois B. Levick's definition:

> A technology platform that merges voice and data services at the functional level to add tangible benefits to business applications.

It was a good definition, but it should be analyzed—just as the little boy's definition of fish was analyzed. What does platform mean in this context? What are the voice and data services, and where are the business applications located? Most of these points were covered in Chapter 1 and subsequent chapters. However, another question now arises. Lois's definition was first published in the late 1980s—does it still have relevance?

The *Association of Computer Telephone Integration Users and Suppliers* (ACTIUS) adopted the following definition for CTI when it was formed in 1992:

> A functional integration of business application software with telephone-based communications.

This is a broader definition, but initially ACTIUS took this as the base definition and pointed to certain things that were not CTI. For example, according to the association, stand-alone voice messaging was not part of CTI.

However, ACTIUS decided to extend its definition in 1995 by adding the following:

> The scope of this definition embraces applications within the call center, at the desktop, and within the workgroup. The integration applies to voice processing and call processing systems and to the media of text, image, and voice.

Clearly, the association was beginning to recognize that CTI extended beyond the call center and that it was about multimedia—not just concerned with telephony. However, note that in the Levick and the ACTIUS definitions the emphasis is on functional integration. This excludes voice and data integration, which is purely a physical sharing of the same transmission channel as in, for example, ISDN. It also excludes two other important physical integrations—that which takes place at the desktop itself and that which occurs at the server. Many people regard physically integrated systems as examples of CTI—regardless of whether any functional integration takes place. Note also the mention of business applications. This excludes applications that are based on a central computer controlling the telephone network; those applications that are managed by a telco. In other words, INs do not form part of the CTI market under these definitions, even though CTI and IN have much in common.

Over the years of CTI evolution I have been regularly challenged over the need for definition. "Who cares?" asks the exasperated supplier, "Everyone knows what CTI is, you are just complicating the whole thing." Generally, it has been the suppliers that have complicated CTI with their strong, and understandable desire to have their own products fall nicely within the definition of CTI—or even to be the definitive CTI products! The suppliers have been aided and abetted in this by some sections of the press. Does it matter? Of course it does. How can you define a market if you cannot define the thing that is being marketed? When people state that the market for CTI will peak at $6 billion, what do they mean? Are they including IN and/or messaging? Exactly what are they including? Who cares?—anyone who has a taste for truth.

At this stage no attempt at a clear market definition will be attempted. This is because the definition of CTI is both broad and nebulous. The important points to remember in examining market figures are summarized in Figure 7.1. These include the different types of implementations that are possible.

When considering a market definition in addition to the characteristics of CTI shown in Figure 7.1, the scope of the market has to be understood. This is best addressed by asking just what is and what is not included. For example, does the market definition contain the following?

- Hardware;
- Software;

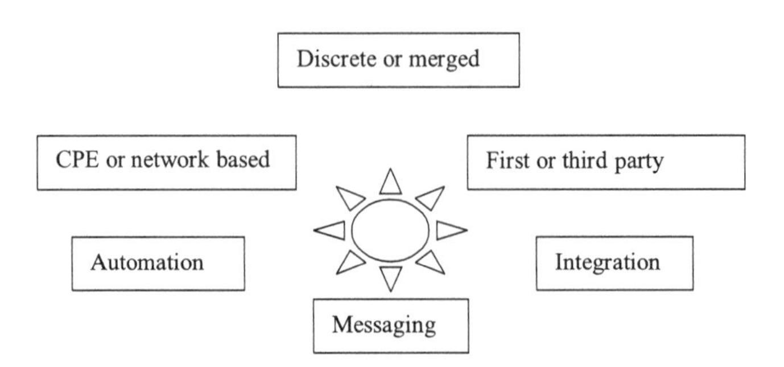

Figure 7.1 Defining characteristics of CTI.

- Wiring;
- Local networking;
- Wide area networking;
- System integration;
- Installation;
- Maintenance;
- Upgrades.

This list could go on—but it is sufficient to make the point. The point is that the scope and coverage of CTI are huge. The scope for misunderstanding is therefore also huge. In some ways, the challenge that Maisie faced in attempting to answer the little boy's question about the number of fish in the sea is less than that faced by the CTI marketer. And the chances are great that answers to the question, how many CTI systems are there in the world, can yield some very silly answers.

7.3. Market Drivers

In Chapter 6, the story of Akio Morita introduced the concept of finding the right application for technology. Also in that chapter the early applications of CTI, dating from as far back as 1969, were recalled. These were genuine applications and their existence raises the obvious question, why did CTI not take off like wildfire then and there? Here, an attempt will be made to answer that question more fully by examining the CTI drivers and inhibitors. First of all, what are the drivers? Figure 7.2 gives some indication of them—according to a survey conducted among CTI suppliers—and indicates their importance. Note that customer service was seen to be the main driver.

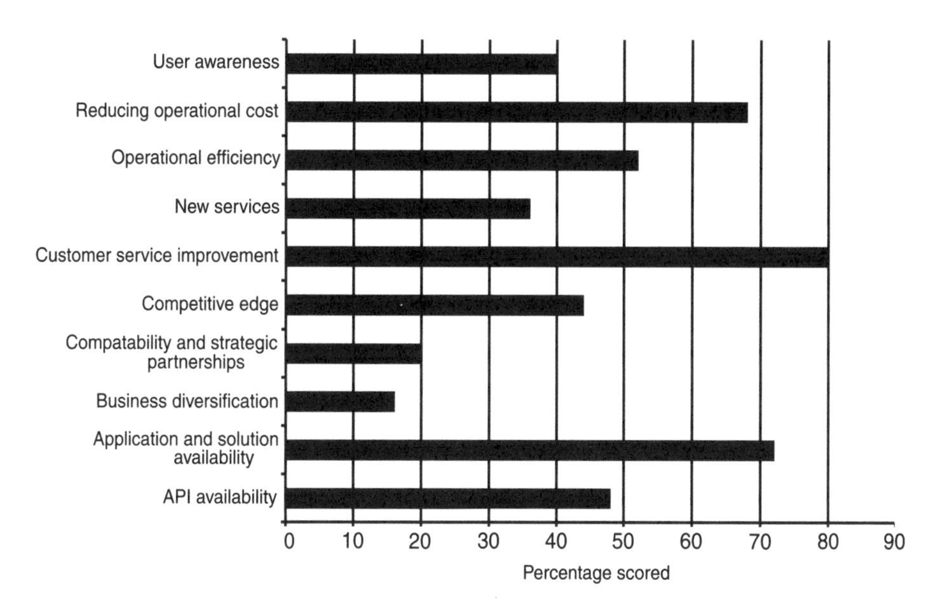

Figure 7.2 Supplier view of the CTI drivers [1].

Here, the major drivers are considered to be the following:

- Expectations;
- Customer service demands;
- Operational efficiency;
- Competitive edge;
- Business diversification;
- Desktop standardization;
- The Internet;
- Application and solution availability;
- Compatibility and strategic partnerships;
- PC card technology.

Each of these will now be considered in some detail.

7.3.1 Expectations

One of the interesting things about CTI is that expectation is a key element in driving this level of integration forward. People just expect to be able to do the things CTI enables them to do—they have no real appreciation of the difficulties involved in doing it. This is analogous to the world of the cellular telephone. I still marvel at the fact that I can use a telephone in my car, on a train, or simply walking along the street—yet I know how it is done. My colleague, who has no conception of how this level of mobile telephony can be achieved (and no interest either), thought it a gimmick at first. Then, far from marveling at the fact that it can be done at all, began to complain of noisy connections, cut-off conversations, and high cost. He did not understand why the ordinary telephone needed wires in the first place, so the availability of wireless telephony was always well within his expectations.

This is also the case with computer telephony integration. The woman who transferred the call to a Spanish-speaking operator in Chapter 1 was not at all surprised that the screen data and the call should simultaneously be presented to her Spanish colleague. The call and the screen are together at her terminal; why shouldn't they stay together? The agent who sees a screen of information on a particular product appear before he or she picks up the call is not surprised—the incoming caller dialed that product line anyway. The user who receives a message from a colleague is quite happy with the reply option—it is obvious that selecting that option will send a message back to the originator. It is no great surprise to the same user that a phone option will place a call to the originator of the message. It is also of no great surprise that voice and fax messages turn up in the in-tray that previously contained e-mail messages, nor that clicking the mouse on a voice mail message causes the phone to ring and that the message is there when the telephone is lifted.

All of these examples reinforce the fact that, on the whole, CTI presents no surprises to its users, sometimes quite the opposite. They may wonder why they were previously unable to transfer the voice and data calls, rather than expressing wonder that it can now be done!

Of course, the people mentioned here are the users, not the purchasers. They do not make the buying decision, nor is it necessarily in their direct interest to make things more efficient. However, if the changes CTI produces were difficult to assimilate or required new skills, these are the users who would form a barrier to introduction. That CTI can be introduced easily is therefore regarded as a driving force for its introduction.

7.3.2 Customer Service Demands

Up to 27% of customers who cannot get through on the telephone will either buy elsewhere or skip the transaction altogether. Unhappy customers return to do more business if their problems are resolved. In fact, 95% of all dissatisfied customers will do business again if their problems are resolved!

Only 40% of dissatisfied customers complain. If their complaints are satisfied, 80% will buy again; if not, 40% will buy again. Of the 60% that do not complain, only 6% buy again. These are the facts that underline a great marketing revelation. Receiving a complaint is an opportunity!

These are just a few of the often quoted remarks concerning customer service. Figure 7.3 demonstrates another pearl of wisdom.

What Figure 7.3 demonstrates is simply that it costs five times as much to acquire a new customer as it does to retain an existing one. The exact ratio will depend upon many factors: it can be much higher. Although this may be an obvious point to make, it is a lesson that we all have to relearn occasionally. There is a tendency to neglect the existing client in the excitement of chasing new prospects. Yet the cost of leafleting, cold calling, developing new relationships, and so on is astronomical compared to the development of existing accounts.

This is a driver for the introduction of CTI, because the generic functions of CTI make it so much easier to provide good customer service—through automatic access to the right person, automatic access to customer data, and the ability to pass customers to someone more able to deal with queries without the customers having to explain their requirements all over again. The affinity routing described in the case studies included in Chapter 6 is a good example here. Regular callers are answered by the same group of people. Taken to its extreme, affinity routing attempts to route a call to the very person who normally deals with that caller.

CTI helps service agents provide better customer service without the technology eroding person-to-person interaction. Hence, for companies wishing to improve customer service, it offers a cost-effective development path.

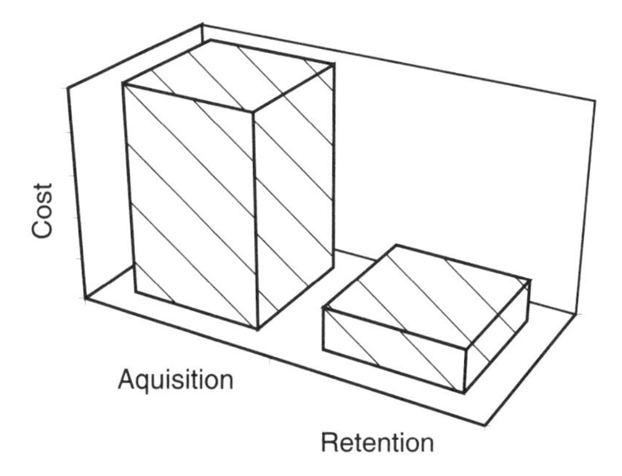

Figure 7.3 The cost of gaining or retaining customers. (*Source:* International Customer Service Organization.)

7.3.3 Operational Efficiency

The end of the 1980s saw a gradual and long-term downturn in the economy. The years of expansion were over, and business began looking inward. It was the era of the accountant—where every business plan was examined over and over again for the least sign of risk, and the operation of the business was under constant scrutiny in the hope of detecting yet more and more savings. Businesses began downsizing—some called it rightsizing. This was hardly the best time to interest business leaders in new technology, but interest was high. What was the driver here—cost saving, of course. Figure 7.4 helps to explain the source and impact of CTI-related saving in the call center. It is often used to explain and underwrite a proposed move to CTI.

Figure 7.4 does not surprise many people. Over the lifetime of a call center, the cost of the equipment is tiny compared to other costs. The true cost lies in the call charges (based in the figure on freephone for an inbound call center) and the cost of staff. Obvious though this may be, it is often forgotten when the initial choice of call center system is being made. The point is that with a ratio of some 10-1 between staff costs and equipment costs, it is well worth spending money on the system to reduce the number of agents, even if the cost of the system is doubled. Moreover, the reduction in staff costs is usually achieved by reducing the holding time of telephone calls, and this has the effect of reducing telephone network costs as well.

Assume, for example, that a call center costs $3 million to operate over a five-year period and that those costs are apportioned as follows:

- Equipment costs (including software): $180K (6%);
- Staff costs: $1.62M (54%);
- Telephone network costs: $1.2M (40%).

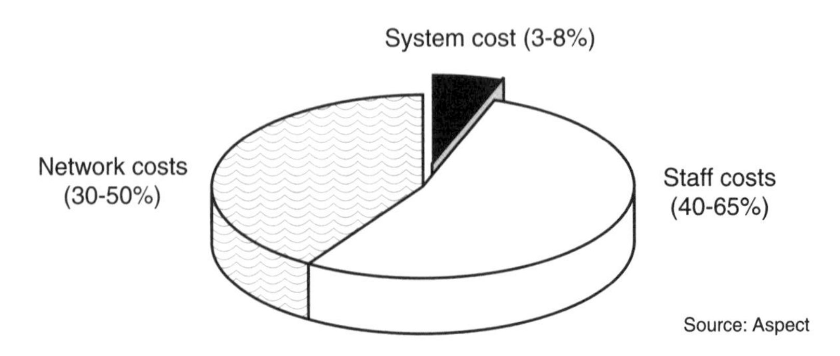

Figure 7.4 Call center costs. (*Source:* Aspect.)

If the introduction of CTI doubled the equipment costs but reduced the staff costs by 25% and network costs by 15%, the total cost of operating the system over five years becomes $2.6M. This is a reduction of nearly 13% on overall running costs, and the actual cost savings are more than double the amount invested to produce them. This is a strong driver. When business managers consider what can be done to call centers to provide savings in staff and call costs, CTI and IVR offer the two main opportunities.

Another well-known statistic in the world of telephone-based marketing is the cost of making a sales call—in person. This cost is estimated as $250[1] in the United States and even more in Europe; it is further claimed that it takes, on average, five sales calls to close a deal—a cost of $1,250. This is astronomical compared to the cost of telemarketing as an alternative sales method; calls made by telephone are going to cost less than $5! CTI reduces those call costs even further.

All the generic functions can be used to increase business efficiency in some way or another, to the advantage of, rather than at the expense of, customer service. Similar arguments can be applied to automation. IVR can extend the hours of operation toward full availability. It also ensures a consistent interface. However, the major driver here is cost saving. In an earlier case study, the cost of providing a transaction at a bank branch versus that of providing the same transaction through a centralized IVR system was shown to be over 12:1. Hence, the capital cost of the IVR installation can be recovered very quickly—but at some cost in the level of customer service. There is a solution to this dilemma. Give the customer the choice, and give those who choose IVR some of the benefit. In many applications, the majority of people will choose IVR.

Note that, though CTI does improve efficiency at the desktop, this area of the market has not proven to be a major driver. The reason for this is simple. Phone utilization at the normal desktop is low compared with that in the call center. Cost justification is, therefore, difficult.

7.3.4 Competitive Edge

CTI is useful here. It can be used to improve the operation in many ways, in particular, by finding innovative ways of selling and of retaining customers. It provides new ways of precision marketing directly from computer databases, and, through the more effective linking of customer data with telephone sales calls, it improves the customer interface. One often quoted example of this is in

1 *Source:* McGraw-Hill Research

cross-selling. Here customers are offered other products and services based on the agent's knowledge of what customers already have or use.

CTI can help to create an aura of efficiency that itself helps to attract customers. Many call centers use the technology of CTI first to save agent time and then utilize that time to improve customer service and hence increase competitive edge. The cycle referred to here is that of inbound/outbound working. Time saved in, say, popping screens to the agents can be used to make calls to any customers that called but dropped out before answer, for example.

7.3.5 Business Diversification

This begins with one of the most well-known functions of CTI, CBDS—popping up screens at the same time that a call arrives. The call centers with the highest requirement for diversification are those that operate as bureaus. The call center capability is leased to various companies that have decided to outsource the telephone handling function. Call center bureaus have two choices: They can allocate a group of agents to each client company or find some method by which the agents can deal with calls from any of their clients' customers. The latter is by far the most efficient in the use of resources, and it inevitably employs some aspect of CTI. The most likely aspect is, of course, CBDS. Here the screen is written with all the relevant client information just as soon as the call center recognizes which client the caller is trying to contact. This is usually based on the called number. Further development may move along the affinity path by using the caller's line identity to route to a particular group of agents, if there is someone available in that group.

Similarly, if a company has established a call center for one product and wishes to expand the portfolio, then CTI enables them to do so within the existing center. Prime examples here are new businesses that begin life by offering automobile insurance directly via the telephone. Having become established, they often move into the related businesses of life and home insurance—using the existing call center.

A further advantage that CTI delivers to diversification is that it can provide meaningful statistics on the various product lines. Managing a multiproduct line call center is naturally more complex than running a single-product center. CCM provides the management with a product-based overview of the operation of the call center.

7.3.6 Desktop Standardization

The 1980s began with a tussle to determine which PC was to become the industry standard. The IBM-compatible PC won that battle, though other architectures such as the Apple Macintosh are still preferred by some users. The IBM-compatible architecture in time became identified as the Intel architecture. Consequent upon that battle was the competition for the software high ground. By the mid 1990s, Microsoft Windows was clearly the nearest thing to a "standard" operating system for the GUI-based desktop. This standardization is achieved by market dominance rather than discussion. However, this does not undermine its importance. The dominance of Windows has gone beyond the desktop. It has made deeper and deeper encroachments into the corporate IT world, gradually usurping IBM and Novell in areas in which these two companies were clear market leaders.

Naturally, the market for desktop applications for CTI is tightly bound up with standardization at the desktop. It is also tied up with the availability of an API, which itself is generally tied to the operating system used. While the desktop market was diverse, only a limited market for desktop integration of PCs with telephony existed. With the extensive installed base of PCs running Windows and supporting TAPI, that market assumed huge proportions. The number of applications available for this environment increased rapidly, although many of these were initially simple dialer applications based on first-party CTI—an inevitable result of the early versions of TAPI supporting first-party CTI only.

In the late 1990s, another wave began to engulf the desktop market in the form of the NC and the use of Java applets. The latter allows CTI applications to be delivered from a server to a client, regardless of the nature of the client. It rids the application world of its dependence on operating systems and PC architecture. Meanwhile, the NC has the capability of reducing desktop costs and simplifying management; both affect CTI implementation.

7.3.7 The Internet

The relationship between CTI and the Internet will be explored in more detail in a later chapter. However, it is significant that the Internet is not mentioned as a driver for CTI in the first edition of this book. In fact, the Internet is not mentioned at all! This itself is an indication of how quickly things change.

The Internet is a driver of CTI in many ways. Though it does not provide the most suitable technology for a multimedia network, it is indeed used as a multimedia network. A fully integrated network capable of carrying real-time and message-based communications in all forms turns CTI on its head. If everything is integrated in the network, what is the problem of integrating it at the user interface? This is a good point—but a slightly misleading one. CTI acts to bring together something more than the bit streams that carry voice, data, and multimedia communications.

However, a fully integrated network should make things easier—and it does. For example, the delivery of CLI across international telephone networks will not be a universal service until well into the 20th century. The delivery of an originator's IP address is an automatic function over the Internet, regardless of country of origin (even though its meaning may not be clear). Another example, the ability to reply to a message in any chosen media or mode (voice, video, real time, etc.) is a tough nut to crack in the telephone world. In the Internet, it is more a matter of possessing compatible software than cracking nuts.

There is a more subtle sense in which the Internet is a driver for CTI. It has helped the spread of information enormously—just as it does with any "new" technology. People talk of Internet years being comparable to dog years. Its ability to spread the word has certainly been a driver for CTI, even though most searches throw up many references to the Computer Teaching Initiative!

7.3.8 Application and Solution Availability

During the early stages of CTI evolution, the cry was often "where are the applications?" The switch supplier provided the link and the computer supplier provided the interface to it, but who was to supply the applications? This was generally expected to be the territory of the independent software vendor. Although such people did exist, the following were not at all obvious to them:

- There was a great need for CTI applications;
- There was sufficient exploitable market in the installed base;
- The CTI world was sufficiently stable to justify the development of applications.

That situation changed dramatically through the 1990s. Chapter 5 and 6 cover just a few of the application programs available. CTI enabling of applications that are used in the call center became a given in most cases. Widely used business application suites such as SAP began to include CTI modules.

Of course, applications are only part of this picture. To drive the market forward, customers are searching for solutions and solution providers. Once again, the situation in the 1990s evolved rapidly. Tools became available to generate CTI applications—all the way from the APIs offered by the computer vendors to fully configurable applications.

In order to deliver integrated solutions, people with integration skills are essential. Initially, specialist system integrators with the knowledge and experience in CTI to deliver working solutions to solve customer problems and satisfy their needs provided these skills. Over time, CTI knowledge has spread in many directions—particularly where call center implementation is concerned. Toward the end of the 1990s, most of the large consultancies and system integrators had grown or acquired the capability to manage the supply of integrated call centers.

7.3.9 Compatibility and Strategic Partnerships

As mentioned earlier, one of the reasons the independent software vendors did not write CTI applications in the early days was that they could not perceive a significant market. One major reason for this was that the interworking of switches and computers was limited. To take an extreme example, the BT Stanza approach mentioned in Chapter 1 would only work with certain models of DEC VAX computers connected to a specific variant of the Mitel SX200 analog PBX, which itself was only available in the United Kingdom. The possibility of finding a customer with all these characteristics was vanishingly small.

Lack of compatibility is a market restraint. Compatibility, often via strategic relationships between switch, computer, and software suppliers, therefore became a driver in this industry. A major area of concern was the interfacing of a switch supplier's protocol to a computer supplier's range of machines. On top of this was the issue of CTI protocol supply by a switch supplier. Application suppliers wanted protocols already supported by the computer/ operating system supplier. Related to this is the testing of these interfaces and protocols to ensure they work with a particular computer and conform to the specifications. This is a necessary prerequisite to the development of applications.

The driver here has been the establishment of accepted methods for physically connecting CTI links together with the widespread use of certain APIs, particularly those mentioned in Chapter 5. However, see Section 7.3.10.

7.3.10 PC Card Technology

Originally conceived as a more flexible basis for creating voice processing systems at the low end, the PC voice card industry expanded in many directions. By the end of the 1990s, it was a key element in the CTI mix and had dominated the voice processing world. Furthermore, it spearheaded the introduction of PC style openness into that world and then began to make serious inroads into the fax and finally the switching sectors. At this point, some wags began to talk about removing the I from CTI. After all, if the whole thing—switch, computer, and voice processing system—could be contained in one box, where is the need for integration?

PC card orientation undoubtedly drove CTI down the road of physical integration in certain market segments. It has also been a major factor in expanding the coverage of CTI from computer telephony integration to complete and total integration.

7.4. Market Inhibitors

To a certain extent, the inhibitors should be a reflection of the drivers, but this is not entirely true, as will be seen. Nevertheless, there is some overlap. Here are some of the things that have repressed CTI market growth—and that, in some cases, continue to do so:

- Standards;
- Regulation;
- Ignorance;
- Culture;
- Integration expertise;
- Cost.

These inhibitors will now be examined in some detail.

7.4.1 Standards

It is clear that the proliferation of protocols, APIs, and enablers that built up as CTI evolved, cried out for standards to be established and used. Furthermore, there is little doubt that lack of standards was an inhibitor to the successful establishment of a substantial CTI market in the early years. It is also clear that the CTI world was without standards for so long it has become established in spite of that lack.

The subject of standards highlights a major divergence of view within the CTI world. Computer-oriented people expect standards to emerge from industry at large and are actually willing to accept a particular supplier's product as a "standard." Telephony-oriented people expect standards to originate with independent standardization bodies and would be horrified at the establishment of a single supplier's product as a telephony standard. Internet-oriented people eschew standards but depend upon them utterly—provided they come from the Internet Engineering Task Force. People oriented towards voice processing share all of these views—sometimes simultaneously. Given this situation, it is hardly surprising that progress toward standards in CTI was slow.

The standards situation will be reviewed in some detail in the next chapter.

7.4.2 Regulation

Whereas standards are international, the regulatory position, especially in telecommunications, is territorial. The regulation situation will be explored in Chapter 8. Sufficient to say here that the use of ANI has been a stimulant to the United States CTI market, and the lack of similar facility in most European countries has been an inhibitor.

There are other facets of regulation that do inhibit the CTI market to some extent. These include the requirements for approval to connect switches to the network and legislation that controls the placement of automatic calls.

7.4.3 Ignorance

I have been actively pursuing activities targeted at removing ignorance in the CTI world. Yet I know that it is still a mystery to many potential users. By the same token, however, so are the workings of the motor car. In the first edition of this book, I wrote that I very much hoped that the book would remove at least some of the mystery. I added that until the CTI industry gets its act (and terminology) together, confusion will prevail. Confusion causes doubts, and doubt is a powerful inhibitor to the development of a substantial CTI market. I used some comments from a survey of potential users that was conducted in 1992 [2].

- "Would like to know more about it. At the moment it all sounds a bit vague, but I am interested in it."
- "I hardly have any information on it, so I can't comment. More information in the future would be nice, as I think it will be a standard within the next 10 years."
- "Lack of awareness or information makes it difficult to justify the need or use of it in the organization—very interested but have no real knowledge of the benefits."

The second user's prediction has, to some degree, been put to the test. Despite many positive comments, the second Schema study conducted in 1996 [3] also revealed a healthy residue of ignorance:

- "CTI is a new technology, and we need to improve our knowledge of it. Pieces of hardware need to be more compatible with one another. There are too many different standards."
- "I'm getting to know the system and its qualities. I'm unsure about it though and apprehensive of so much technology—things can go wrong and then really wrong."
- "CTI sounds interesting, and we would like to know if it will work with existing elements—there may be a lot of adaptation."

In fact, the survey showed that, of those with no interest in CTI, the major reason for that lack of interest (45%) was "need more information about CTI." Furthermore, a survey amongst suppliers conducted in 1995 [1] showed that the respondents rated confusion as the strongest impediment to the adoption of CTI, as shown in Figure 7.5

Some ignorance of CTI will always remain, but the late 1990s did see some of the veils removed and some of the hatchets buried. The entry of Microsoft and Novell into the CTI world in the early 1990s certainly helped to raise awareness levels. At the same time, however, it probably raised the confusion levels. The sheer amount of literature on the subject has helped, as has the establishment of some excellent magazines. As ignorance of the law is no

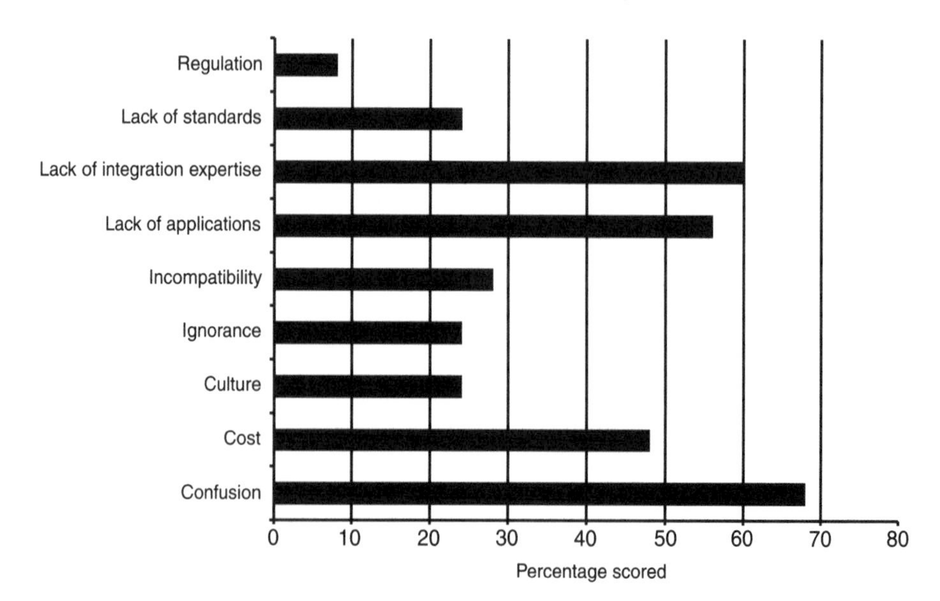

Figure 7.5 The impediments to CTI implementation [1].

excuse for breaking it, ignorance of CTI is no longer a valid excuse for not implementing it.

7.4.4 Culture

This goes right back to the introduction to CTI in Chapter 1. There is a telephony culture and a computing culture, and that is often reflected in company organization (and company politics). Where this division exists, selling CTI solutions always meets barriers that do not exist in either camp. For example, to whom does CTI belong in the bigger organization? Is it the telecommunications department, or perhaps the information technology department or maybe the business department within which it is to be implemented? Generally, the telecommunications skills have been absorbed by the IT department, but pools of telecommunications specialists still exist.

When one department becomes a CTI solution champion, it has to sell the concept to other departments; if not, the other departments will resist the activities of the champion. In some organizations, the internal departments battle for control of CTI, viewing it as a threat to their own existence or a means of subjugating someone else's existence. Organization-wide acceptance of the technology and support for its introduction is a necessary background to successful CTI implementations. Cultural and organizational dissension is an inhibitor to its introduction.

7.4.5 Integration Expertise

As the CTI industry grew, more and more existing integrators began to develop CTI expertise. However, the rate of growth in the call center world certainly outpaced the growth rate in CTI expertise. As time went by, more and more role shifting occurred. Switch suppliers and voice system suppliers leaned toward solution supply—as did some computer suppliers. This led to an unbalanced distribution of skill levels. The main lack of skill—and the main impediment to CTI implementation—tends to have centered on the large corporation and is more concerned with legacy working than CTI, per se.

In other areas, the emergence of PC-based call center systems has removed the need for some skills that were lacking.

7.4.6 Cost

CTI is as difficult to cost as it is to define. However, there is little doubt that perceived or actual cost has been an inhibitor throughout the evolution of CTI.

CTI implementations in large call centers can literally cost one million dollars and yet still be cost-justified over time. Such an expenditure can result in an implementation cost of $10,000 per agent seat. Clearly such costs are not justifiable at the desktop; it may be difficult to justify a cost as low as $100.

Many CTI implementations are complex—in the sense that CTI is an integration rather than a system or product. Particularly in third-party CTI, there are many components to be costed before the overall price can be assembled. It is this total cost that is often unacceptable. Fully integrated systems do not have this problem. Pricing for a PC-based call center is fairly simple. However, prospective users often compare the "all-in" price with, say, the cost of a basic telephone system. The comparison is unreasonable, since the merged solution does a lot more than the basic PBX or ACD—but there is a perception problem here.

Voice processing costs were reduced significantly during the early 1990s—primarily under the competitive pressure created by the introduction of PC-based systems. As the technology has become a commodity, the major suppliers have moved into the more profitable solutions and value-added market.

The entry of Microsoft and Novell into the integration world certainly created an expectation of reduced prices. Both companies took some time to deliver their solutions, and there were some disappointments with regard to capability and robustness. There is little doubt that TAPI and TSAPI have reduced CTI costs; nevertheless, cost remained an inhibitor. Some people claim that CTI should actually be free at the client end and that suppliers should take their income from CTI server sales. This business model is particularly suited to JTAPI implementations.

7.5 CTI Benefit Analysis

Business benefit is the chief driver of the CTI market—as it is of most other markets. Before attempting to quantify the market, it is therefore desirable to quantify and analyze the benefits CTI can offer.

7.5.1 CTI Benefits

There are a number of areas of benefit. The advantages, which derive almost directly from the drivers previously described, are the following.

- Competitive edge;
- Cost savings;
- Customer service;

- Enhanced profile;
- Increased revenue;
- Job satisfaction.

The first question considered here is, which of these are most important to the user? Fortunately, the study conducted by Schema in 1992 [4] offered some early answers to this question. The results of rating customer benefits from this study are summarized in Figure 7.6 The ratings were based on European customers; however, this tends to make the findings more, rather than less, surprising. The largest group of respondents thought the key benefit of CTI was customer service! Generally, the United States is well in advance of Europe with regard to attitudes towards customer service. It was fully expected that this study would indicate that cost savings would be the highest rated benefit in Europe. However, in the survey, cost savings were rated as a poor second to customer service. Job satisfaction and competitive edge followed at much the same rating; revenue increase and enhanced profile were similarly paired.

The Schema study was repeated four years later. This time the categorization of benefits was slightly different. The results are shown in Figure 7.7. Apparently the conclusions are much the same: Improved customer satisfaction is the prime reason for introducing CTI. However, looking more closely at these

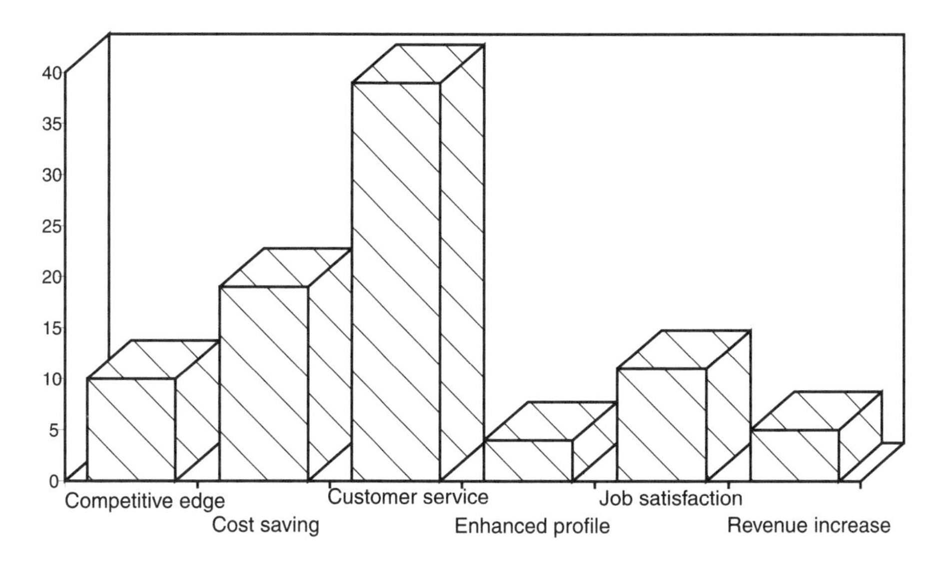

Figure 7.6 Rating of CTI benefits.

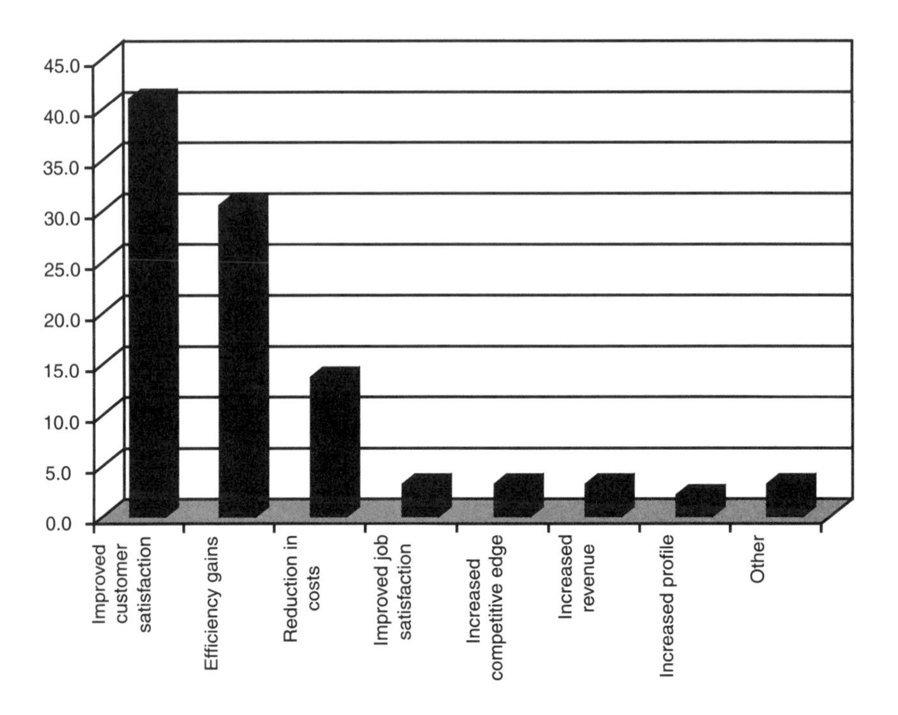

Figure 7.7 Rating of CTI benefits—four years later.

two figures shows two things. First, if efficiency gains and reduction in costs are taken together from Figure 7.7, then they amount to approximately 44%. These two can be considered as the equivalent of cost saving in Figure 7.6. Looking at the figures in this way indicates that in the later survey, cost saving is the most important benefit from CTI. There is another trend here. The other benefits of CTI—job satisfaction, competitive edge, etc.—are shown to be of almost negligible interest in the second survey.

Here we have an interesting comparison. The survey results on CTI awareness are of particular interest. In 1992, 62% of those interviewed had heard of CTI (or at least CST, as it was then called). In the second survey, 70% had. This does not seem a significant increase over four years! However, in the second survey, 59% of the respondents had a potential interest in using CTI in the call center, and 34% were interested in using it at the desktop and workgroup. In 1992, only 4% were planning to implement a CTI system at the time the survey was carried out; in four years, this has risen to 13%—over threefold. It can therefore be concluded that the first survey examined a very immature market and the second, a market that was actually implementing. The polarization exhibited in Figure 7.7—in comparison to Figure 7.6—is therefore explainable.

Note that these surveys are directed at end users and potential end users—most surveys rely upon the suppliers to form a picture of the market.

An alternative way of looking at the prime benefits of CTI is to consider which are tactical and which strategic. This grouping is shown in Table 7.1.

Table 7.1 actually categorizes the benefits into short-term benefits and long-term benefits. Combining this with the results of Figure 7.6 indicates that over 60% of respondents favored the strategic group. This is an interesting conclusion, given the ease with which short-term benefits can be demonstrated and achieved for CTI.

This book has often drawn attention to the generic functions of CTI. It is therefore important to know which of these are the most important. By mapping the generic functions defined here with the terms used in the original Schema study it is possible to show a clear preference for the functionality provided by VDCA. This is one of the most demonstrable of the CTI-generic functions and the one that most impresses users who are new to this technology. Significantly, it is the most difficult to cost-justify but is the one that has a direct effect on customer service. This also reinforces the conclusion that users are looking for strategic benefit, not simply tactical cost savings and revenue increases.

Table 7.1 can be taken one stage further by examining each generic function against the list of benefits. The results of this examination are contained in Table 7.2. It is interesting to observe that in Table 7.2, cost savings tend to require the most functions.

Fortunately, a direct survey of the generic functions was undertaken in the *Financial Times* survey of 1995 [1]. The results are shown as Figure 7.8 and include automated call handling, which can, in this discussion, be interpreted as IVR. Note the surprisingly low value allocated to CCM. This may be accounted

Table 7.1
Classification of Benefits

Tactical Benefits	Strategic Benefits
Cost savings	Competitive edge
Revenue increase	Customer service
Job satisfaction	Enhanced profile

Table 7.2
Allocation of CTI-Generic Functions to Benefits

Benefit Category	Cost Savings	Revenue Increase	Job Satisfaction	Competitive Edge	Customer Service	Enhanced Profile
Benefit class	Tactical	Tactical	Tactical	Strategic	Strategic	Strategic
SCB	✓		✓	✓		
CBDS	✓	✓	✓	✓	✓	✓
ACRI	✓	✓		✓	✓	
ACRO	✓	✓	✓			
VDCA			✓		✓	✓
DT	✓					
CCM	✓	✓			✓	

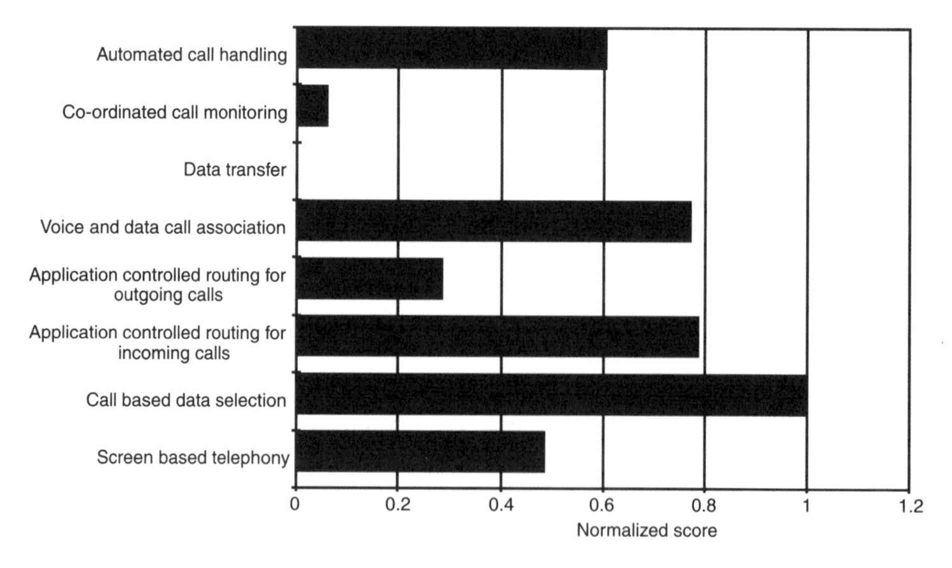

Figure 7.8 Normalized view of the popularity of the generic functions [1].

to the fact that this survey is supplier-based. Here CBDS is the most popular function. Of all the integration functions, this is the one that is most likely to provide cost savings—though automatic call handling can yield more. This is closely followed by ACRI and then VDCA. ACRO does not score highly considering that it is the basis for predictive dialing.

7.5.2 Quantification of CTI Benefits

There are plenty of installed CTI systems from which the real benefits of this integration can be gleaned. Unfortunately, it is not always possible to obtain quantitative feedback for many reasons; one reason is that users often regard such figures as confidential. The following examples have been derived from many different sources. The experiences reported provide an overview of what has been achieved and an indication of the level of benefit that can be realized.

The reports are tabulated by benefit category, and in each case the industry application is named, together with descriptive notes, where relevant. Finally, a subsection is devoted to gains that can be made from the use of power dialers, since this is a major source of short-term return.

7.5.2.1 Competitive Edge

Quantitative evidence of the competitive-edge benefits provided by CTI is impossible to obtain. Other benefits, such as faster answering times and better-informed agents, tend to fall under the customer-service heading—but, of course, better customer service naturally attracts customers and therefore adds competitive edge. Table 7.3 provides some qualitative feedback on competitive edge from various industries.

7.5.2.2 Cost Savings

As Table 7.4 shows, there is much more in the way of quantitative information here, as one might expect. Some of the savings are very impressive, demonstrating very short payback periods on the initial investment. Many of the larger savings are based on the use of power dialing, and the source of those savings is described at the end of this section.

Table 7.3
Competitive Edge Benefits

Industry Application	Competitive-Edge Benefit	Notes
Banking	More complete service	Banks compete through customer service
Health product supply	Easier and quicker to order	
Transport	Faster repair times	Through increased agent productivity, autodialing, VDCA, and routing of incoming calls to the correct agent

Table 7.4
Cost-Saving Benefits

Industry Application	**Cost -Saving Benefits**	**Notes**
Financial call center	26% saving in the number of agents required	This reduction was made possible by integrating an existing call center with the database computer
Inbound debt collection	The number of agents handling incoming calls reduced by 45%	First level of debtors dealt with fully automatically (60%)
Outbound debt collection	18 agents reduced to 4	This represents a 78% reduction
Hotel management	Reduction of fraud by 50%	E.g., telephone calls from unoccupied rooms
Hotel management	Two man days per week saved by automatic transfer of billing information	200-room hotel, transfer of lobby, room service, and telephone billing information
Health product supply	10–15 seconds off the length of an incoming call	Phone bills 13% lower
Inbound insurance call center	Average inbound call duration of 4 minutes reduced by 26 seconds	Saving 41 days/annum
Outbound insurance call center	Average outbound call duration of 5 minutes reduced by 40 seconds	Saving 270 days/annum
Medical	Call set-up time reduced by 10–12 seconds	
Multi-user desks	Savings in accommodation costs of 40%	A saving of £2m/annum for a 700-person office

Power Dialing

Power dialing was described in Chapter 6. It can provide major benefits in outbound applications. Predictive dialing in particular can increase agent productivity quite spectacularly by removing all automatically generated calls that are not answered by a live person. Figure 7.9 illustrates the savings that can be made. The computer application places the call and so removes the preview time. The call progress detector within the power dialer allows the detection of tones and removes any calls that meet a busy tone or ring tone with no answer.

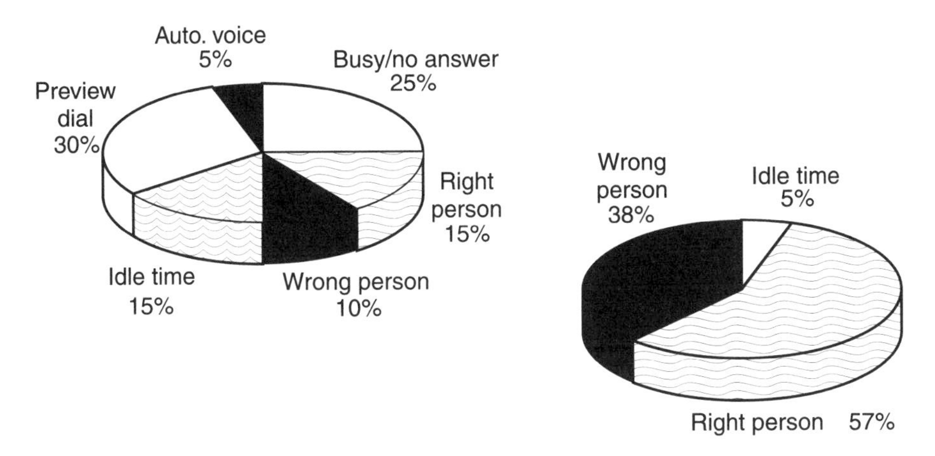

Figure 7.9 The benefits of predictive dialing.

It may also be possible to detect whether the voice at the other end is from an answering machine or a person.

The predictive dialing software only launches sufficient calls to ensure that an operator is available when a call is answered. All of this can produce productivity increases of 200% to 300%. Of course, this is dependent on the nature of the task. If the agent holding time on each call is long in relation to the time saved by automated placement, the benefits are much lower. Table 7.5 illustrates where and when the predictive dialer is most efficient.

Table 7.5
Relative Efficiency of Predictive Dialers [5]

Characteristic of Called Parties	Relative Dialer Efficiency
Consumer daytime (2-min. call)	Very high
Consumer daytime (5-min. call)	High
Consumer evening/weekend (2-min. call)	High
Consumer evening/weekend (5-min. call)	Medium
Business-office at home	Low-medium
Other businesses	None

7.5.2.3 Customer Service

Much of this experience relates to speedier and more responsive call handling. The benefits that occur because agents are better informed is less easily unearthed because it is a secondary effect in speeding the transaction with the customer. Table 7.6 provides some examples.

7.5.2.4 Enhanced Profile

It is very hard to find any real examples of companies benefiting from enhanced profile, but Table 7.7 gives a few. This is also the lowest on the list of priorities for potential users.

7.5.2.5 Increased Revenue

One area that can realize large increases in revenue is debt collection. Once again, this is primarily due to the use of power dialers. Table 7.8 shows that the other sources lie in increasing the calls made or answered and increasing the closure rate on those calls.

Table 7.6
Customer Service Benefits

Application	Customer Service Benefits	Notes
Insurance call center	Calls answered within 20 seconds increased from 48% to 86%	
Health insurance	Reduced customer wait time by 75%	
General	High level of customer acceptance	
Hotel management	Quality of service to guests	E.g., speed of check-out
Insurance	Establishes ownership of calls	
Health product supply	Calls handled more quickly and more accurately	Reduced number of call transfers
General	No need to repeat information already supplied after transfer	
Insurance call center	Speedier and more active response to callers	
Inbound debt collection	30% reduction in abandoned incoming calls	
Finance investment	Abandoned call rate reduced from above 7% to below 1%	Target was 2%
Health insurance	Reduced lost calls by 50%	

Table 7.7
Enhanced Profile Benefits

Industry Application	Enhanced Profile Benefits	Notes
Retail/wholesale	CTI will enable us to have a higher profile and be more competitive, so everyone benefits—clients and suppliers [3]	
Finance investment	Yet to show a prospective customer our center without winning the business	Powerful endorsement of the technology

7.5.2.6 Job Satisfaction

Once again this is difficult to quantify. Though there is a known danger of burn-out in telemarketing agents, there is no evidence that CTI increases the rate of staff turnover at call centers; rather the opposite is observed. Table 7.9 provides qualitative feedback.

Table 7.8
Increased Revenue Benefits

Industry Application	Increased Revenue Benefits	Notes
Call intercept	Increased call revenue over investment (2:1)	Automation of calls to numbers that have changed
Debt collection	Paid for itself within six months	
Debt collection	228% increase (585 to 1920) live connects/agent/month	
Sales appointments	Number of agents to produce 3,000 appointments reduced from 41 to 15	
Health insurance	Handles 20% more inquiry calls per day	
Health care	Increased outgoing call attempts by 250%	
Telemarketing	25% increase in VCR sales	
Health product supply	Recovering callers that hang up	Caller's number is registered then presented as an outgoing call
Telemarketing (pharmaceuticals)	Number of outgoing calls/day increased by 80%	Conversion rate is 23%

Table 7.9
Job Satisfaction and CTI Benefits

Industry Application	Job Satisfaction Benefits	Notes
Vacation broker	Job enrichment	Avoids negative customer service; removes menial tasks.
Health product supply	Enables agents to do their work more easily	Removes some of the frustrations of the job
Insurance call center	Improved agent performance and job satisfaction	Via automation of repetitive tasks

7.5.3 A User's View

In 1996, Norwich Union, a U.K. insurance company, won the annual ACTIUS award for best integrated call center of the year. One year later, Mike Holmes, the IT manager responsible for this Centrex-based distributed call center wrote an article entitled "CTI—Why bother?" [6]. The article was clearly written from the heart and is reprinted here with permission from the *CTI Informer.*

> "I have to admit to a level of amusement at the contortions exhibited by companies justifying the business benefit of CTI against the cost investment. I dispute none of their figures and applaud their imagination. Thankfully I never had to justify our implementation. As a learned sage of two years standing I can now assess the myths and realities of CTI in my particular organization. Taking some common claims in turn:
>
> - CTI will enable you to pop up caller information and save lots of time.
>
> This may well be true in installations supporting predominantly regular callers. But for our inbound sales teams the number of times CTI is usefully matched is very small (10%) and the saving in time is not great. The value of CTI could never justify the cost of implementing CTI in this organization.
>
> - Synchronized call and data transfers makes your operation slick.
>
> This can be extremely useful, but if a customer is being passed from pillar to post something is probably wrong. One transfer may be acceptable to the

customer but operationally it is expensive. We strive to improve our processes to reduce transfers to a minimum.

- Interactive voice response is essential to any modern call center.

Really good for the call center but what does it do for your customers? I could easily fill many pages with the good, the bad, and the appalling. Basically if the customer knowingly calls a computer it is good. If a customer is confronted with unwanted automation it needs to be very, very good. We prefer to use real people to provide a service to real customers.

- Improves management information.

In our case this proved to be the key benefit. Being able to integrate business information with telephony statistics helped to shape the call center operation. We are able to analyze the type of business and success level of every call from a business perspective and input this to training to pricing and marketing. *This information almost certainly justifies CTI on its own.*

- A predictive dialer increases productivity by 300%.

Our call center was initially almost entirely inbound. When we hit peaks we took call back messages on scraps of paper but this was not effective. We built a simple outbound dialer which managed the callbacks during quieter periods. Suddenly we had control and we could manage our peaks and troughs. When we needed to manage outbound sales campaigns the dialer was there. This outbound dialer alone justifies CTI.

In summary, and in my experience, the CTI hype has got most things wrong. There are very real advantages in using the technology but they are not the obvious ones. It all depends on how CTI is utilized to support the business objectives. Just popping up screens does very little. We manage a virtual call center of 1,000 agents split between four cities. CTI does not do that for us. It is however, a key tool in supporting the good management processes that have enabled the business to flourish. It may never have been cost-justified but it would cost us a fortune if it was taken away."

7.6. CTI Market Quantification

To quantify a market, one needs to know the following.

- What is being sold;
- Who buys it;

- What it costs;
- Who sells it.

CTI is not an easy topic as far as market quantification is concerned. An attempt to describe what is being sold was made in the market definition established at the beginning of this chapter. No clear definition could be established. The problem is, everyone has a different definition, so market figures might not address the same thing. Figure 7.10 makes the point. It was assembled in 1995—but the year makes little difference. The point is this: What were Tern, Schema, Ovum, and the FT talking about when they made their estimates? In fairness to them, and the many other sources of market figures for CTI, the definition will be in the report. Given the impossibility of arriving at sensible market figures, it is more practical to look at the other items on the list at the beginning of this section.

7.6.1 Who Buys CTI?

The computer market is a source of constant surprise. I did not dream in the 1970s that my local paper would have an advertising section devoted to computers; that there would be at least six shops dedicated to computer and com-

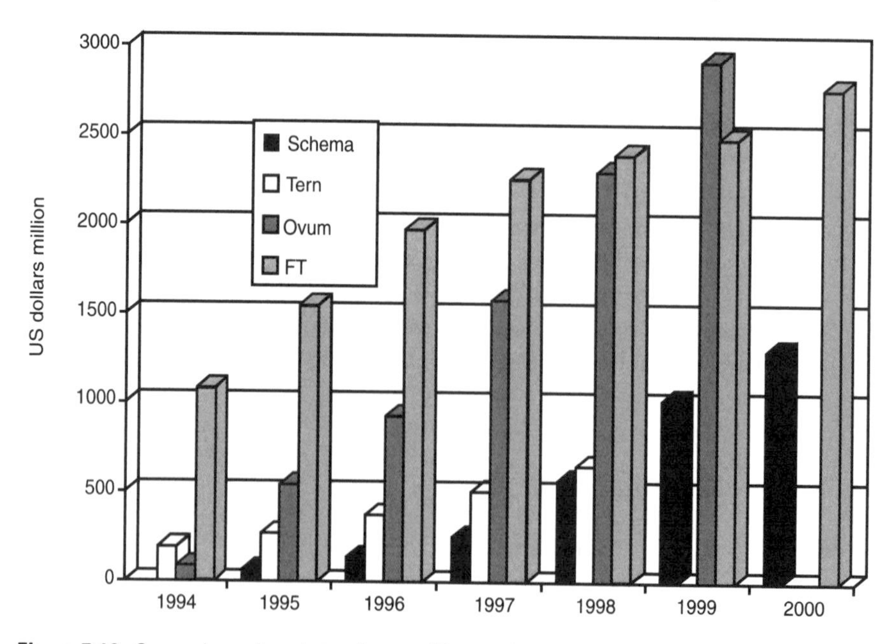

Figure 7.10 Comparison of market estimates. (*Source:* Various.)

puter peripheral supply in my nearest big town; or that news-agent display shelves would be crammed full of fat, colorful magazines with titles like *Computer Shopper*. Nor did I dream that those same bookshelves would be crammed with fat colorful magazines on the Internet. Though my powers of prescience failed me, I was fairly confident that CTI would never be a consumable in the way the PC has become—simply because CTI is an integration, not a product. However, the component parts of CTI have become consumables—and the day of the prepacked CTI solution is clearly visible.

In the early 1990s the buyers of CTI included the following.

- Call center managers looking to improve margins and customer service;
- Large organizations looking to improve overall efficiency and reduce staff costs;
- Companies that wish to do most of their business through the telephone;
- Help desk managers wishing to improve the service they provide.

However, the potential application areas for CTI were almost unlimited. Anywhere a telephone and terminal coexist, there exists a potential CTI application. Any business that has automated its major customer processes and does a significant amount of business over the telephone is a potential CTI user. Even the lone PC user who keeps a telephone directory on the PC is a potential user. As the 1990s progressed, the list of purchasers grew. The arrival of TAPI brought forth a long list of simple CTI applications for the lone PC user and others. This began the march of the CTI from the call center onto the desktop and into the workgroup. Perhaps the word march is too strong here. A dignified toddle may be a more suitable description. During most of the 1990s, the true crock of gold for CTI has remained firmly within the walls of the call center.

In the call center, the CTI buyer can be somewhat distant from end users. In CTI, end users can be any of the following.

- An incoming caller;
- An agent;
- A supervisor;
- A call center manager.

Only the latter is likely to be directly involved in the purchase of the system. The type of customer is often defined by the cost and complexity of the CTI implementation. Naturally, as cost decreases, the level of the buying decision falls.

In general, the customers for CTI are most likely to come from the manufacturing sector—because it is the largest. Figure 7.11 demonstrates how the market segments [1].

7.6.2 What Does CTI Cost?

I am often asked how much an average CTI implementation costs. My usual reply is that hackneyed phrase, "How long is a piece of string?" There are many variables involved, including the following.

- Number of users;
- Complexity of the application;
- The switch involved;
- The computer involved;
- Whether it is a new system or an upgrade;
- If an upgrade, whether the existing computer/switch is to be retained.

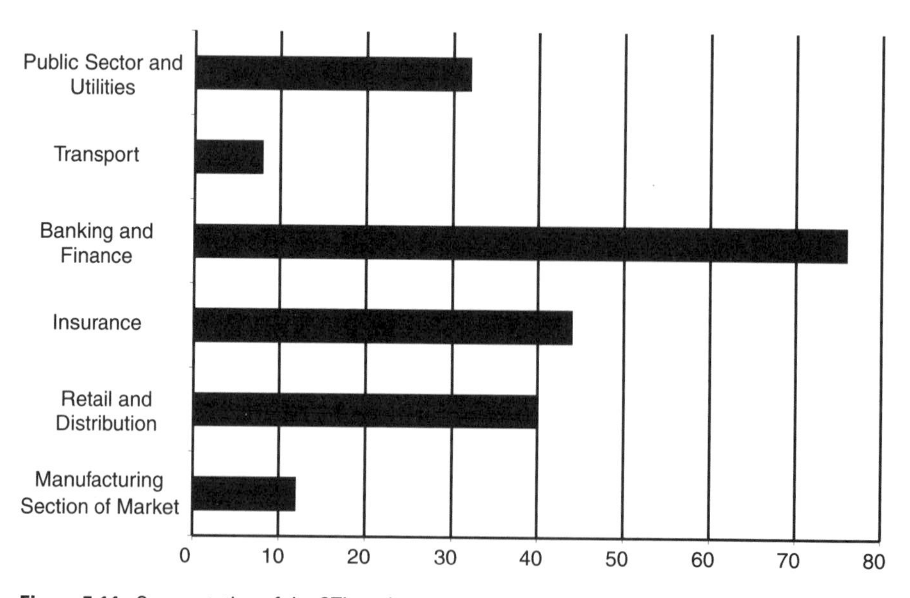

Figure 7.11 Segmentation of the CTI market.

If the implementation is an integration of existing items then the cost will lie predominantly in software and services. This does not simplify things much. Within that cost there are a number of items:

- Writing new applications, if any;
- Interfacing to existing applications;
- System integration—making the whole thing work;
- The basic API;
- Higher-level enablers;
- Application generation tools.

If there is a choice between a first- or third-party CTI implementation, then there are two cost models in operation. First-party costs start at zero and increase on a per-user basis. Third-party usually involves a major initial cost and a lower incremental cost. Figure 7.12 makes the point. It is highly dependent on the assumptions made in its construction. It is included here only to make the

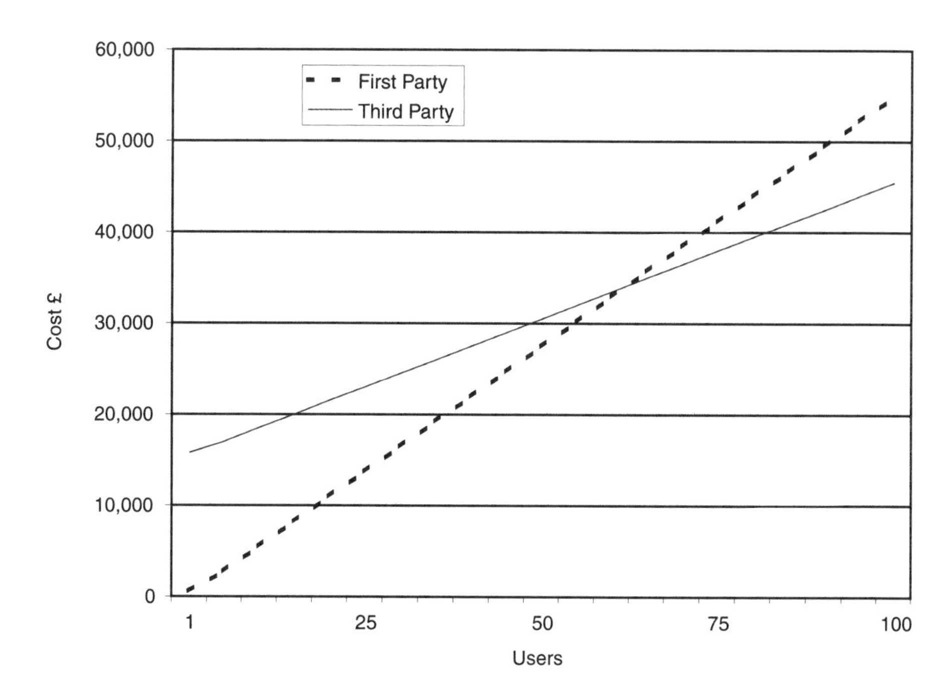

Figure 7.12 First- and third-party CTI cost example.

point that there can be a changeover point at some stage—even though third-party CTI has the higher functionality.

7.6.3 Who Sells CTI?

The simple answer to this is nearly everyone. Directories of CTI suppliers have become huge. Everyone with the slightest connection with the computer or telephony world wants to be considered a CTI supplier. One of the many Web sites devoted to CTI (www.ctinet.co.uk) used the categories shown in Table 7.10.

Suppliers that access this Web site can request that their details are added to the directory contained there. The distribution of suppliers found is given in

Table 7.10
Categorization of CTI suppliers

Supplier Category	Supplier
Equipment and software	Computer system
	Switch
	Component
	Voice processing
	Independent software
	Integrated systems
	Peripherals
Service	System integrators
	Telcos
	Consultants
	Bureaus
Value added	System developers
	Value-added resellers
Distributors	CTI distributors
	Switch distributors
Enabling technology	Speech recognition
	Speech synthesis
	Speech general
	Software tools

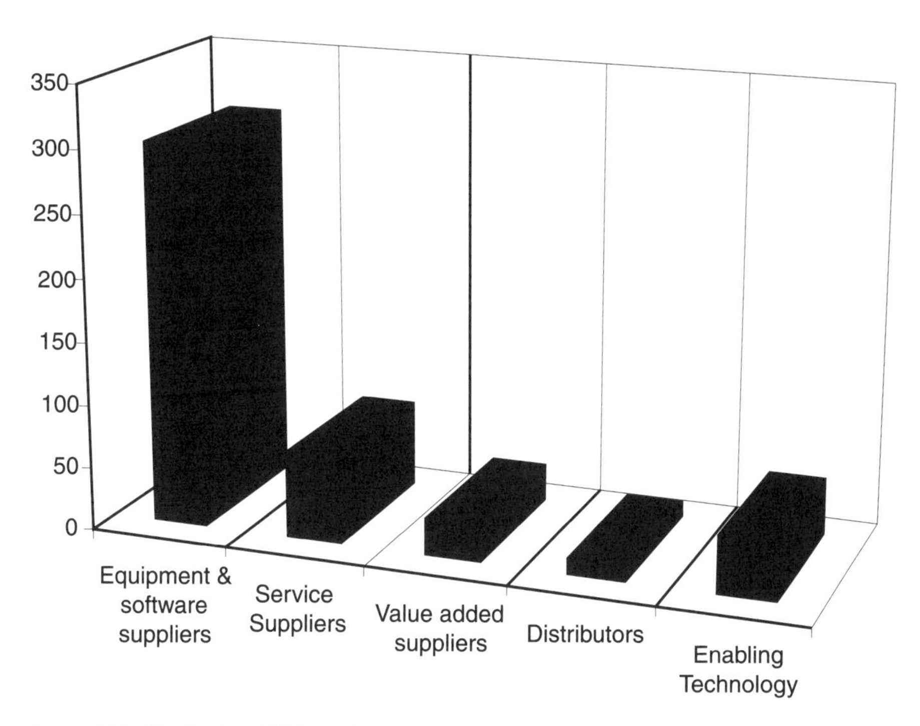

Figure 7.13 Distribution of CTI suppliers.

Figure 7.13. In 1997, there were nearly 800 companies registered—a high number and, fitting figure on which to end this chapter on the CTI market.

References

[1] "Computer Telephony Integration," *Financial Times Management Report,* ISBN 1 85334 400 1. 1995.

[2] *Computer-Supported Telephony in Europe Market,* London: Schema, 1992.

[3] *Computer Telephony Integration in Europe Market: An Update,* London: Schema, 1996.

[4] *Computer-Supported Telephony in Europe Market,* London: Schema, 1992.

[5] Communications Series, McGraw-Hill, 1992.

[6] Holmes, M., "Members Spout—CTI Who Cares?," *CTI Informer,* Issue 3, October 1997.

8

Computer Telephony Integration System Engineering

8.1 What is System Engineering?

Engineering is all about getting things to work; therefore, system engineering is all about getting a system to work. A CTI solution is a system like any other; it therefore requires the same discipline that system engineering of a chemical plant or any major software implementation requires—fine words and fine intent. This section attempts to lay down some guidelines that lead to good system implementation. It focuses primarily on the larger CTI implementations, those that require the skills of a system integrator. However, the guidelines are of value in all areas of CTI and can be easily applied to product and system development.

The starting point for good system design has to be a clear statement of customer requirements. Consequently, the failure of a number of CTI implementations can be directly traced to a lack of clear requirement definition. A number of system integrators have coined the phrase, *managing customer expectations.* CTI holds great promise, although delivering those promises may be impossible without a major upgrade of the customer's telephone system, computing system, or both. Such an expenditure must be identified at the very beginning of a CTI project. To accomplish this, the system integrator must have the necessary knowledge and tools to do the following:

- Assess compatibility and fitness for purpose of the various elements involved in the system;

- Select the correct architecture for the system;
- Size the system elements for correct performance.

This chapter takes a very pragmatic view of computer telephony integration—as is necessary in system engineering. It expands the topic of requirement definition, introduces the relevant measures of performance, considers the problems of testing CTI systems, expands the topic of CTI management, and addresses the compatibility issue with a detailed consideration of the existing and evolving standards available within the CTI world. It is by no means a system engineering manual. Many of the concepts and approaches considered here apply to the implementation of any computer-based system. I believe it was President Eisenhower who said, "Plans are nothing, but planning is everything." System engineering is very much concerned with planning, but it does not regard plans as nothing.

The plan is the mechanism for detecting when system implementation may be going astray. Early detection provides the signal to the project manager that corrective action may be needed. So now for the "motherhood and apple pie." Read it if you need to.

8.2 Implementation

Implementing CTI systems is often very much concerned with the management of customer expectations. What does this mean? Surely customer expectations are to be met, not managed. Of course customer expectations are to be met, but, unfortunately, experience has shown that they are rarely met in the IT world. Reports of long overdue delivery milestones and of projects that vastly exceeded budget are available from all sectors of the marketplace. Expectations were high, and they were not met.

Of course, it is quite possible that if expectations were low, projects would never start in the first place. Chapter 7 explains that the cost of acquiring a new customer is five times that of retaining an existing one. Thus, creating a dissatisfied customer is far worse in the long run than persuading a potential customer not to proceed because the risks are too high. At least the latter is still a potential customer.

The risks in CTI implementations are greater than the risks in many IT projects. After all, a CTI implementation is an IT project but an IT project with extra ingredients—telephony, voice processing, messaging, and so on. So, CTI projects have all the usual pitfalls, plus a few more, provided by the extra ingredients. Naturally, potential customers must be exposed to, and convinced of, the benefits of computer telephony integration, but the benefits must not be oversold. The customer must have a clear vision of what can be achieved, how long it will take, and how much it will cost. This is the sense in which customer expectations must be managed.

Good implementations of CTI follow the general rules of good implementation. The guiding principles include the need for a formal methodology for all stages of the design and implementation of a CTI system. The methodology will deliver the following.

- An architectural overview that contains all the relevant protocols and interfaces;
- An agreement from the suppliers covering the availability and release level of the protocols and interfaces;
- Clear and agreed-upon design objectives for the system—in functional and performance terms.

8.2.1 Formal Implementation Methodology

Whenever I have to plan anything that is serious, I make notes, usually in the form of mind maps and lists. I think all serious planners do something similar. Whilst we are doing so, we secretly envy those that seem to be able to manage without such props. Dream on. A formal methodology should contain the following steps.

1. Establish project management and tracking procedures;
2. Agree on user requirement definition and validate;
3. Define component specifications;
4. Design and configure system;
5. Develop and customize implementation plan;
6. Perform component and end-to-end testing;
7. Perform acceptance testing against user requirements;
8. Establish contractual relationships.

Most of these steps will be familiar; a good methodology provides an effective balance between the efficiency of the production team and the bureaucracy of managing the process. Some of the points listed deserve further attention from the CTI standpoint.

8.2.1.1 User Requirement Definition and Validation

The collection and validation of user requirements is just as important in a CTI environment as it is in any IT project. There are, however, special factors that need consideration because the telephony ingredient has been introduced. Vanguard Communications Corporation lists the following items as key to this phase:

- Establish specific, quantitative objectives;
- Obtain data on the performance of the existing arrangement;
- Establish a preliminary call-flow sequence (i.e., a description of the task);
- Perform a busy-hour traffic analysis, which should include the following:
  - Call durations;
  - Call handling time saved by automation;
  - Daily, weekly, and seasonal variations:
  - Computer traffic levels and response time.

At a detailed level, it is essential that inbound and outbound calls are analyzed. This entails breaking the calls down into their constituent parts and examining each part for its integration, automation, or workflow potential. The parts of an incoming call might be those shown in Table 8.1 [1], which also contains suggested CTI methods that can improve each part.

It is at this early stage that a cost justification should be performed—assuming that it has not already been carried out. This requires that a preliminary systems design be undertaken based on these implementation steps

Table 8.1
Call Segmentation and Analysis

Call phase	How CTI can be used
Ascertain reason for call	Routing to product/service group and/or screen pop
Obtain caller identity	Routing to affinity group and/or screen pop
Deliver caller information	Automatic delivery of customer data, etc.
Communication	Use information to provide greeting and announce service offered to customers
Wrap-up	Auto completion of contact logs and workflow information

and on the call center application. Application cost justification should assess the following.

- Number of agents;
- Switch dimensioning;
- Computer dimensioning;
- Voice processing requirements;
- CTI requirements.

Quantitative benefits achieved within various applications are listed in Chapter 7. Here, some more general figures are provided as rules of thumb. These should be used only in first-cut cost justification. Table 8.2 shows the range of agent productivity improvements that can be expected from various CTI-based functions.

Productivity gains have to be translated into real cost savings to provide a meaningful cost justification. Table 8.3 provides some examples of cost savings that can be achieved. The present value of savings presented in the last row of Table 8.3 uses a discount rate of 12% and can therefore be compared directly with the actual investment required to implement the system. Naturally, these figures are very sensitive to the loaded pay rates of agents, which vary both within and between countries. The figures used are from the United States and are provided for 15% and 40% productivity improvement. These may arise from screen popping or automated fulfillment or a combination of functions.

Cost justifying automation is generally a simpler matter—the decision to automate is cost driven but may well be rejected because it may reduce

Table 8.2
Some Rules of Thumb for Estimating Productivity Improvements

Agent Function	Generic Functions	Productivity Improvement
Basic incoming call management	CBDS, ACRI	15–20%
Voice and data transferring	VDCA	5–10%
Voice and data conferencing	VDCA	5–10%
Automatic call-back	CBDS, ACRO	10–20%
Predictive dialing	ACRO	100–300%

Source: Based on data from Aristacom, etc.

Table 8.3
Cost Justification Analysis

	Productivity Improvement	
	15%	40%
Time saving per agent per annum (hours)	290	774
Loaded savings per agent per year ($)	3,020	8,054
Equipment cost savings ($)	281	748
First year savings ($)	3,301	8,802
Annual savings($)	3,020	8,054
Present value of savings over five years ($)	11,168	29,780

Note: Based on data from Aristacom.

perceptions of customer service. Table 8.4 provides some insight into the dramatic savings that can be made by providing automated banking through IVR (see 6.6.17).

Another means of automation is to bypass the call center—partially or entirely—by providing access to data via the Web. Actual costs of providing Web access are often lower than those of providing IVR access.

In a hypothetical example, *Business Communications Review* compared a mortgage application using freephone access to a call center versus mixed Web and call center working. The call center-based transaction took 15 minutes per call and attracted a call charge of between £0.07–0.14 per minute. The alternative required 12 minutes of Web access and then a three-minute phone call. The overall transaction costs on this basis were £4.30–5.35 for the call center-based version versus £2.00–£2.66 for mixed working.

Table 8.4
Comparative Transaction Costs

Access Method	Cost per Transaction
Traditional clerk	$2.93
Call center agent	$1.82
IVR	$0.24

Source: Periphonics.

Though cost justification is subject to a number of assumptions, it is the CTI benefit most amenable to quantitative analysis. Improved customer service (often a prime motivation for CTI implementation, as detailed in Chapter 7) can be more difficult to justify. It is therefore even more important that customer expectation be managed effectively in this area—and this requires clear and validated customer requirement specification. A formal approach to requirement gathering, specification, and validation is recommended here. Most system integrators have a well-defined method for this.

8.2.1.2 Component Specifications

In Chapter 3, the components of a discrete third-party integration were listed as follows.

1. Application programs;
2. Switch;
3. Computers;
4. Telephone;
5. Terminal;
6. Physical link;
7. Communications stack;
8. CTI protocol.

All of these components do need to be specified—and in considerable detail. Item 1, the application software, needs to be closely specified at the requirement level, as defined previously. The switch needs to be specified, not only in terms of model number and overall dimensioning (trunks, agents, supervisors, traffic capacity), but also in terms of software release and interface ports. Often, the software release determines the availability of a CTI protocol—and the functionality of that protocol. Beyond the basic specification comes the issue of configuration. The switch has a database through which the switch controller knows what is connected to each port. All of this has to be programmed. For a new switch, this is often carried out by the supplier prior to delivery. In the CTI environment, it is sometimes necessary to provide *virtual* terminations, which have no equipment connected to them. Calls are connected to the virtual terminations pending some other action—in much the same way that calls are queued against an ACD agent group.

The computers—both server and clients—need to be specified in terms of model and, of course, in terms of physical CTI link capability for the server. The major criteria in specifying the computers are identical to those pertaining to supplying computers in nonintegrated applications. Where the differences lie are in the performance and reliability of the machine. Both are more exacting in the CTI environment.

Item 4, the telephone, includes any of the many possible telephonic terminals—from an agent turret to an integrated voice and data terminal to a digital keyphone. Similarly, item 5 can include a range of terminals, from the dumb to the very intelligent, passing thin client and NC along the way. The last three items are particularly relevant to third-party discrete CTI. The main object of clear specifications at this level is compatibility—between the switch and the server hardware at the lower levels and between the switch and server software at the higher levels.

A similar analysis can be applied to the components of a first-party CTI implementation. The major differences will concern reliability and performance, both of which are a lesser issue in first-party CTI because it is a distributed approach, there is no central server here. Meanwhile, a physically merged solution, a CT-server, presents reliability and performance challenges but reduces the component count to one and therefore simplifies component specifications and the entire implementation process.

The components of automation and messaging are primarily computers that are performing a very specific task. The majority of IVR systems and messaging servers are PC-based. Specification must ensure adequate performance, scalability, and reliability. The former is primarily directed to providing adequately powered processors and sufficient storage for the specified task. Reliability can only be guaranteed by investing in some degree of redundancy in the hardware, and in thorough testing of the software. Systems that are not PC-based may provide superior scalability.

8.2.1.3 Component and End-to-End Testing

CTI systems consist of a large number of elements, some of which are subsystems in themselves and some of which are software modules. Following the usual good practice of system engineering, each module will have a specification, and each module will be tested against its specification.

Test harnesses capable of generating messages that will exercise all parts of the software are necessary and are usually produced as part of the component testing strategy. End-to-end testing is sometimes problematic. Many system integrators cannot afford to have every model of every switch at their premises so interworking tests with the switch can only be performed at the installation site. For a very large company—IBM, for example—the provision of a *switch farm* to test compatibility may be practical. However, most system integrators partially overcome this problem by developing switch emulators. Usually PC-based, the emulators are flexible to the extent that the message sets they generate and receive are programmable. However, the majority of emulators cannot test the overall performance of a CTI application; they usually emulate only one call at a time. Perhaps the most demanding systems to test are those that use a large number of different switch types networked together to provide a virtual call center.

Using the switch as a basis for performance testing is not necessarily the best solution, even though this approach makes it very easy to generate test message sequences simply by lifting the telephone! However, it is not so easy to provide high loadings of repeatable traffic. Fortunately, the switch suppliers have already had to meet this problem in testing their own call processing software. Call generators are available that can generate multiple calls and can be scripted to test such telephone features as transfer and conference.

As the CTI industry has developed, it has nurtured the growth of specialized testing companies. One of the earliest of these called itself Hammer. It provided script-based testing tools capable of exercising voice automation components in a repeatable way. Such companies in turn gave birth to others that specialized in the preparation and maintenance of test scripts and schedules.

The story never ends, but end-to-end testing must take into account the full complexity of the CTI system components. It must test the synchronism of the system in delivering calls and data to the telephones and terminals as well as to ensure that voice responses are correct. The latter is sometimes tested by using speech recognition.

8.2.2 Risk and its Containment

The major areas of technical risk in CTI system implementation concern the following.

- Reliability;
- Performance;
- Portability and interworking with existing applications;
- Development and integration cycle;
- Flexibility.

These risk areas and ways to minimize risk will be examined in more detail.

8.2.2.1 Reliability

Basic platform reliability is usually determined by the computer hardware. The required reliability level can be achieved by a judicious choice of a computer system. Reliability is usually specified by quoting a required Mean Time Between Failure (MTBF). Very high MTBF figures can be obtained by employing a redundant computer architecture. Reliability embraces repair. Telephone systems are built on the basis of *hot-plug.* This means that a faulty circuit board can be changed without powering down the system and without disturbing the rest of the system. Computer systems were not traditionally built this way. CTI systems do demand hot-plug, so platforms that offer this must be selected. This applies to integration and automation servers—the latter have more points of failure as the plug-in card count rises. Fortunately, later systems placed the synchronous bus on the backplane. Earlier implementations used the less reliable ribbon cable.

Software reliability is not so easily specified or improved. Here, the concern is to minimize the number of software errors in the release. Minimization is achieved during the initial design phase by employing application generators and/or pretested objects. It is further reinforced during the development and field trial phases through rigorous testing and in the field through the use of dynamic recovery mechanisms. Reliability is discussed in more detail later.

8.2.2.2 Performance

There are two performance concerns: delay and throughput. Delay can occur in the following areas:

- The switch;
- The links from the switch to the server;
- The server itself;
- The client workstations;
- The voice server;
- The fax server;
- The database server.

Performance of CTI systems in terms of both delay and throughput is explored in more detail later.

8.2.2.3 Portability and Interworking

Portability is an important issue for some end users. It defines the ease with which an application can be moved from one platform to another. There are three areas of concern here:

1. The ability to run business application software on a variety of computer platforms;
2. The ability to connect the computer platform to various telephone switches;
3. The ability to move business applications between different computers and switches.

The problems encountered in moving business applications between computer platforms are minimized to some extent by using an open operating system as a platform. Openness is a key issue for many companies, and it is as important in the CTI context as it is in any other. Many of the approaches available from the larger companies did tie the customer to that supplier's computer hardware. However, as IBM moved its platform to cover a range of operating systems, including Microsoft Windows, and as Digital's CIT was taken over by Dialogic and ported to Windows and UNIX, such issues became an irrelevancy in the CTI world.

The lack of standardization in CTI protocols and functionality has produced serious portability problems. This particularly affects the larger corporates that have inherited a range of switch types from previous mergers, takeovers, and other business reorganizations. Over the years, a degree of switch normalization has been introduced into most APIs—shielding the application from this variability. Normalization cannot cause a switch to perform functions that are not supported. API requests for such functions will be rejected, and applications will therefore be aware that these functions are not supported and will have to take appropriate action.

8.2.3 User Concerns

CTI is dangerous because it is often uniting two disciplines that have previously been entirely separate. Potential users justifiably have a number of reservations with regard to undertaking an implementation of this technology within their businesses. Figure 8.1 locates those concerns as they relate to first- and third-party CTI systems.

The major overall concern is that CTI represents unfamiliar and untried technology. Following this, the area most likely to cause problems as far as users are concerned is component integration. Many of the items shown in Figure 8.1 are ingredients of these concerns. The most difficult to deal with are compatibility issues and management, though both have eased as implementation technology has improved.

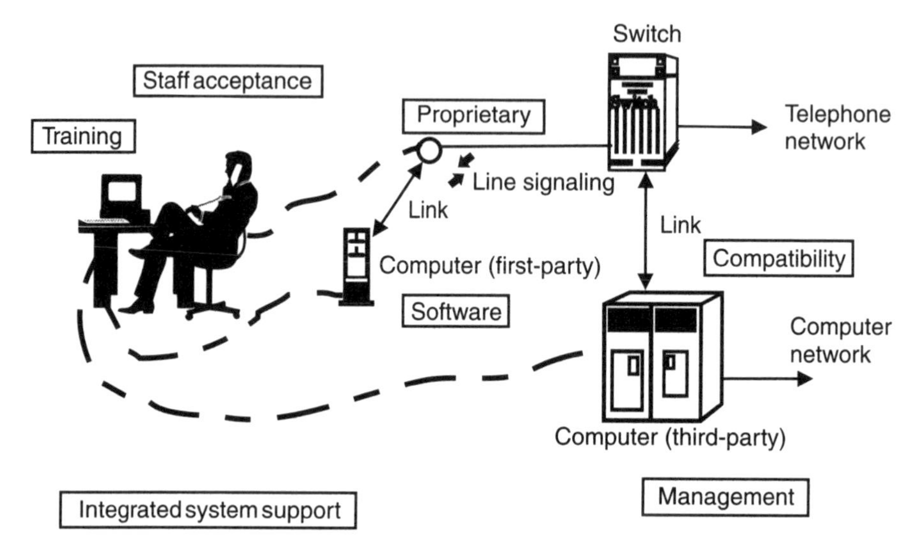

Figure 8.1 CTI user concerns.

The extreme issue of automating a process entirely is a difficult one to resolve for any business, though there is little doubt that the acceptance of voice automation increases with time in all cultures. The sensible answer is to give customers the choice—and reflect the cost savings made by the company in the charges made for the automated option.

Integrated-system support is a genuine and practical concern. Who do you call when things malfunction? Is it for the user to determine whether or not the computer system or the telephone system is at fault? There are many examples of tears spilled in this area—and of users needing to call the system integrator back into the disaster area to pinpoint a problem so that it can be dispatched to the relevant supplier for resolution. Fortunately, some maintenance organizations are becoming adept at supporting integrated systems. Also, some of the larger computing companies have established links between their own and the switch supplier's design areas so that design faults can be handed over if necessary. Naturally, as more systems have been installed, the skills to support them have become more plentiful. However, where there is no prime contractor, it is natural for subcontractors to point their fingers at the other contractors when problems arise.

Merged systems do not inherit the finger-pointing problem. They present a single point of failure—but also a single point of resolution. True, the problems experienced may be traced to a component of the merged system—an interface card from one of the card suppliers, for example. Nevertheless, it is still clearly the system supplier's job to get the problem fixed.

8.2.4 Networking Options for Third-Party CTI

In first-party CTI, the connection between the computer and the switch is simple. Though the computer itself may be provided as one per line or as one computer connected to many lines, the switch connection is always on a per-line basis. This is implied in the definition of first-party CTI. This is not the case for third-party CTI, for which there are many options.

Take the simple case of a two-site call center. Each site has an ACD, and each site has LAN. The LANs are interconnected via a private WAN and may appear as one network—no problem moving data around here. The ACDs are linked by a private network, so there is no difficulty in moving calls between them (though there are plenty of options here). The prospect of the two call centers becoming CTI-enabled poses some interesting possibilities:

- Will there be one CTI server or two?
- If just one, how does it pick up the CTI link from the ACD at the other site?
- If two, then how do they interact?
- If a call arrives at one site but is moved to an agent group at the second site, will the second server pick up the necessary information to pop screens?
- If a call is transferred from one site to another how are the call and data associated?

These questions are easily posed—but not easily answered. It was hoped that the ECMA standards group would provide guidance on CTI networking in the third release of the CSTA specification. It did not do so. Perhaps this subject is still too complex to allow for the adoption of a single approach.

The issues here cover function, performance, and reliability. A single server will provide good functionality but also presents a single point of failure. In addition, the CTI link from the remote switch has to be extended in some way. The most cost-efficient method is to use the LAN; this presents performance challenges since the LAN's loading will vary and the WAN may present a bottleneck. The two-server solution seems preferable even though it may be more costly. The major area of server interaction will occur when calls are transferred from one site to another.

The problem is that, though the switch can redirect a call to the other switch at the instigation of the computer, the switch receiving the call may not be able to discriminate between the transferred call and a newly arrived one. However, if the switches have an intelligent connection, the origin of the call can be detected, and the call can be associated with its data. This requires the private network to carry some form of call identifier and therefore depends upon the private network signaling system in use. What if there is no private network? Can calls still be transferred across the public network? Surprisingly enough, some novel solutions have evolved to allow this to work. Any solution usually involves the establishment of a DDI/DID line at each switch. This is reserved for CTI transfers only so that the receiving server can identify the call and associate its data.

There are many other configuration options, including the connection of multiple switches to a single computer and so forth. All of these introduce architectural and implementation issues that must be separately analyzed. In practice, many switches limit the number of CTI links that can be supported, and many API implementations limit the number of links open for communication.

Interestingly, CTI has achieved a star role in the routing of calls to networked ACDs. In implementations such as that provided by GeoTel, IBM, and Genesys, the CTI server is linked to the public network *Signaling System 7* (SS7) and to each of the ACDs in the network. The server collects statistics from the ACDs and can therefore make decisions on where best to deal with a newly arrived call. The CTI server can then work with the public network's CTI server (the service control point) in order to direct incoming calls to the appropriate call center across the public network itself. This is sometimes called network-based routing as opposed to switch-based routing discussed above.

8.3 Performance Constraints

As previously observed in the consideration of risk, there are two major performance concerns in integration: delay and throughput. However, there are also general concerns over the performance of media processing devices that are related to processor loading and the speed with which signals can be processed. Media processing systems are generally processor- and memory-intensive. This may affect the time taken to recognize speech or the time taken to generate a phrase from text, for example. Merged systems do not escape entirely from the performance check-up. Here the central processor often has a wide mixture of activities—from call processing to media processing. It is therefore difficult to predict the behavior of such systems in all possible conditions.

8.3.1 Delay

The switch is often said to be a real-time system. However, telephonic responses are not excessively demanding. In general, the maximum delay allowed is one second for the return of dial tone and for connection of a call. However, these responses must be met, and the switch is designed such that, except at times of extreme overload, they will be.

When the computer makes requests of the switch, overall response times for call set-up should be as previously specified. This includes any time taken in transferring the request from the computer to the switch across the CTI link. Meanwhile, the time from the occurrence of a telephonic event to the presentation of the relevant status message to the computer will be determined by the scheduling constraints of the switch operating system plus any delays in transferring these messages to the computer via the CTI link. In normal circumstances, the computer should be presented with status messages within 100 ms of the occurrence of an event.

Performance problems are more likely to occur in situations in which the switch is the client and the computer is the server. When a switch needs to request routing information from the computer, the computer becomes part of the real-time system. For connect times of less than a second, it must respond in much less than one second. This is not an arduous target—but it may well place strict limits on the loading of the computer system. For this reason, CTI implementations often exist in a dedicated server that communicates with, but does not run, the main business applications. The situation is exacerbated where routing depends upon an access to back office customer data and the back office system is not designed for fast response time.

What of call set-up time itself? One of the most demanding tasks in integration is that presented by power dialing. As the control computer cycles through its "calls to be made" list, it is waiting for the first call to be answered—by a person. The moment this occurs, the computer has to bring an agent into the call. Otherwise, the called party is going to say, "hello, hello" a few times, then hang up—vaguely suspecting that a nuisance call has been received. This is called overdial in power-dialing parlance and is to be avoided. The power dialer software is supposed to ensure, within defined bounds of probability, that an agent will be available when a caller answers. However, what if the switch cannot react quickly enough?

The sequence of events begins when the called party answers. This produces an event message to be transmitted to the computer dealing with power dialing. This will cause a connect message to be sent to the switch that will request that the answered call be sent to the turret of the agent selected to deal with the call. At the same time, the computer must pop the relevant data onto the agent's screen. The whole transaction from call answer to screen pop and switch-through must take place in less than one second. It is worthwhile making an example of this transaction to study the timing. The example will be based on a number of assumptions:

- Assume that the switch can deal with 3,600 busy-hour call attempts (BHCAs);
- Assume a CTI link speed of 9.6 Kbps;
- Assume an event-message length of 30 bytes;
- Assume a connect-message length of 40 bytes;
- Assume each message has an overhead (to manage the link) of 10 bytes;
- Let the delay between call-answer detect and presentation of the event message at the CTI link be . . . *ta* ms;
- Let the response time of the computer from event-message reception to connect-message transmission be . . . *tc* ms;
- Let the time from when the connect message is received at the switch to the point when the call is connected be . . . *ts* ms.

On the basis of these assumptions, the time taken for the event message to travel across the link is

$$(30 + 10) \times 8/9.6 = 33.3 \text{ ms}$$

and the time taken for the connect message to travel across the link is

$$(40 + 10) \times 8/9.6 = 41.7 \text{ ms}$$

Hence, the delay between detection of the call-answer condition and connection being established is

$$td = ta + 33.3 + tc + 41.7 + ts \text{ ms}$$

When the system is not busy, it can be assumed that

$$ta = tc = 100\text{ms}$$

in which case

$$td = 275 + ts \text{ ms}$$

Hence, the overall response time between detection of call answer and connection of the call to the agent must be less than 0.725 seconds. However, this assumes that the switch and the link are dealing with only this one call. In fact, there will be other messages on the link and other calls to process in the switch. The latter will lead to increases in ta and ts, the former will increase link occupancy and cause delay, because messages have to wait while others are using the link. The end result of this is that less and less time is available for the computer to respond to the event message.

The time allocated to the computer *tc* is also load-dependent. The computer has to deal with screen popping, but it is also dealing with other events and other applications—one of which is the generation of calls from the predictive dialing list.

8.3.1.1 Link Speed

Early link speeds were often as low as 9.6 Kbps, though some implementations used ISDN-like links, which allowed faster speeds. As CTI links developed, it became conventional to use a 10-Mbps Ethernet link—usually dedicated to CTI linkage. On the whole, the information rate on a CTI link is not high, because calls have fairly long holding times and the actual number of messages per call is not high. Assume, for example, that a call holds for two minutes. Also assume that all messages pertaining to that call emanate from and are transmitted to the computer and that the total number of messages for the call is eight, each averaging 50 bytes. The total average bit rate for that call is then only 26.7 bps. Even if the computer is handling 100 very busy agents, the call rate is unlikely to exceed one per second. So, using the same assumptions, the composite average bit rate will be 3.2 Kbps—a loading of only 33.3% on a 9.6-Kbps link and negligible loading on an Ethernet link.

There will be other messages on the link, such as management messages. Nonetheless, it can be seen that the traffic-related bit rate is not likely to determine link speed.

8.3.2 Bus Capacity

PC-based media processing systems are usually dependent upon a synchronous bus to link the cards together. The buses have limited, though reasonable, capacity—usually stated in terms of the number of 2 Mbps channels supported. Each channel is divided into 32 time slots, each having a capacity of 64 Kbps and each usually used to carry PCM voice or data. The original MVIP specification included eight channels, and Dialogic's SCbus has 16. These two industry standards therefore supply sufficient capacity for 256 and 512 simultaneous connections, respectively. These are large numbers and unlikely to be limiting for any but the most demanding of applications—residential voice messaging, for example. It is these enormous systems that may require a multichassis implementation. It is then necessary to check the performance of the interchassis links.

8.3.3 Throughput

Throughput and delay are obviously related. Delay will increase, usually exponentially, as the number of calls being processed increases and as the number of active processes in the computer increases. Figure 8.2 provides an indication of this increase and shows that delay increases very rapidly as full loading is approached.

The line in Figure 8.2 indicates the point at which one-second delay occurs. This indicates a maximum loading of 78%, which determines the maximum throughput for this response requirement. Figure 8.2 applies to all components in the CTI system: the switch, the link, and the computer system. In reality, however, one of the components usually limits throughput—that is, one of the components presents a bottleneck. Assume for a moment that Figure 8.2 describes the performance of the switch alone. If each processed call consumes *C* seconds of processor time, the loading will be determined by multiplying this figure by the number of calls processed per second. Note that the

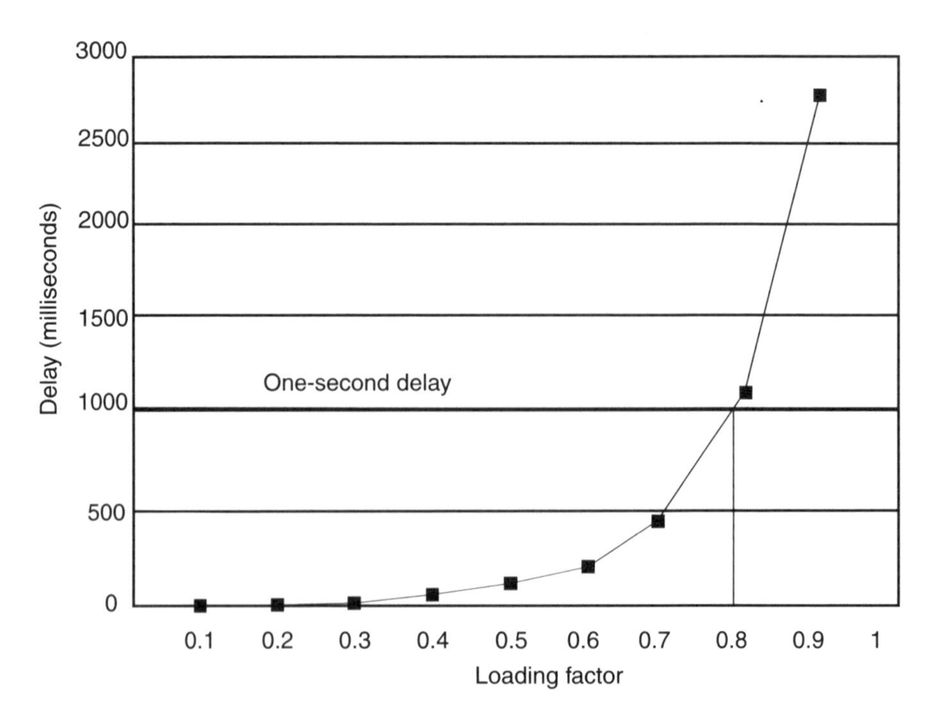

Figure 8.2 Increase of delay time as loading increases.

time taken to process a call initiated from the CTI link may be more than that taken to process a normal call (one initiated via a telephone). Usually the rate at which the switch processes calls (B) is indicated by the number of busy-hour call attempts. Thus, the switch processor loading (L) is

$$L = B \times C/3600$$

So, using the maximum loading of 0.78 from Figure 8.2

$$L = 0.78 = B \times C/3600$$

Hence

$$B = 2808/C$$

C is determined by the call processing software and the switch processor hardware. Assume it to be 0.25 seconds.

The maximum switch throughput is then

$$2808/0.25 = 1{,}123 \text{ busy hour call attempts}$$

Similarly, assuming Figure 8.2 applies to the link alone, the 0.78 maximum loading time determines maximum allowable throughput. That is, to find the maximum usable link speed, simply multiply the raw link speed by 0.78. This reasoning can also be applied to the computer system. Unfortunately, the calculation is not so simple. The variety of applications and of tasks making up applications make it difficult to translate a maximum loading figure into a meaningful throughput figure for the computer (e.g., maximum number of transactions per second). However, where the computer is solely providing a server function, transaction times are not so varied, and an estimate of throughput in terms of messages per second can be calculated. In these circumstances, performance is deterministic—in circumstances in which the computer's load is unpredictable, it is not.

The loading on media processing servers should be predictable. The task is not generally onerous, provided that the media processing function is carried out within media processing cards rather than on the central processor. Bottlenecks are most likely to occur at the disk interface. Reading or writing speech files to the disk rapidly becomes a major challenge in large systems. It is for this reason that many of the larger end designs are proprietary and not based on PC technology.

8.3.4 Reliability Implications

Some of the reliability concerns that attach to an implementation of computer telephony integration are the following.

- What happens if the CTI link is severed?
- What happens if the computer fails?
- Can the computer cause calls to be misrouted or lost?
- What happens if the media processing system fails?

In general, if the CTI link fails, the switch will eventually return to its normal mode of operation. With the exception of dumb switches that do not contain a call processing algorithm, a switch can never be entirely dependent upon the computer for call-routing instructions. Every action taken by the switch is protected by a time-out. Time-outs ensure that a reasonable action will be taken when the computer or link does fail.

However, the computer application can potentially cause calls to be misrouted if it has not been adequately tested. Virtually all implementations of CTI on switches analyze request messages to ensure that the parameter list contains acceptable values. If it does not, the request is rejected before any action is taken. Even so, the request may be asking for something that is not possible, connecting a telephone to itself, for example. Switches have to deal with such problems all of the time. It is generally accepted that users—human users, that is—will always do the most bizarre things, dialing their own number, for example! Switches are programmed to deal with such behavior, always failing the call in an appropriate manner and returning the relevant tone to the deviant caller. In integrated systems, requests from the computer will be dealt with in just the same way. For example, illegal requests are rejected at the outset, and requests that ask for something that proves to be impossible (because of the state of the switch's network, for example) cause an error message to be returned.

The switch will misroute calls if the computer requests it to do so (assuming that the request specifies legal telephone numbers), just as it would if a human user made a similar request.

Despite the fact that CTI systems are secure against loss of calls and can continue operation if the computer system fails, there is an expectation of application reliability from users once CTI is introduced. Generally, our expectation of the telephone system is that it should never fail, whereas our expectation of the computer system is that it will fail—that's why we do backups! Put the two together and reliability expectations do not fall to the lowest common denominator; they rise to the highest—that is, they rise to telephone system reliability levels. In order to achieve this, there are many applications that require nonstop computing systems in the CTI world.

Migrating the switching function and call processing to the server to create a switching server has special ramifications from the reliability viewpoint. In the CT-server everything is placed in one basket. Any failure that affects the basket is likely to bring down the computer and telephone system.

The problem here is the software—not the hardware. Ensuring reliability on the hardware front is primarily a matter of investment. If high reliability is required then hardware must be duplicated and fault detection and changeover mechanisms supplied—just as they are as a matter of course in the telephony world. The software is a different matter. Desktop operating systems are not generally reliable. They do not usually provide methods that prevent one process from intefering with the memory used by another. This is essential in reliable CTI systems, so the choice of operating system is usually constrained to those that provide a reliable operating environment.

The applications themselves should undergo the level of testing that is applied to telephone systems. However, this may prove unrealistic—simply because of the flexibility of operation provided by merged systems. Hence, the system is likely to be less reliable but more suited to a changing business environment.

8.4 Management

Management is of critical importance in CTI implementations. What follows will not do the topic justice. If space had not been at a premium, a whole chapter of this book would have been devoted to management. There are management issues at all levels, from the CTI link to the business application level. Because integration is concerned with linking two intelligent systems, the switch and the computer, it automatically leads to a sharing of databases; in particular, the switch database becomes of great relevance to the computer system. Uncoordinated changes can lead to disastrous failures that, in turn, can be very difficult to track and remedy.

Ease of management may be a major reason for selecting first-party CTI over third. It may also provide a good reason for selecting a merged system rather than discrete. In merged systems the database that describes the telephony world is merged with that describing the computer world. Though users are naturally interested in functionality, they are also deeply concerned about management of the total IT system.

8.4.1 Link Management

The CTI link is a data link and requires the same sort of management that data links usually require. Meridian Link from Nortel, for example, provides the following.

- Flow control;
- Message recording;
- Message monitoring;
- Link maintenance.

The link management functions may form part of the protocol or be provided as a separate facility. The management functions provided by Meridian Link are quite comprehensive, but proprietary protocols vary considerably in their management capabilities. Many do not provide message recording and monitoring functions, presumably leaving such functions to standard protocol analyzers.

In order to communicate, the computer and switch have to establish a session. Establishing a session between computer-based applications and the switch is a higher level function than those mentioned so far. Some protocols cover this field; others leave it to the API or the application program. Similarly, from the computer's viewpoint, protocols vary considerably in the manner in which telephony devices are given visibility. Some implementations assume that all devices are of interest unless the application indicates otherwise. Others assume the opposite—that no devices are available to the application for monitoring or manipulation until the application registers an interest in them.

Similarly, facilities for selecting certain message types are considered to be desirable—though this could be dealt with by the application rather than at CTI protocol level. Filtering can be accomplished in two ways: by indicating the events that the application wishes to receive or by indicating those that it does not. Both approaches are found in the existing protocols.

8.4.2 Database Synchronism

In addition to the basic management features already considered, there are higher level management functions in a CTI environment. One of these concerns the synchronism of databases within the various parts of the system. How can a system ensure that the various databases contain compatible data? Take a simple example: Most CTI applications require the establishment of an association between a telephone and a terminal. Telephones are managed by the switch and terminals are managed by the computer. Thus, if the number of a telephone is changed at the switch, the computer may not be aware of it. If that telephone is associated with a particular terminal for incoming calls, the lack of tracking will cause the screen of one agent's terminal to fill with call-related data while the call itself is placed with another agent!

This is a simple example and can be dealt with by establishing the association at the time a user logs on. That is, the computer can ask the user for the telephone number. However, the number could still be changed at the switch during the session, which, though it is not likely, would certainly cause a good deal of confusion.

If the association between a terminal and a telephone is not established each time the user logs on, there must be some database that associates the two on a semipermanent basis. If such a database exists, it must be managed. It must be possible to establish new associations, delete old ones, and modify existing ones. The availability of a single management position covering both the telephone and the computing system becomes increasingly attractive.

If a voice messaging system joins the fray, this too may have a database that references telephone numbers—to enable the system to route an incoming call to a mailbox and for message delivery. Similarly, a voice response system may have the telephone number of the accounts department or the sales department embedded within its database. In these cases, there are multiple systems functionally linked through CTI, each with knowledge of telephone numbers and each with a separate management terminal. Things can get very complex in such an arrangement, especially since a similar action at each of the various management terminals may not produce the same result.

Once again the superiority of a merged system from the management viewpoint is clear.

8.4.3 Coordinated Call Monitoring

One of the generic functions of CTI mentioned throughout this book is CCM. CCM is the ability to keep a single log that contains a history not only of the calls made but also of what the agent did during each call. This requires establishment of a log that collects and time stamps all call and business application events and that is able to sort these events by call reference. The management of such a log requires allocation of sufficient storage; definition of the events to be entered; and a means of reading, sorting, and building reports. At first the implementations were fairly basic—but the log becomes a more important management tool than telephone call analysis or computer resource reports can ever be. Distributed CTI systems can pose particular challenges here since it can be difficult to track calls that move from switch to switch within the enterprise.

Solutions such as IBM's CallPath Enterprise Server attach an enterprise-wide identity to each call, allowing birth-to-death logging.

8.4.4 The CTI Manager

A number of potential problems and opportunities in CTI management have now been considered, and there are many more. Taken together they present a clear argument for unified management. A single point of CTI system access would ensure that there are no conflicts and also assure consistency and synchronism of databases. The computer system seems the ideal location for the CTI manager; it would deal with all management administration and collation of logging outputs. These are tasks to which the general purpose computer is best suited.

Although solutions for central collection of data from the many devices that may be used in a call center did evolve, the centralization of management control developed more slowly. What did evolve were tools that allowed the graphical creation of call center call flow and applications that could be configured to control call handling. In this way management control began to drift toward the computer without any positive step taken to remove it from the switch. A much stronger trend is demonstrated by the evolution of merged systems; here the establishment of a CTI management point is intrinsic to the design.

8.5 The Regulatory Environment

There is little or no regulatory legislation that clearly addresses CTI. However, there are many areas of regulation that impinge on the CTI industry, particularly concerning systems in which the use of CTI might be regarded as intrusive. Such systems cover the whole area of inbound and outbound telemarketing; in fact, they cover almost any sector in which intensive business is conducted via the telephone. There are two points of interest here. The first point covers the entire world of privacy and protection of individual rights. The second is entirely rule-bound, particularly in Europe, and covers the approval of equipment for connection to a telephone network.

In the regulatory field, this book considers North America and Europe only. In other geographical and political groupings, the situation might be very different. Even the areas chosen have major differences. North America is more advanced in the adoption of telephone-based marketing and less restrictive than Europe with regard to connection approval.

8.5.1 North American Regulation

In the United States a fairly liberal view is taken of the use of the telephone as a sales and marketing device. However, there are restrictions in place and there may be more to come. A federal law prohibits certain telephone debt collection procedures, particularly, in relation to the use of the telephone with intent to annoy, abuse, or harass a person [2]. The use of repeated ringing or engagement in long telephone calls and the abandonment of calls or placing of calls in long queues could all be considered harassment.

Of course, there are also state laws. Many of these attempt to protect the individual requiring that the consent of the called party be obtained prior to a message being played, restricting the hours in which automatic calling systems can be used, barring the use of automatic systems in selling goods and services, and so on.

8.5.2 European Regulation

Each European country has its own regulatory stance, though the *European Economic Community* (EEC) is providing some commonality within those countries that belong to it. The general restrictions in Europe include limitations on the following.

- Automatic generation of calls;
- Transmission of unwanted telemarketing calls;
- Release of CLI.

The ANI facility, commonly available in the United States in the late 1980s, was not generally available in Europe in the early 1990s, except for use in nuisance-call prevention. Some say that this inhibited the growth of the CTI market. Its introduction began in 1992 in the United Kingdom and quickly spread throughout Europe, the personal privacy concerns being dealt with by various schemes for suppressing the transmission of CLI, if required.

The release of CLI together with other aspects concerning the use of the telephone for selling began to attract the regulatory eye of the European Union. In the United Kingdom, for example, automatic generation of outgoing calls was clearly regulated by the general license that applies to all telephone users. Calls that are automatically generated and connected to a recorded or synthetic voice may only be made to persons who have consented to receive them—in writing. In Denmark such calls can be made providing there is some specific signal to indicate that a recorded announcement is being made. Meanwhile, in

Germany, there are no specific restrictions on the playing of recorded messages, but unsolicited calls to sell goods or services are prohibited.

The overall solution favored by the direct marketing industry was to have the industry regulate itself—through the creation of telephone preference schemes. Here, the onus is on the recipient to request that no further approaches be made by calling his or her telco. That person's number then becomes part of a blacklist of numbers that must not be called. The scheme provides the blacklist to all telemarketing companies, which must then stop bothering the person.

8.5.3 European Approval Issues

European approval issues constitute a difficult area. Europe has been fairly restrictive from a connection-approval standpoint, though this has changed in some areas. From the integration-only perspective, a CTI implementation has no direct connection with the telephone network. It is therefore free from the many technical requirements for connection approval placed on switches (safety, electrical, performance, and so on). On the other hand, the connection of a computer to a switch can be likened to a brain transplant and can potentially negate many of the performance checks that form a fundamental part of switch approval.

The software in a switch is considered part of the switch itself. When the switch is first approved, all the items that make up the switch gain approval. If the supplier wishes to change any item, some mechanism of re-approval has to be negotiated. This includes the software of the switch and, normally, each new software release is approved by the relevant authority. Often the approval authority requires that strictly limited numbers of sites are provided with a new release. These are regarded as trial sites and carefully monitored before the new software is generally released. This procedure can take many months.

Some regulators wished to regard the whole thing as one system, that is, to approve the computer, CTI link and switch as one. With this approach, no change can be made to the computer's software without prior approval, alpha and beta trials, and so on! This position could have made the introduction of CTI very difficult because it freezes the software in a computer that needs to be reprogrammed and updated regularly. However, the trend was away from government involvement in regulation, so this position was not adopted.

By far the most sensible approach, and the one that is generally taken, is to approve the CTI port as part of the switch approval. This approval includes safety requirements and examination of, and perhaps exercise of, the protocol. Once approved, any suitable device could be connected to the port. If the protocol were to be changed, this would require a change to the switch software and re-approval. In this scheme it is unnecessary to gain approval for the computer's software.

One of the biggest boons to the CTI industry in Europe was the introduction of common standards and common testing. The other was the approval of a CTI card for connection to the telephone system, rather than a complete system. This has enabled system developers to build systems from pre-approved boards without the need to gain approval for their systems. To a certain extent this has bypassed much of the regulatory process.

8.6 Standards

It is clear from Chapter 1 that CTI is by no means a new idea. It is also clear that many users do appreciate its potential benefits without the aid of persistent and convincing salespeople. Given this, why did the market not develop quickly? One reason is contained in that particularly pertinent adage: "It takes two to tango." This is particularly true of third-party CTI.

A switch with an excellent CTI protocol is not very valuable if there are no computers to connect to it. Worse still, if all the computers already support a number of different protocols, computer suppliers will not be keen to add yet another. In the early 1990s, the CTI industry was characterized by each switch manufacturer having no protocol at all or else one of its own specifications. Each computer or API supplier supported those protocols that it considered to be strategic. Hence, the CTI software market became a segment within a segment within a sector. For any particular application it was necessary to do the following:

- Identify companies likely to take up a CTI option;
- Remove those who did not have suitable equipment;
- Further reduce the numbers by isolating those with equipment that supports compatible protocols.

All of this conspired to suppress application production, yet the key players in CTI are the independent software vendors. These people needed a common platform upon which to base their software to provide the confidence that justified the massive investments that software development requires. In the computing environment, the IBM-compatible PC and Windows are examples of the availability of common platforms. A similar base was needed to encourage the production of CTI-compatible software. Two candidates appeared to fill this gap—Microsoft TAPI and Novell TSAPI. Both were flawed in different ways, and as the decade draws to a close, attention has turned to JTAPI, which offers portable CTI applications at the client level and greater scope than its predecessors.

To a certain extent, the APIs shield the programmer and the application purchaser from the viscitudes of the various protocols used in interfacing to switches. However, each new API development requires that a new set of drivers be written to interface to the switches. Wouldn't it be nice if, at a minimum, the switch interface was standardized?

Perhaps, however, the importance of standards in CTI can be both underestimated and overestimated. The lack of standard protocols unarguably checked the growth of the integration market, but many people argue that there are other factors that would have limited growth even if standards had been available and implemented in the 1980s. Note the statement "and implemented." Standards are not necessarily the panacea that many potential users believe them to be. Suppliers only implement standards if it is in their best interest to do so. Those interests might be best served by retaining a proprietary protocol if, for example, the market is growing well. In other circumstances, the growth of the market may require that suppliers help its formation by implementing standards.

There is a well-known quotation from an anonymous standards cynic: "The nice thing about standards is that there are so many to choose from." One system integrator in the CTI world went one step further. He said, "If standards are so good, why don't we all have one?" When the standards situation is confused, suppliers are not likely to commit the significant resources necessary to implement them. When there are many possibilities to choose from, the supplier is more likely to retain its proprietary solution—a solution that is probably more feature-rich than any of the standards. When confusion exists, users are more likely to resolve it than the suppliers—simply by specifying a particular standard in all procurements. However, forcing the use of standards in this way can have its problems.

In Alan Bennett's play "Forty Years On," one of the leading characters puts a very pertinent question to the headmaster.

Franklin: "Have you ever thought, headmaster, that your standards might be a little out of date?"

Headmaster: "Of course they're out of date. Standards are always out of date. That is what makes them standards."

The headmaster has a point. For some years the problem for CTI implementers was the sheer paucity of standards. In 1992 and 1993 the standards began to emerge; the problems that emerged with them have already been highlighted by Franklin and the headmaster. Standards by their very nature are slow to evolve and difficult to change. For this reason, they tend to offer a solution that is functionally poor compared to a proprietary solution.

Until the release of the first standard in 1992, attention focused entirely on the standardization of a protocol—the definition of the messages that flow between the switch and computer. Various people have pointed out that, in the CTI world, it is more relevant to standardize the application programming environment than the communications link—a very good point. In the wider world of data processing, this is comparable to the Open Systems Interconnection (OSI) and operating system controversy. Is it more important to standardize the communication architecture or the operating environment? In this world, OSI dominated the scene, and the issue of an open operating system took second place. That seemed to change throughout the 1980s, as more and more users wished to achieve hardware independence—something that a standard operating system can give them but open systems interconnection cannot. In any event, the Internet's TCP/IP now has an almost complete stranglehold on the data communication scene, and Microsoft Windows enjoys a similar position at the desktop and, to a lesser extent, as the server operating system. The original arguments and players are all but forgotten.

As the coverage of CTI has expanded, then so too has the relevance and scope of standards. The inclusion of media processing brings with it a whole compendium of standards—from those dealing with interfacing and compression to those dealing with media conversion. Fortunately, the standards for this world have clear owners and are already largely in place, though still evolving. Relevant standards include the G.7XX series from the ITU, which cover the compression of speech, and the H.3XX series, which covers the transmission and compression of video. The merging of networks brings with it an even larger treasure chest of standards. These range from the specifications that underlie the creation of LANs (courtesy of the IEEE) through to those that define ATM technology from the ITU and the ATM Forum. Here the only area that this book spotlights is that which links the packet and circuit networks most closely—voice over packet.

In the integration world the need to link computers and telephone systems is unique to CTI. It is for this reason that CTI link protocol standards are so important and therefore a primary focus here. It is probable that the standards world is led by technical rather than user considerations. This could explain why the CTI world focused on the protocol issue and entirely neglected the development of a standard API, at least for third-party CTI. In the mid 1990s, cries for a standard API grew strong. The ECMA committee laid aside its work on refining a protocol for one meeting and considered these demands. It concluded that API standardization was not feasible and that their protocol work should aid the existence of APIs rather than replace them. API standardization is not easy. It is argued that APIs offer a partial solution only and that there are many levels at which APIs can exist. APIs require language and operating system-specific implementations. Also, in order to support a number of different switches, they tend to address the lowest common denominator as far as functionality is concerned—or leave it to the application to sort out the differences. JTAPI does at least address the problem of operating system independence, but it leaves the other points untouched. However, there is one approach that can achieve operating system and language independence; this is the simple computer telephony protocol (SCTP), which is considered at the end of this chapter. In the media processing world there is an API standard—S.100 from the ECTF. First released at the beginning of 1996, it was not immediately embraced by the CTI industry. The S.100 standard and others from the ECTF will be considered later. Video standards were summarized in Chapter 4; they are covered here under collaborative working standards.

8.6.1 CTI Linkage

At present, the *computer supported telecommunications applications* (CSTA) protocol is the most important independent standard in the integration world. The global nature of that standard was emphasized when the European Computer Manufacturers Association, which created the standard, changed its name to ECMA in the mid 1990s. This reflected the fact that half of its members were United States- or Asia-based. However, coverage will also be given here to the *American National Standards Institute* (ANSI) work on the *switch computer application interface* (SCAI) standard, which forms the basis of some Centrex implementations.

Note that the standard protocol activities described concentrate entirely on third-party CTI. First-party protocols are determined by the line-signaling definition between the telephone and switch. The most likely basis for standardization here is the ITU definition of signaling for the ISDN. This is defined in recommendation Q.931, and the various APIs that are being defined to provide program access to that protocol. In the analog world, line signaling standards tend to be country-specific—though similar for basic implementations. Where first-party enhanced is used in the analog world, the signaling system is almost invariably proprietary.

8.6.1.1 Background and History

In the early 1980s, various initiatives began to explore opportunities for standardizing the links between switches and computers. AT&T was at the forefront of many of these activities, both in North America and in Europe, formulating a *digital multiplexed interface* (DMI) protocol in 1984 that initially provided a HP-AT&T linkage [3]. However, DMI was defined for data transfer rather than control and status and was therefore not capable of providing CTI capability as defined here. Northern Telecom offered a similar solution in the *Computer PBX Interface* (CPI), but there is no record of CPI entering the standards scene. DMI had its European launch in 1984 and provided a major input to the ECMA standard for combining 30 data channels on an E1 multiplex provided between a PBX and a computer. Though this standard was implemented by some PBX suppliers (Plessey provided the interface on its iSDX switch, for example), and some computer suppliers began to experiment with it (DEC had an implementation working at its Annecy facility in France), the concept was virtually ignored by the market—in common with most attempts to provide data transport across PBXs.

DMI and related protocols did, however, pave the way for command and status protocols to some extent. AT&T (subsequently Lucent) sponsored a vendor forum called the ISDN/DMI Users' Group. This attracted 170 switch and computer suppliers, and, in 1987, a special interest group was formed to study the technical considerations required for the implementation of a switch/computer application interface. The outcome of this work was the definition of ASAI, a protocol that can be used with a PBX or a public exchange. Though ASAI was open—in the sense that the specification could be purchased for a minimal sum from AT&T—the linkage with AT&T was too strong to allow this protocol to become a universal standard. It is, of course, still used and supported by Lucent across its switch range.

Meanwhile, in the United Kingdom, British Telecom arranged a series of meetings aimed at providing a national standard for CTI. This built upon the success story that had been achieved in the private signaling area with the *digital private network signaling system* (DPNSS). These meetings were based primarily on the use of the Mitel HCI protocol as a starting point for the definition of a national and, ideally, later an international standard. In the event, Mitel found difficulty in releasing its HCI specification for this purpose, and the initiative was upstaged by the commencement of the ECMA CSTA activity.

ECMA took up the task of providing a European standard for a CTI protocol late in 1988. The activity was stimulated by a contribution from DEC, which also supplied the original chairperson of the standardization group. Time frames for producing a report and then a workable standard were ambitious and, ultimately, unachievable. The group initially labeled the activity Computer-Supported Telephony Applications, but later changed telephony to telecommunications to broaden the scope.

The North American activity began in 1989 under ANSI. It took advantage of some of the ECMA work, but later diverged from it in certain areas. The ANSI activity is labeled *switch computer application interface* (SCAI). Both

CSTA and SCAI are protocols for interchanging messages between switches and computers—neither ECMA nor ANSI define an API. The two protocols evolved slowly but surely from the initial commencement of the standardization activity, and both groups released formal standards. However, only the ECMA work was continued beyond the first release.

8.6.1.2 ECMA: Computer-Supported Telecommunications Applications

Background

As stated earlier, work began within ECMA in 1988. The committee was always fairly small, with less than 20 active members, but well-balanced. Representatives from the three major supply areas—computer suppliers, switch suppliers, and telcos—were present throughout.

During the important period in which the first issue of the CSTA standard was produced, the committee was chaired by computer suppliers—first DEC, then IBM. The input the committee lacked during this period was from users, software suppliers, systems integrators, and voice processing suppliers. It is, of course, very difficult to get positive input from users in standards making. Though users have a very important role to play in the process, their interest usually cannot be sustained over many years because the standardization issue is peripheral to their business interests. However, in later years, the balance of suppliers became more evenly distributed as companies recognized the importance of the ECMA work.

Though the standardization activity commenced in 1989, it was not until 1992 that Phase 1 was released as ECMA 179 and 180. The first document is a readable specification; the second is much more rigorous, presenting the protocol in a formal language (ASN1). Interest in the standard continued as suppliers such as Ericsson, Philips, and Alcatel implemented links based upon it and as, coincidentally, Novell decided to base its API (TSAPI) on the protocol definition.

Phase 2 was released in 1995 as ECMA documents 217 and 218. This new release contained various extensions to the basic protocol, including the following:

- A service to associate data with a call;
- A complete set of services to deal with voice processing;
- A new service set to deal with data interchange.

Phase 3, released in 1998 was expected to include specifications on networking and management. However, it happened that much of the ECMA group's activities were devoted to aligning the specification with the work of Versit and the specification embodied in the Versit Encyclopaedia [4]. This work provided a stronger base for the protocol and therefore a good base for implementing switch-independent APIs. The coverage in this book relates to Phases 1 and 2.

During the early evolution of the standard there were some clear areas of dispute. Contention arose during the evolution of the ECMA technical report on the definition of a call model to describe the switching function. In fact, a definition of CTI states and events was included in the technical report—but for "tutorial purposes only." Evidently, telephone switch suppliers were worried that the manner in which calls are processed in their products might be standardized. Not surprisingly, a supplier would be particularly unhappy if the standard model did not describe the way in which its own systems currently behave.

Another area that caused considerable contention concerned the number of CSTA services that a telephone system supplier needed to implement before compliance with the standard could be claimed. Some telephone system suppliers thought that one was sufficient. However, application and computer suppliers thought that either all, or at least a minimal set, of services should be implemented. The switch suppliers triumphed here: It is only necessary to implement one service to the specification in order to claim CSTA compliance.

For all of this contention, the ECMA work started well; it began with applications.

Examples of Applications

The ECMA group, well aware of the importance of the user's view of CTI, began its work by examining the market for applications and selecting a number of typical examples. These were included in the group's first output, a technical report that formed the basis of the standardization activity [5]. The list, edited somewhat for clarity, is as follows:

- *Personal telephone support:* Provides improved human interface and support for users;
- *Telemarketing:* Provides enhanced monitoring of telemarketing activity;
- *Outbound calls:* Assist agents in making calls to customers;
- *Inbound calls:* Assist agents in handling incoming calls from customers;
- *Customer support environment:* Provides support for handling customer inquiries;
- *Integrated message desk:* Integrates telephony message desk facilities with computer-based electronic mail;

- *Emergency call applications:* Includes industrial alarm system and fire service control examples;
- *Data collection/distribution:* Exploits the voice terminal for the collection and distribution of data;
- *Data access:* Supports the switching function accessing data held on a computer and vice versa;
- *Hotel application:* Identifies typical hotel telephony functions;
- *Switched data application:* Provides support for the connection and configuration of data ports.

The first example is generally referred to in this book as SCB. The description includes the display of call progress information and the status of the user's terminal. The second example, telemarketing, is one of the most significant applications of CTI as far as revenue potential is concerned. The third and fourth examples are of interest since they provide applications in which a computer system and a voice system are integrated with the switch. The fourth example has been reprinted here with minor modifications to demonstrate the level of definition provided by the ECMA report.

Integrated Message Desk

Typically, a user invokes the service by providing information about projected absence. Incoming calls are redirected to an attendant or to the answer point; then, the attendant handles the call on behalf of the called user. These functions are described here in detail.

Service invocation. This can be via the user's voice or data terminal. The application conducts a dialogue with the user to ascertain such information as:

- The user's scheduled whereabouts at various times during the day;
- The preferred message attendant to handle incoming calls;
- If there are any callers the user would like to speak to urgently (rather than having a message taken);
- Any messages the user would like given to particular callers;
- An emergency number (perhaps the number of the user's manager).

If the service was invoked via the voice terminal, the switch affects the necessary call forwarding and informs the computer application of the details. If the service was invoked via the data terminal, the computer requests the switch to effect the necessary call forwarding. In either case, it may be necessary to ensure that the data held on the switch is consistent with that on the computer. Typically, this would be done by one side (for example, the switch) requesting an update of all stored information held by the other side.

Message forwarding. This function handles the redirection of incoming calls in the following manner:

- The application requests notification of incoming calls for the selected attendant(s).
- The switch forwards the incoming call to the attendant and notifies the application:
  - That a call has arrived (i.e., is ringing on an identified attendant's voice terminal);
  - That the call has been forwarded from another extension (together with the number of any calling or called line identification information that is available).
- The application will use the supplied information (particularly the originally called extension number) to update the computer screen of the attendant with the appropriate data for the called user (i.e., the information originally entered by the user). Any data available about the caller would also be displayed.

Attendant actions. These actions will be based on the caller- and the user-supplied data (displayed on the attendant's screen). The attendant will often be attempting to make internal calls and would benefit from being able to invoke call completion features on encountering busy or absent users. The attendant's telephony operations (such as transferring the call) should ideally be invoked only at the computer terminal. Depending on the caller and the user's instructions, the attendant might do the following.

- Give the caller a specified message;
- Take a message and forward it to the user (for example, via electronic mail and including the caller's number in the mail header). The user could subsequently access items (in any order) from a single menu of e-mail and telephone messages, giving the time received, the called number, and the calling number. When a voice system is available, the application may be able to provide notification of the voice message. The user may have the ability to place telephone calls directly from e-mail messages generated by the attendant;

- Arrange to make the user aware of messages via the message waiting lamp on the voice terminal, if required;
- Transfer the incoming call to the user, as indicated by the user's schedule;
- Transfer the call to the supplied emergency number.

Architecture and Services

In common with all standards bodies, ECMA endeavors to use existing standards. To this end, the CSTA group did not attempt to create a new architecture for the protocol. The group based it on the existing architecture and on the evolving work of other international standards groups, particularly ISO and ITU. The aim is to ensure that the protocol is completely independent of the underlying transport layers. Hence, the request and status messages may be transported across wires, coaxial tube, or fiber and may use lower-layer transport protocols, such as X.25 packets, or the Internet protocol datagram service—IP. The choice is of no interest at the CSTA protocol level, providing that a message path does exist.

The general basis for the CSTA protocol architecture is provided in Table 8.5.

Overall model. Having established the foundations upon which to build a standard protocol—the sample applications and the architecture—the next step is to define models to portray the behavior of the two worlds, computing and telephony. It is the latter that is most critical, since the switch acts as a server for most of the services required in implementing the sample applications. Agreeing on a suitable call processing model became very difficult. Finally, though the objects and states for such a model were defined, the detailed sequence of operations to establish calls, clear calls, and generally manipulate calls were not. The elements that were defined are covered in Chapter 3, in the section that introduces the concept of a switching model. The ECMA work was taken as a basis

Table 8.5
Basis of ECMA CSTA Architecture

Architectural Area	Based On
Functional	Open distributed processing
Interconnection	OSI ROSE
Operational models	ISDN

for that introduction. Table 3.5 of that chapter lists the basic objects to be manipulated—devices, connections, and calls together with attributes and examples of operations.

The attributes of a CSTA device are its type and its identifier. Figure 8.3 shows the various device types defined by ECMA.

Each device will also belong to at least one of the defined classes—data, image, voice, other—and each device will have an identifier. ECMA defines two possibilities: static and dynamic. The static identifier remains stable over time—a directory number provides an example. The dynamic device number may be a short form used for the duration of only one call. Thus, the device corresponds roughly to a physical thing that the computer can identify. Devices can be associated with calls, and, if they are, they will have a connection to that call.

Both connections and calls have two attributes: identifier and state. Both calls and devices can have multiple connections. The call state is defined as the set of connection states that comprise the call. This is fairly straightforward for a two-party call. If you lift the telephone, a call is created and a connection to that call is established. Your telephone will have a directory number; that is its static identity. It might also be given a dynamic identity of "one," simply because it is the first device to connect to the call. Your connection with the call is in the initiated state; there is no other party yet because you have not dialed the number, so the other connection is in the null state. The call state is then

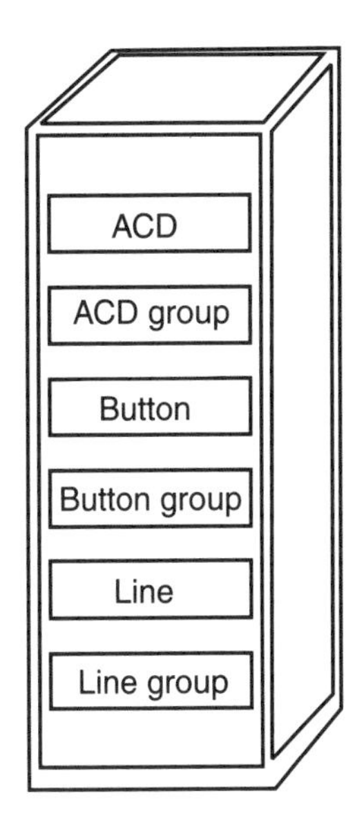

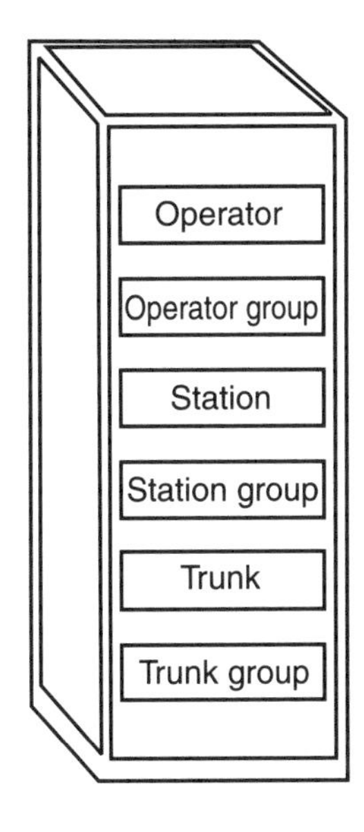

Figure 8.3 ECMA CSTA device types.

defined as pending. After you have dialed the number, a connection can be established to the other party. Your connection state then becomes connected. If you are listening to a ringing tone, the connection state of the other party is alerting. The call state is then defined as delivered. Every time a call changes state, an event is generated to indicate that change. This event will be transmitted to the computer if the computer has registered an interest in the call. The sequence of connection state changes and resultant events for a simple call setup is illustrated in Figure 8.4.

Note that much of this area of the ECMA work is related to ITU ISDN standards. Also note that there is a correspondence between the connection states and ISDN call states, though a direct relationship does not exist. Figure 8.4 might not look simple—but in telephony terms it is. Remember that devices can have multiple connections and so can calls. There are seven states possible for each connection, so for a two-party call there are 49 possible combinations, or 49 possible call states. For a call involving three parties, there are 343 possible call states, assuming that all combinations are possible. However, not all combinations are not possible. There are rules that apply to telephony that

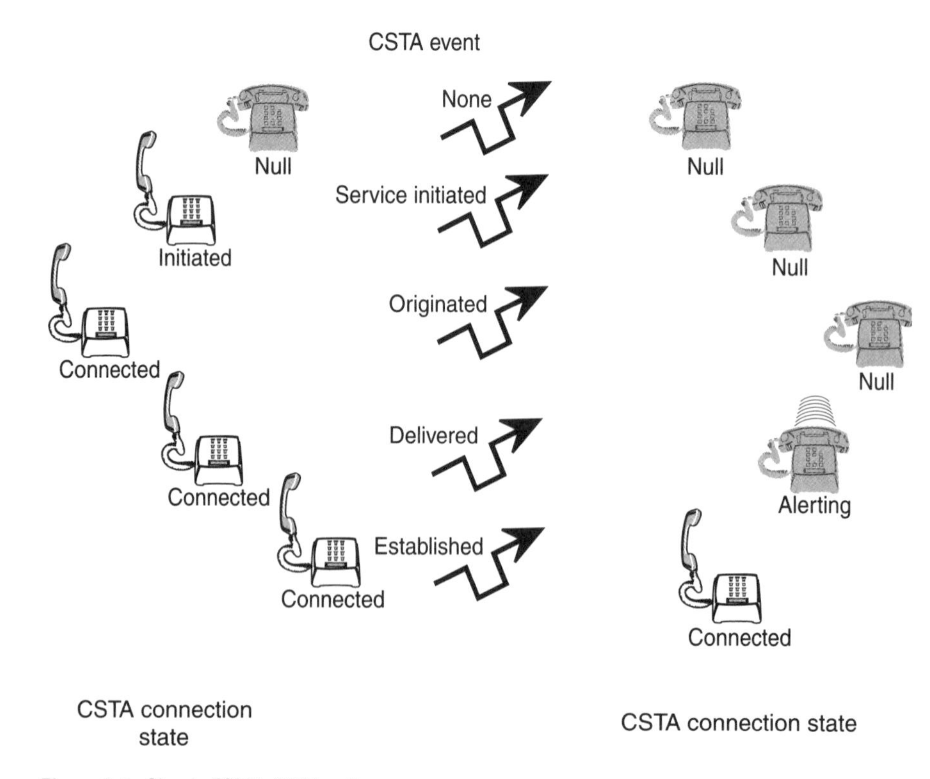

Figure 8.4 Simple ECMA CSTA call sequence.

constrain the number of combinations and the freedom to move from one state to another. The ECMA standard does not place powerful constraints on what can be done; it does not impose a call model. However, it does suggest a simplified connection state model, which is illustrated in Figure 8.5.

The circles in Figure 8.5 represent connection states and the arrows represent typical transitions. The transitions cause events and are therefore the basis for letting the computer application know what is going on.

Even with constraints, there are still many perfectly reasonable call states possible, and the ECMA standard does not define them all. The call state is simply the list of connection states. However, most calls do only exist in a small number of widely recognized states. These are known as the simple call states, and they are defined in Table 8.6.

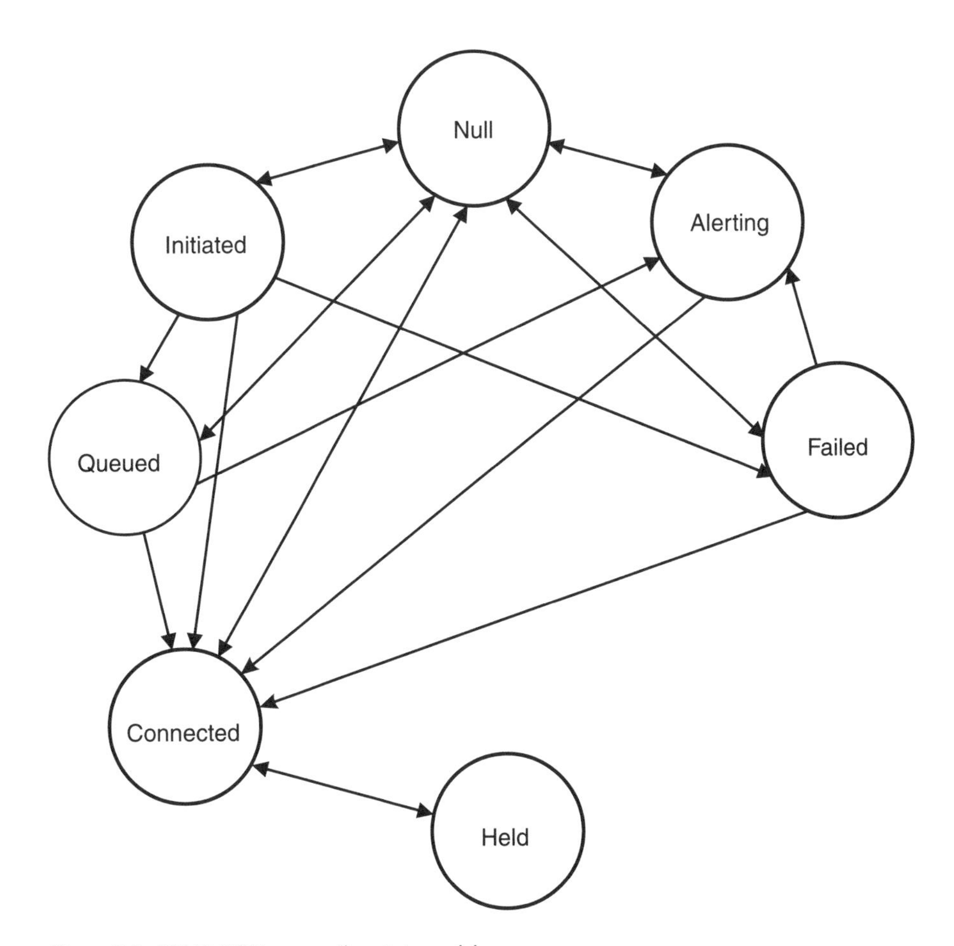

Figure 8.5 ECMA CSTA connection state model.

Table 8.6
ECMA CSTA Simple Call States

Local Connection State	Other Connection State	Simple Call State
Alerting	Connected	Received
Alerting	Held	Received-on-hold
Connected	Alerting	Delivered
Connected	Connected	Established
Connected	Failed	Failed
Connected	Held	Established-on-hold
Connected	Null	Originated
Connected	Queued	Queued
Held	Alerting	Delivered-held
Held	Connected	Established-held
Held	Failed	Failed-held
Held	Queued	Queued-held
Initiated	Null	Pending
Null	Null	Null

To summarize, before putting this view of the operation of the switch to work, the ECMA CSTA standard defines devices that have known identities and types. These devices are associated with calls via an object called a connection. Both the connection and the call can exist in a number of states, and the state of the call is determined by the states of all of the connections to it. As connections change state, event reports are generated, and it is through these that the computer application knows what is going on.

The standard services. Having established the way in which the computer sees the switch through the ECMA standard, the next step is to examine what the standard allows the computer to do to the switch and vice versa. These operations are called services by ECMA. They are, in fact, request messages that pass mostly from the computer to the switch, though there are a few in the other direction.

The second release of the ECMA standard defined six categories of services; they are listed in Table 8.7. As is usual in these protocols, many of the requests gain an immediate response. Generally, the service response is an indication of acceptability—that is, the request is legal—it does not imply that the request has been carried out.

Table 8.7
ECMA CSTA Services

Category	Service
Switching functions	1. Associate data
	2. Single step transfer
	3. Send DTMF tones
	4. Single step conference
	5. Park call
	6. Divert call
	7. Set feature
	8. Make predictive call
	9. Call completion
	10. Answer call
	11. Conference call
	12. Transfer call
	13. Reconnect call
	14. Alternate call
	15. Consultation call
	16. Retrieve call
	17. Hold call
	18. Clear connection
	19. Clear call
	20. Make call
Status services	1. Snapshot call
	2. Snapshot device
	3. Query device service
	4. Change monitor filter
	5. Event report
	6. Monitor stop
	7. Monitor start
Computing	1. Route used
	2. Route select
	3. Route request
	4. Route end
	5. Reroute

Table 8.7 (continued)

Category	Service
Voice unit	1. Query voice attribute
	2. Set voice attribute
	3. Review
	4. Suspend
	5. Resume
	6. Stop
	7. Reposition
	8. Synthesize message
	9. Delete message
	10. Concatenate message
	11. Record message
	12. Play message
Input/output	1. Send multicast data
	2. Fast data
	3. Send broadcast data
	4. Resume data-path
	5. Data-path suspended
	6. Data path resumed
	7. Suspend data-path
	8. Send data
	9. Stop data-path
	10. Start data-path
Bidirectional	1. System status
	2. Escape

The call handling services, or switching function services, to use the ECMA nomenclature, are listed in Table 8.7. As one might expect, the protocol allows the computer to do most of the things that a human operator could do.

The status services of Table 8.7 allows for the tracking of events generally or those following a request. In order to receive events, a monitor needs to be started. This can be associated with either a specific call or a device. In setting the monitor, a filter is established that determines which event reports will be returned. In the ECMA protocol, the settings of this filter can be adjusted at any time. The protocol contains a range of status reporting services, which are listed in Table 8.7. Included here are the establishment of permanent monitors and the ability to take snapshots of current status.

Monitoring an object causes event reports to arrive when a change of state takes place. There are a number of event report types. Table 8.8 lists some of the possibilities. Note that the ECMA definition does include the concept of an ACD "agent."

Though the bias of the protocol definition is toward treating the computer system as the client and the switching system as the server, some computing functions have been included. These are entirely concerned with providing the switch with new routing for a call—at the switch's request. The routing services are listed in Table 8.7. The voice processing functions are also listed there. These are primarily directed to messaging applications but can be used for IVR.

The input/output section of Table 8.7 lists the data transfer functions. These are extensive even though this set of services is only rarely used.

The final set of functions defined in the ECMA protocol are truly bidirectional. The escape service is a particularly useful compromise between rigid standardization and the use of proprietary extensions to improve functionality. Such extensions need to be agreed on from each end of the link, of course.

8.6.1.3 ANSI: Switch Computer Application Interface

The activity began in 1990, and the first release of the standard came in 1993. Though forming the basis of the Centrex CTI link protocols used by Nortel and AT&T, no further work was devoted to the standard after this initial release.

Table 8.8
States Returned in the Event Report Messages

Event Report Type	**Possible States**[1]
Agent state events	Logged on, logged off, not ready, ready, work not ready, work ready
Call events	Call cleared, conferenced, connection cleared, diverted, established, failed, held, network reached, originated, queued, retrieved, service initiated, transferred
Feature events	Call information, do not disturb, forwarding, message waiting
Maintenance events	Back in service, out of service

[1]Event reports include a "cause code" parameter which may amplify these definitions

Background

The North American standards activity, SCAI, progressed under the T1S1 committee of ANSI. The SCAI group was part of the T1S1 committee that also included work on intelligent networking and related topics. The SCAI group had approximately 40 active participants. Key players included Bellcore and the regional Bell operating companies, AT&T, NT, Rolm, IBM, Harris, Hewlett-Packard, and Mitel. The SCAI committee was chaired by Bellcore.

The activity commenced somewhat later than the ECMA work and proceeded on a parallel course. There was liaison activity between the two groups, but it proved impossible to persuade them to converge. The dominance of the public network carriers within SCAI led to fundamental differences in approach. There was much greater emphasis on alignment with emerging IN standards and on the need for interworking between private and public network systems. There was therefore a distinct emphasis in the SCAI activity on the connection of local exchanges (central office switches) to CPE-based computers, an emphasis that naturally encompasses Centrex working. The major application driving the activity was telemarketing, primarily inbound.

Architecture and Services

ANSI, just like ECMA, endeavored to use existing standards wherever possible, and this is reflected in the SCAI approach; however, there are differences in emphasis and in the choice of standards with which to align. The general basis for the SCAI protocol architecture is provided in Table 8.9.

Overall model. The SCAI switching model is based on a view of switching objects and their behavior. The SCAI view of the telephone world is said to be similar to that held by an end user using an advanced telephone. The switching model objects are calls, devices, parties, and agents. Calls are subdivided into two classes: two-party calls and conference calls. Devices are similarly sub-

Table 8.9
Basis of ANSI SCAI Architecture

Architectural Area	Based On
Layering	Application layer protocol within the OSI model
Interconnection	OSI ROSE
Association control	Association control service element
Encoding rules	Basic encoding rules for ASN.[1]

divided into individual devices and device groups. The relationship between the SCAI objects is shown in Figure 8.6.

The multiple lines that emanate from the call symbol in Figure 8.6 simply indicate that a call can have many parties. Similarly, the lines emanating from the device indicate that a device can be associated with a number of parties. A party is an associative object linking a single device to a single call. It is similar to a CSTA connection but by no means identical. A party has an identifier and a state; the latter is termed a condition within SCAI. As shown in Figure 8.6, only three conditions are allowable for a party: null, active, and held. Thus, the party condition determines the current relationship between a particular device and a particular call. The call identifier and device identifier are also attributes of the party; the party identifier is a combination of these two.

The basic call object in SCAI has only one party and exists only when a call is partly formed; this object has only one attribute: the call identifier. Two-party and conference calls have an identity and a state in the SCAI definition, together with the identities of the parties involved in the call. SCAI splits the process of making a telephone call into two halves: originating and terminating. It also defines the allowable transitions between states, that is, the definition imposes a call model, as shown in miniature in Figure 8.6. The

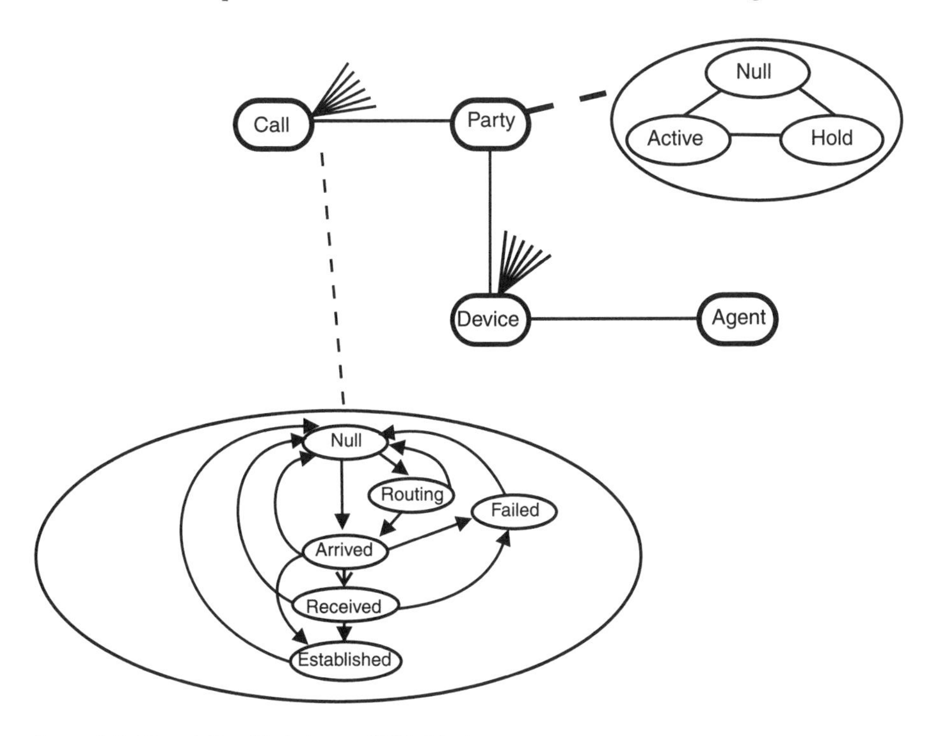

Figure 8.6 The relationship between SCAI objects.

simple call model in Figure 2.8 shows a call in set-up. It was derived from the SCAI definition but omits the FAILED state and a few of the more obscure transitions, for clarity. The full model for the SCAI originating call is defined in Figure 8.7.

A normal call proceeds through the steps illustrated in Figure 8.7. A telephone that is on-hook is in the IDLE state. Lifting the receiver will move the call into the PENDING state and generate a "service initiated" event report. In this state the switch is collecting the digits dialed from the telephone. When all the necessary digits have been sent, the call moves to the ORIGINATED state and a "call originated" event report is produced. In this state the destination address for the telephone call is determined, as is the route the call will take. The call can then be moved to the DELIVERED state, in which case the called telephone is ringing and the originating telephone hears a ring tone. This transition is indicated by a "call delivered" event report. Finally, the called telephone is answered, and the call state becomes ESTABLISHED, indicated by a "call established" event report.

As Figure 8.7 shows, there are a number of other routes a call might take through the model—especially if something goes wrong. However, the point about the SCAI call model is that only the transitions shown in Figure 8.7 are allowed during the set-up of an outgoing call. A separate call model exists for incoming calls, defining the terminating states and transitions. Thus, if the call

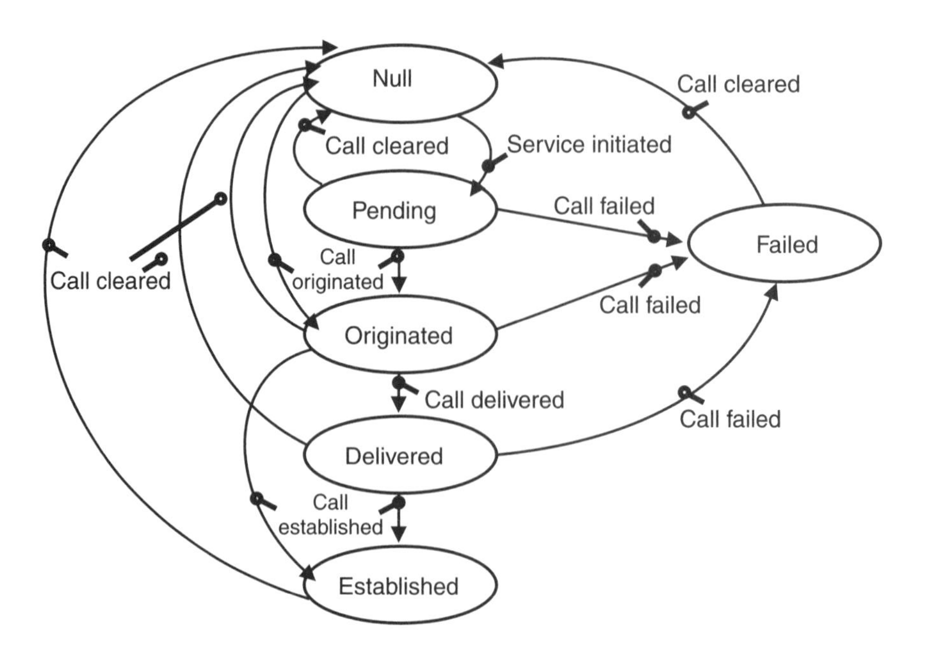

Figure 8.7 SCAI call model for originating call.

was viewed from the point of view of the called telephone, the states transmitted for the simple call would be the following.

- NULL;
- ARRIVED;
- RECEIVED;
- ESTABLISHED.

The ARRIVED state is entered when the call arrives at the switch. The RECEIVED state is entered when the called telephone is ringing, and, finally, the ESTABLISHED state indicates that the called telephone has been answered.

Naturally, there is a correspondence between the call state and the party condition. Returning to the states shown in Figure 8.7 for the originating call, Table 8.10 shows the corresponding party condition for each call state.

It is interesting that the SCAI protocol identifies an agent as a specific object. It does so because, unlike a telephone user, agents can log on to and log off from the call distribution mechanism. Hence, an agent object has a state as well as an identifier. The operations an agent can perform that are of interest in a CTI context are allocated event reports in SCAI. These are defined in Table 8.11.

The states an agent can exist in are the following.

- NULL;
- READY;
- NOT READY;

Table 8.10
The Correspondence of Party Condition and Call State in SCAI

Call State	Party Condition (Originating)
NULL	NULL
PENDING	ACTIVE
ORIGINATED	ACTIVE or HOLD
DELIVERED	ACTIVE or HOLD
ESTABLISHED	ACTIVE or HOLD
FAILED	ACTIVE

Table 8.11
Agent Operations and Event Reports in SCAI

Agent Operation	Agent Events
Log on	Agent logged on
Log off	Agent logged off
Ready	Agent ready
Not ready	Agent not ready
Working ready	Agent working ready
Working not ready	Agent working not ready

- WORKING READY;
- WORKING NOT READY;

The meaning of these states is self-evident and their relation to Table 8.11 is defined in the SCAI definition via an agent state model.

The standard services. SCAI services are defined as *functional elements* (FEs). The SCAI FEs are not defined in detail (see [6]), but Table 8.12 lists each beside the CSTA Phase 1 service it most resembles. The mapping is not exact. SCAI and CSTA are not compatible, even though the table indicates an overlap of terminology. Similar mapping is performed for event reports in Table 8.13.

CSTA and SCAI: The Differences

The difference in emphasis between the two standards produced by the public network dominance within SCAI has already been mentioned. Within that context there were many attempts aimed at aligning the two approaches. Some of these were successful, and the output of the two groups does have a lot in common. However, there are differences. Perhaps at the heart of these is the definition of a rigid call model within SCAI. The SCAI call model goes considerably further than the connection state model defined by ECMA. The state transitions to set up and clear calls are defined and are part of the standard. Attempts to standardize at this level were strongly resisted by the PBX manufacturers within the ECMA group. However, CSTA Phase 3 should be more rigid.

There is a difference in scope; the ECMA activity was based on a range of applications, whereas SCAI used inbound telemarketing as the prime base ap-

Table 8.12
Mapping of SCAI and CSTA Messages

Message	SCAI [6] Functional Elements	CSTA [7] Service
Request (basic)	Make call	Make call
	Answer call	Answer call
	Clear call	Clear call
	Predictive make call	Make predictive call
		Call completion
		Divert call
Request (conference)	Conference existing calls	Conference
	Drop conference party	Clear connection
Request (transfer)	Single step transfer	Transfer call
	Consultation transfer	Consultation call
Request (hold-retrieve)	Hold call	Hold call
		Alternate call
		Reconnect call
	Retrieve call	Retrieve call
Request (feature)	Query feature	Query device service
	Set feature	Set feature
Request (routing)[2]	Route request	Route request
	Routing toggle	Reroute
		Route end
Monitor	Initiate monitor	Monitor start
	Cancel monitor	Monitor stop
	Change monitor filter	Change monitor filter
	Query monitor	
		Snapshot call
		Snapshot device

[2]These messages emanate from the switch whilst all others are from the computer.

plication for its work. There are also many detailed differences, including differences in the definition of object identities. These variations around a common theme can, and do, lead to significantly different standards. Clearly, the two standards are not compatible; the operation of services can differ significantly even in situations in which low-level compatibility is achieved.

Table 8.13
Mapping of SCAI and CSTA Events

Event Report Types[3]	**SCAI**	**CSTA**
Call set-up	Service initiated	Service initiated
	Call originated	Originated
	Call delivered	Delivered
	Call arrived	Queued?
	Call received	?
	Call established	Established
	Call cleared	Call cleared
	Call failed	Failed
	Call transferred	Transferred
	Calls conferenced	Conferenced
	Conference party dropped	Connection cleared?
	Call held	Held
	Call retrieved	Retrieved
	Call diverted	Diverted
	Network reached	Network reached
Routing responses from the computer	Route selected	Route select
		Route used
Information on established monitors	Monitor report	
Agent reports	Agent logged on	Logged on
	Agent logged off	Logged off
	Agent ready	Ready
	Agent not ready	Not ready
	Agent working not ready	Work not ready
	Agent working ready	Work ready
Feature events		Call information
		Do not disturb
		Forwarding
		Message waiting
Maintenance events		Back in service
		Out of service

[3]Difficult Mapping. Most protocols have a single status message with parameters to define the event.

Harmonization meetings were held between SCAI and CSTA proponents in the early 1990s. The original aim of these meetings was to explore the potential for producing a single standard rather than two. The outcome of the meetings was disappointing. The official reason for not bringing the SCAI and CSTA protocols together was timing. Both protocols were near to the completion of their first issues when the harmonization meetings commenced. Making changes at this stage would have resulted in a significant delay. In addition, both groups displayed an unwillingness to change. At one meeting a delegate was heard to say, "Why don't we compromise and do it my way?"

During the harmonization meetings, a great deal of time was spent endeavoring to explain and justify the different approaches taken by the two groups. SCAI is biased towards public network implementations (Centrex), and CSTA is biased towards private networks. As a result of its bias, SCAI is a more constrained protocol, whereas CSTA has more functionality and fewer restrictions. However, significant though these differences are, in the final analysis the differences do not of themselves justify the need for two protocols.

8.6.1.4 Related Standards Activities

ITU

Study Group XI of the CCITT (now ITU-T) had a work item on *telecommunications applications for switches and computers* (TASC). The work began in September 1992 within a subgroup that specializes in INs. The first meeting established a plan for achieving an international standard and drafted some very interesting definitions.

TASC was defined as a bidirectional client/server protocol for use between a telephone network and a computer network that facilitates their integration. The scope of the TASC work extended to public, private, and hybrid networks. It was to be a network access technology, whereas IN is an internetwork or intranetwork technology. TASC was to complement IN and be able to invoke IN services. TASC and IN were to adopt a common call model to avoid interaction conflicts with IN based services.

Though the work started with some enthusiasm, this did not continue. Attendance at the meetings was very low so the activity was frozen after the production of some early reports.

Japan

TTC in Japan began work on a standard in June 1990. Progress was slow and the standard that was agreed on in 1995 was almost identical to that produced by ANSI.

Simple Computer Telephony Protocol (SCTP)

SCTP is a means of presenting call control messages to and from a TCP/IP socket in HTTP compatible form. Since most links are now TCP/IP-based, this seems to make sense. HTTP is the basis of the World Wide Web. The CTI link messages are not normally in HTTP form, of course. In a conventional link the form and meaning of the messages depend on the switch. The message set might be proprietary, e.g. Meridian Link, or standard CSTA as described above.

The CTI link protocol is usually terminated in driver software in a server. The driver will then present the protocol as an API within the server or, via TCP/IP at the clients. The API might be defined by IBM's CallPath or Microsoft's TAPI or Sun's JTAPI. This is normal CTI architecture—just where does SCTP fit into this world? It does not really fit at all; it replaces. Basing this new protocol on HTTP does have advantages, not the least of which is that it is easy to extend messages. It also has the strong advantage that it is operating system and programming language independent. Pitted against these advantages are the facts that there are no applications for SCTP as yet—and there are no drivers. Also, though claimed to be simple, it is not particularly so. If it achieves the same functionality as, say TAPI, it will be equally complex.

Here are the goals of SCTP according to its creators [8]:

- To provide a simple and flexible interface that requires no third-party development tools or middleware besides standard TCP/IP SDKs. A competent programmer should be able to build a working prototype of a client or server application within a few days.
- To provide first- and third-party call control services typically offered by digital telephone handsets—and by client telephony APIs such as TAPI and TSAPI—and to facilitate multipoint conversations (i.e., conference calls).
- To enable clients to handle multiple calls simultaneously.
- To enable clients to connect to multiple servers simultaneously.
- To perform common voice mail management tasks (i.e., check for messages, indicate waiting messages to a client).
- To be capable of interacting with POTS, ISDN, and proprietary digital phone line services and of treating virtual telephone circuits as real circuits.
- To enable users to modify their telephone settings (i.e., do not disturb, call waiting, call forwarding, etc.).
- To enable agents to log into and out of ACD groups.

- To be scaleable from small key/intercom systems to large central offices and telephone networks.

Messages are relatively simple; they are composed of ASCII characters, and there is no predefined limit to message length. Messages are terminated with the carriage return, line feed characters. Here is an example in which a call with the label 1234 is to be transferred to extension 301. It should be self explanatory:

```
action=transfer&cid=1234&destination=301<CR><LF>
```

Table 8.14 provides some examples of the call-related actions that can be requested via SCTP.

Currently, SCTP is yet another stone cast into the CTI protocol pool. It has lots of technological advantages—but may be providing too much too late. It is proposed by Pacific Telephony Design and supported by Nortel, Altigen, and others. If it finds a standards-based home, it could become the enabler of all CTI applications.

8.6.2 Voice Over Packet

This topic is dealt with in Chapter 9 under converging networks. There are many standardization issues here: from the coding used to compress speech to the number of samples per packet and the choice of buffer sizes.

The central choice is the coding algorithm. At present, the trend is very much towards the adoption of ITU G.723.1. This algorithm runs at 6.3 or 5.4 Kbps and uses linear predictive coding and dictionaries that help provide smoothing. Smoothing is necessary to deal with the inevitable loss of packets, which will occur in any network that does not guarantee quality of service. G.723.1 is specified for this environment and is therefore appropriate to the Internet. It is part of a group of standards existing under specification H.323, which deals with video and audio communications across nondeterministic networks. Other implementations are based on the standards produced by ETSI for GSM. This a well proven approach that works at 13 Kbps in the error-prone cellular environment.

There are plenty of industry standards here. VoxWare, for example, is a proprietary encoder that has been bundled with the Netscape browser. It delivers 53:1 compression and very low jitter. VoxWare requires very low network bandwidth but has lower speech quality than G.723.1.

Table 8.14
SCTP Messages

Message	Description
Accept	Instructs server to accept call, create voice path
BlindTransfer	Instructs server to blind transfer call to another destination (local or remote)
Call	Instructs server to place call to a destination address
CampOn	Instructs server to camp call to a destination
ConfCreate	Instructs server to create a conference resource to which calls can be joined
ConfDelete	Instructs server to destroy a conference resource
ConfJoin	Instructs server to add a call to a conference resource
ConfLeave	Instructs server to remove a call from a conference
Fwd_RNA	Instructs server to treat call as ring no answer
Fwd_Busy	Instructs server to treat call as busy
FwdCall	Instructs server to forward call to specified target without answering
Hangup	Instructs server to hang up/terminate call
Hold	Instructs server to place a call on hold
Park	Instructs server to park a call on hold for retrieval by a user; either server or client can specify parking address (behavior will vary by system)
CallState	Server uses this event to notify listening clients of a change in state in a particular call
TransferStart	Instructs server to place call to destination, place current call on hold; this is used to set up a supervised transfer
TransferDrop	Instructs server to abandon transfer attempt, return to original call.
Unhold	Retrieves a call from hold.

8.6.3 Messaging Standards

In the e-mail world, the two global standards are X.400 from the ITU and SMTP from the Internet Engineering Task Force. The latter is the most used because of its association with the Internet and membership of the TCP/IP suite.

The ITU's standard is a multimedia messaging specification. It can carry voice messages, for example, as tagged voice message bodies. However, it is not much used in this manner.

The *simple message transfer protocol* (SMTP) is simple. It is designed simply to carry ASCII encoded text messages and that's all. SMTP is tricked into carrying other message media by encoding. A binary message is encoded to appear as a string of ASCII characters and decoded at the receive end. *Multi-*

purpose Internet messaging extensions (MIME) improves this trick by embedding an indicator of the message type so that it can be opened by the correct application when it is received—the spreadsheet, for example.

Voice profile for Internet mail (VPIM) takes this one stage further by providing a specification of how voice messages can be transferred via SMTP and MIME. It has been evolved by the European Messaging Association and is based on an earlier specification from the Information Industry Association of the United States. Called the *audio message interface specification* (AMIS), this standard was penned for use between networked voice messaging systems. It had two versions, one analog, the other digital. The digital version was itself based on X.400, and it is this that is associated with VPIM.

VPIM messages carry a spoken name and spelled name field. They are normally based on ITU G.721 encoding (32-Kbps ADPCM) for the voice itself together with a specified encoding to present the file as an ASCII string.

8.6.4 Standards for Collaborative Working

Though slow to follow the progress made in the provisions for voice and video communications, collaborative standards were established in the mid 1990s. These exist under the ITU T.120 umbrella. They cover anything that allows two or more people to collaborate across a communications network as if they were sitting in the same room. The T.120 set includes the standards listed in Table 8.15.

Table 8.15
Collaborative Working Standards

Recommendation Number	ITU Description
T.120	Transmission protocols for multimedia data
T.121	Generic application template
T.122	Multipoint communication service for audiographics conferencing
T.123	Protocol stacks for audiographic teleconferencing
T.124	Generic conference control
T.125	Multipoint communication service protocol specification
T.126	Multipoint still image and annotation protocol
T.127	Multipoint binary file transfer
T.128	Audio-visual control for multipoint multimedia systems

References

[1] Walters, R., et al, *CTI In Action,* Wiley, 1997

[2] Computer-Telephone Integration: Markets, Products, and Suppliers, Tern Systems, 1992.

[3] Foard, C. F., et al., "Switch to Computer Networking in the Nineties," *AT&T Technical Journal,* Vol. 70, No. 5, 1991.

[4] Computer Telephony Integration Encyclopaedia, Available from www.versit.com, 1996

[5] "Computer-Supported Telecommunications Applications," ECMA TR152.

[6] "Switch-Computer Applications Interface (SCAI)," Draft ANSI Standard T 1.626, 1992.

[7] "Services for Computer-Supported Telecommunications Applications (CSTA)," ECMA-179, June 1992.

[8] Simple Computer Telephony Protocol (SCTP) version 0.11., Pacific Telephony Design (brianm@phonezone.com), September 1997.

9

Merging on All Fronts

9.1 Review Of CTI

Simple astronomy indicates that all of the stars of the universe are moving away from us. A simple conjecture follows: We are at the center of the universe. Of course this is not the case. The explanation is simple enough—since the whole universe is expanding, we are receding from the stars that follow us as the stars in front recede from us. The words "following" and "receding" assume a center—and for the universe that is the location of the big bang.

CTI is also expanding, but the center of the explosion is more difficult to locate. There has been no big bang here. Growth has been slow—almost painful. However, as the millennium approaches, the red shift is plain for all to see. CTI's expansion is not traceable to one big bang; its progress has been driven by a series of smaller bangs. These in the end have led to an industry that is almost awesome in its extent.

The odd thing about the CTI universe is that it is fueled by convergence. Convergence suggests compression and therein lies a conundrum: CTI's expansion is driven by compression. As the worlds that now constitute CTI have collided they have lost more and more of themselves to CTI. The process is by no means complete. It is ongoing.

In this manner CTI gets fatter at the expense of the telephone world, the computer world, and the messaging and multimedia world. What is more, those worlds are themselves expanding rapidly. As a result, we have an expanding universe here, created by convergence and fueled by an expansion of the converging parts. No wonder it is such fun.

At the beginning of this book, CTI was introduced by the story of the debt collection agent. It is a very revealing tale: Although it is quite clear to those who understand the complexity of call and data transfer that CTI is doing something tricky, to those who do not appreciate the complexity, the functions performed are only to be expected. And those who do not appreciate the complexity certainly make up the majority of people who will, and do, use CTI. It is important to realize and to retain that knowledge. No one introduces the fundamental changes that are required to introduce CTI without good reasons; there have to be palpable and quantifiable business benefits. Achieving something that impresses technically oriented people is not sufficient. Some sectors of the market might say "wow," as the latest CTI trick is performed. Sadly perhaps, they represent only a small portion of the market. It needs something simpler than "wow" to get to the mass market.

CTI's gentle giant was the call center—it still is. Now, however, another giant, nowhere near so gentle has arrived to supplant it. This giant, like CTI itself, has its origins in the 1960s. The decade that saw the invention of sex, drugs, music, and CTI also saw the birth of the Internet. Growing organically in the gardens of academia for many years, it suddenly shot to prominence as the vice president of the United States announced the information superhighway. Suddenly this thing that many of us had struggled with for years began to flower. With that flowering, the alternative meaning of CTI as complete and total integration has become more apposite. The Internet drives many things as the decade of its discovery moves towards the millennium. Its impact on CTI will be covered in this chapter as integration at the server moves to the desktop and then the network itself.

The aim of this chapter is to tidy up the subject of CTI. It links to most of the chapters that you have already read. First, the history of CTI is re-examined, this time looking for the major milestones that have driven this new industry on since the late 1960s. You will then be reminded of the basics of CTI through a little song in which a few outstanding topics are tackled—terminology and openness. The market for CTI is revisited and that rather neglected area that can cripple market growth, the ownership of intellectual property, is examined. The book reaches its finale by reviewing the merging that is taking place at the desktop, server, and network levels and then peers into the future.

9.2 History and Evolution

The beginning of this book introduced the history of CTI. Most people are quite shocked to find out that there were viable CTI applications in use in the late 1960s. Hence, in subsequent chapters, one question is raised repeatedly: "If CTI was as good then as it is claimed to be now, why didn't it spread like wildfire in the 1970s and 1980s?" This is a good question and one worth reflecting on further in this review. However, before that, it is worth considering this: On the wall of the office in which I first typed these words there was a small cartoon. In it a man stood in the Spartan surroundings of an office with nothing in it at all except bare floorboards and a bare light bulb. He is speaking to a young woman who hovers at the open doorway. "Ms. Thompson," he says. "Get me a desk, a phone, a client, a commission, and a cold glass of beer." Like this gentleman, CTI had to start somewhere.

One of the first points to remember is that the installations of the 1960s and 1970s were provided by one supplier—IBM. Powerful as IBM was in those days, telephony was a new discipline for the company and one that was not fully embraced by the U.S. portion of the corporation. Furthermore, those early systems were introduced largely to circumvent a regulatory obstacle—an obstacle placed in the path of IBM by the German regulator and PTT. The circumstances were therefore distorted. The imaginative efforts of the IBM staff were geared toward fighting the regulatory restrictions rather than answering a market need. The real issues of those days were not concerned with the concepts of linking telephones and computers. The real issues of the 1960s and 1970s were the introduction, for the very first time, of computers into businesses, or—for the more sophisticated users—the debate over the use of minis versus mainframes. These were the days when computer programming was an esoteric art and those who did it were regarded as geniuses. These were the days before computer communication and networking became essential to most businesses. These were hardly the days to introduce a new concept that linked together a world that was still in the formative stage (computing) and a world that was for the most part still based on electromechanical systems(telephony).

In those early years of CTI, the prophets wandered in an infertile desert. As the years passed, green shoots began to make their way to the surface of this desert. Table 9.1 plots some of the major relevant events in CTI from 1969 onwards. Through this, some picture of the evolution of the industry and market can be garnered.

Clearly, the decade of the 1980s was the period for early product announcements—most of these were enabling products, though some applications were also introduced. On the basis of events recorded in Table 9.1, it is now possible to answer the question, "Why didn't CTI happen in . . . ?" The answer is that the environment has to be right. The enabling products have to be in place, and the goods must be available to purchase. Furthermore, awareness has to increase, and that is a people-related problem. It is people-related in two senses: First, the people who will carry the message have to be made aware of the benefits of CTI, and second, they have to carry this message to the potential customers. The 1980s were the "Gee whiz, look what we can do!" years. The missionaries were about, but they were few in number and spoke the wrong

Table 9.1
Some Historical High Points in the Evolution of CTI

Year	CTI Development
1969	First known CTI application (IBM)
1970	Applied Data Research installs first call center supervisor
1973	Rockwell installs first ACD
1979	Call answer and clear detection patent granted (c.f. predictive dialing)
1981	Northern Telecom ran CTI laboratory prototype on SL1 PBX
1983	First predictive dialer introduced
1983	Northern Telecom defines command and status link (C&SL)
1983	Northern Telecom introduces an in-house CTI server
1984	AT&T introduces DMI (data transfer only)
1985	Aspect Telecommunications formed to produce ACDs
1987	AT&T defines adjunct switch application interface (ASAI) protocol
1987	DEC announces computer-integrated telephony
1987	DEC displays working CTI applications at Geneva
1987	IBM patents intelligent call transfer
1987	Norite formed to design and develop a fully integrated system
1988	IBM announces CallPath Services Architecture
1988	Mitel introduce HCI protocol
1988	Rolm introduces an early first-party PC-based CTI solution
1988	Work begins on a European standard protocol (CSTA)
1989	Northern Telecom introduces ISDN/AP (later called Meridian link)
1989	NPRI installs telemarketing software with CTI functions (Humana)
1989	Multivendor integration protocol (MVIP) introduced
1990	Number of ACDs in the United States exceeds 5,000
1990	Rolm provides CTI protocol (CallBridge)
1991	Alliance of Computer Telephony Application Suppliers (ACTAS) formed in the United States
1991	GPT announces CTI protocol on iSDX switches
1991	Mitel introduces MiTAI (Mitel telephony applications interface)
1992	AT&T and Novell announce client server CTI solution
1992	First CTI protocol standardized (CSTA)
1992	First public telephone link (Centrex) to IBM's CallPath installed
1992	Northern Telecom introduces central office-based protocol (Compucall)
1992	Rolm and AT&T introduce Windows-based first-party products

Table 9.1 (continued)

Year	CTI Development
1993	ACTIUS formed in the United Kingdom
1993	First book on CTI appears (first edition of this book)
1993	Signal computing system architecture (SCSA) introduced by Dialogic
1993	Second CTI protocol standardized (SCAI)
1993	Work begins on international protocol (TASC)
1993	Novell and AT&T announce CSTA-based API
1993	Microsoft announces telephony API
1993	Internet discovered
1994	First CT-servers released
1994	First CTI middleware applications appear
1995	Microsoft releases Windows 95—with TAPI in the box
1995	Dataquest Inc. predicts that CTI market to reach $7.9B by 1999
1995	Enterprise Computer Telephony Forum created
1996	First Internet telephony gateway
1996	First release of Java TAPI
1996	Mitel demonstrates the first universal serial bus (USB) telephone
1996	Early speech understanding applications installed
1997	ATM-based voice and data systems appear
1997	Microsoft includes TAPI2 in Windows NT
1997	Draft of the simple computer telephony protocol published
1997	TAPI3 announced by Microsoft
1998	Phase 3 of ECMA CSTA released
1998	Switch suppliers release CT-servers
1998	Voice over IP comes of age
2000	All CTI systems pass—into the next millennium

language. They told customers how wonderful integration was and spent a long time talking about CTI links and protocols and DSPs. In the 1990s, the missionaries were replaced by the marketers. There were more of them, and they spoke a language that was much closer to that of the customer. They spoke of needs and solutions, applications, and benefits. The missionaries were still around, but they became increasingly redundant, spending their time speaking at conferences and writing books on CTI.

In the early 1990s, everything began to fall into place—the protocol availability, the standards, the training courses, the software, and so on. Witness to this is the formation of industry and user groups that specialized in promoting understanding of CTI. These were the *Alliance of Computer Telephone Application Suppliers* (ACTAS), later to become part of the Multimedia Telecommunications Association in the United States and ACTIUS in the United Kingdom.

Further witness to the growing level of support was the sheer number of companies involved in CTI. In 1992, Tern Systems [1] listed 180 CTI software suppliers, and in 1993 ACTAS listed over 140 companies in its CTI Business Catalogue [2]. By 1997, CTInet had over 750 suppliers listed in its Web-based directory, and in the following year the CT Expo event attracted over 400 CTI exhibitors.

Of course the major events that helped to put CTI on the map, if not into the office, were two announcements from software companies. The companies concerned were Novell and Microsoft. Announcing their intention to provide enablers for CTI as part of their product portfolio immediately drew more attention to CTI itself. Though neither company added anything really new to the CTI world, they gave it credibility and immediately attracted the interest of a host of smaller software developers. By 1995, there were a growing number of TSAPI and TAPI applications. Those for TAPI were predominantly simple dialers for the desktop since it was not until later in the decade that Microsoft began to seriously address the corporate CTI marketplace with client server TAPI implementations. It was also in that time frame that the first "portable" enabler became available in the form of JTAPI.

None of this activity succeeded in drawing CTI very far out of its traditional home, although some of the more imaginative middleware products such as FastCall and Phonetastic did create some movement. However, there was other action. The voice processing world began to actively embrace CTI in the early 1990s. It also began to change the nature of its products—beginning a strong drift towards PC-based systems. This strengthened the role of the voice card suppliers, particularly the then market leaders: Dialogic, Natural Microsystems, and Rhetorex. These companies needed to become part of the burgeoning CTI world, yet were trapped in the media processing world. The solution arrived as a compromise term, computer telephony, heavily promoted by the Flatiron publishing company through magazines and exhibitions. This initiative was underwritten by the formation of the ECTF in 1995. Many of the company initiatives promoted as standards moved to the ECTF and gained credibility through its pan-industry membership.

Openness became the mantra and the CT and CTI terms began to be used interchangeably. Voice processing extended into integrated and then unified messaging, continuing the steady expansion of things that are CTI. Toward the end of the decade, the switch suppliers were themselves beginning to launch systems based on PC-based technology, and the convergence of industries took another leap forward.

Also, in the late 1990s the CTI newcomers began to look inside the telephone networks. There they found applications for CTI and CT—in the form of the IN. Software experience in routing and system experience in IVR constitute very useful skills in this environment. Accordingly, the CTI suppliers began to move in force. Meanwhile, another initiative was gaining steam under the CTI banner. Internet telephony began to threaten the very survival of the telephone networks themselves—in word if not in action. IP gateways began to appear everywhere—almost all of them made from that ubiquitous PC-based technology.

If CTI is a battlefield, then the armies are assembled from the computer world on the one side and the telephone world on the other. At present, there is little doubt as to who is gaining most ground. It is the computer world, and its success is almost entirely traceable to that amazing little development of the early 1980s—the PC—and to the genius who decided to make the thing extendable via the expansion bus.

9.3 The Basics of CTI

The basics of CTI are contained in those generic functions that have been referred to so many times in this book. Here they are again:

- Screen-based telephony (SCB);
- Call-based data selection (CBDS);
- Application-controlled routing for incoming calls (ACRI);
- Application-controlled routing for outgoing calls (ACRO);
- Voice and data call association (VDCA);
- Data transfer (DT);
- Coordinated call monitoring (CCM);
- Interactive voice response (IVR);
- Messaging exchange (ME).

If you find you cannot remember them, you might find it easier if they were embedded in a song. I can still remember the equation for solving quadratics from my school days—simply because it was taught to me as a song.

Unfortunately, there are not too many people writing songs about CTI at present. I'm sure that will change as the CTI market accelerates. In the meantime, however, I have repeated here some relevant lyrics in the first edition of this book. When singing it, you may find that they do not quite fit the tune—just draw out some of the syllables, where necessary. I do not suppose this song will ever be very popular, but I can assure you that it has seen public performances in London, Stockholm, Dubai, and Singapore. It has been slightly extended here to encompass the broadened coverage of CTI.

The Computer Telephony Integration Song

(Sung very approximately to the tune of "On Top of Old Smokey.")

Oh, the telephone system
Worked well on its own.
Who needs a computer
To talk on the phone?

So what is this good thing
For you or for I
Of providing a linking
That they call CTI?

Well the old screen-based dialing
With screen popping and all,
Yes, that's what you're getting
From a CTI call.

Application-based routing
And monitoring and all,
They're the benefits you're getting
From a CTI call.

Associating the data
With the telephone call
This is the magic
Of a CTI call.

Oh, day and night service
And messaging and all
These are automating
Your CTI call

So let's integrate them,
The computer and phone,
So the old telephone system
Is no longer alone.

9.4 Convergence Problems

The benefits of CTI lie in the bringing together different worlds—usually computing and telephony—and so do the problems. After reading Chapter 2 there should be little doubt of the real differences that exist between those two worlds, not the least of which are those directly related to the people concerned—their backgrounds, attitudes towards change, and aspirations. The people factor is relevant in all areas of CTI though no more so than in the computer and telephony worlds.

Problems can occur in the procurement phase, during which a customer's department may be involved in power struggles focusing on the CTI system. The telephony department may feel that their existence is threatened by the insurgence of a technology that appears to be taking control of the telephony infrastructure. This can cause difficulty because the telephony department's reaction is often to attempt a take over of the CTI procurement and implementation activity—so that they control it rather than it controlling them! This can work, but there are dangers here. Often, the telephony people do not really know enough about the application side of the business, nor do they usually have the skills to select and implement solutions that require the interconnection and interworking of a range of computers, LANs, and WANs.

Telephony is a horizontal market. Obviously, the telephony requirements of one business do differ significantly from another in terms of geographic spread, size of offices, and so on. This is also true of computing at the hardware level. However, it is at the application level that the two worlds diverge significantly. The telephone application is much the same for any business. People want to make calls, receive calls, have calls dealt with while they are not at their desks, and so forth. It is rare for a special version of call processing software to be written for a particular business. The different size and interconnectivity needs are dealt with by choosing the correct switch and configuring it for use. The software is not changed—in some countries the software cannot be changed without the permission of the local regulatory authority. Call processing is a generic application. It may seem that ACD provides an exception to this rule. However, the ACD application is itself generic and is rapidly becoming a standard function for any type of switch. This is not to say that all switches are the same; some are better than others at certain things, without doubt.

Nevertheless, the call processing software within all the different switches is very similar in function if not in form.

In contrast to this, business applications vary significantly according to the customer's trade. Application software may well be altered quite frequently—new applications will be purchased and modifications carried out by the customer or a subcontractor. This is not the world of the telephone system engineer. Though some will adapt to the new, integrated environment with relish, others will try to impose constraints on the development of business applications similar to those that exist in the telephony world.

Fortunately, many large companies have amalgamated the telephony department with IT department and the parochial problems outlined above should not occur. However, enclaves of telephony people still exist. They are still needed, given the importance of basic telephony to the operation of most businesses. Hence, cultural clashes still occur. Perhaps CTI will be the vehicle of change here.

This people business is not constrained to the customers alone. It applies to the industry as a whole where the expansion of CTI has made it more and more difficult to find the few people who have the necessary background and attitudes to span the worlds of CTI. Recruitment by all elements of the industry—computer supply, switch supply, system integration, software supply, voice processing supply, and messaging application supply became increasingly competitive. As early as 1995, one prominent voice processing supplier was heard to say, "Things have gotten so bad that we are now considering training people rather than recruiting from the other CTI companies."

The people problem extends into the standards area. Where conflict arose in the production of the ECMA CSTA protocol, it was often between the computer suppliers and the switch suppliers. Though some of these conflicts were commercial rather than cultural, many had their source in culture. Computer people wanted a full range of functions from each switch; the switch suppliers argued that implementation of just one standard function was sufficient to allow the supplier to claim conformance to the standard. Similarly, the computer people wanted a fixed model for the operation of the switch. The switch suppliers said that every switch may operate differently. They maintained that it was sufficient to be able to describe the telephonic operations, that it was not necessary to constrain the order in which operations can take place. In each case, the switch supplier's view won the day. Interestingly, the ANSI SCAI people came up with different conclusions—but that committee was dominated by yet another group of people, the telcos, and the developments in CTI have shown that a strong call model is indeed a highly desirable thing for the application program suppliers.

These are detailed conflicts, but there is a fundamental one that affects all areas of CTI, including voice processing and messaging. This fundamental conflict lies in a different attitude to standards. In the computer and allied worlds, the standards are generally set by the industry—often by one company. In the telephony world, standards are set by independent bodies—the ITU, for example. Though this conflict is deeply ingrained, the latter part of the 1990s began to see sensible compromises arising. Signs of flexibility within the telephony world were matched by an increasing willingness to adopt ITU standards as the computer world became more and more involved in communication. Nowhere is that more marked than in the adoption of video, voice, and collaborative standards by Intel and Microsoft (H.320, H.323, T.120, etc.).

It is interesting to observe the effect of repeated attacks on the traditional switch suppliers by the newcomers with their PC-based solution. The switch suppliers have been called dinosaurs and accused of restrictive practice, proprietary practices, and so forth. Though there are many theories as to why the dinosaur became extinct, none of us would be very confident if challenged by a Tyrannosaurus rex one dark night. Our superior intelligence may not be sufficient to deal with the giant lizard. In fact, it might turn out to be brighter than we think—even capable of using our own technology against us.

The switch suppliers are not dinosaurs. They have teams of bright developers working for them. They may be constrained by the very market that they have created—but they can turn. At the time of writing, I have just returned from a major CTI trade show. Most of the switch suppliers announced PC-based solutions at the show. The dinosaurs had suddenly become the competition—and extinction is much more likely to occur amongst the myriad of new players than among the established suppliers.

Similar things have developed in the unified messaging market. Unified messaging is not a stand-alone product and easily becomes an option within e-mail systems rather than a special from the voice mail suppliers. The client and server software seems to be universally based on Microsoft Exchange so differentiation in this world becomes increasingly difficult. However, the e-mail suppliers have established markets and distribution chains—often the clincher in significant market development.

Similar in some ways is the world of collaborative tools. Here there was no industry as such—simply a number of players with incompatible collaborative solutions ready. The market was set to grow when the ITU finalized the T.120 set of collaborative standards and a number of suppliers did produce T.120-compatible applications. However, the one solution that quickly became the de facto standard was NetMeeting from Microsoft. Its popularity was partly due to

Microsoft promotion but chiefly due to the fact that it was provided free on the Internet.

In the voice processing world, the traditional suppliers began a gradual transformation from box supplier to system integrator. Over a period of five years, Syntellect, one of the leading IVR suppliers, dropped manufacture and began the slow march toward system integration [3]. It started with the wrong staff balance, and its implementation team was a cost, rather than a profit center. Over the five years, all that changed so that it achieved a position where half of its revenue came from supplying service—including customized development and maintenance.

In the early 1990s, extending an IVR system was just a matter of adding ports. Over the decade, extensions became more complex, and extra applications were often required—and that meant development and therefore billable services. In this way, Syntellect began to obtain much more revenue from a system in the long term than it used to, which was good business. Part of this change is the move into integration software. Syntellect produced CTI middleware, called VistaLink, which was based on Dialogic's CT-Connect. This gave it access to a wide range of switches. To this it added more software: a predictive dialer, a rules engine for call routing, and a special API for controlling the IVR. This was how at least one IVR company became a fully fledged system integrator to the call center market.

9.5 Technology and Terminology

The technology of CTI can be daunting. This book goes to great pains to try to convince the reader that the topic is not a particularly technical one. When I am giving a presentation on CTI, I often display a slide that is crammed full of very detailed instructions that are quite incomprehensible to most of the attendees of my seminars, but not to all. I ask the audience if they think this slide is technical, and everyone agrees that it is, everyone that is except those people who knit. The slide is in fact an extract from the instructions for knitting a sweater—by hand. It looks very technical because most of us are not knitters and therefore do not know the semantics or syntax of that world.

The most technical part of CTI was its terminology and, perhaps, the CTI protocols that enable the two worlds to speak to each other. The first edition of this book included a number of protocol descriptions. People who read the draft found the protocol descriptions very hard going. Not many people would have a simultaneous interest in a number of protocols. If interested in that level of detail they are usually only interested in their own particular switch. It is similar with knitting patterns. Even devoted knitters do not, as a rule, pour over the instructions for knitting a particular garment unless they have already committed themselves to producing it.

However, the scope of CTI has evolved in many directions. Encompassing media processing has certainly brought some vibrant technology into the fold. This includes the software algorithms that perform speech recognition and those that create speech from text. Most people in the CTI world do not get involved with this detailed technology. Even the application programmer is pretty well-shielded from it by enabling layers that simplify interactions down to a few commands and responses. The most noticeable trend in CTI has been its modularization. In the search for more openness the systems have become more "Lego-like." They are often built from off-the-shelf components—of software and hardware. Thus, although the technology has increased in complexity, building systems has become simpler. Nowhere is that more true than in the CT-servers that were developed in a widely available server hardware, using the Windows NT operating system. The cards were often purchased from a third party and linked using a standard bus that also provided the switching function. To this, driver software was added in order to offer a constrained world through well-defined APIs and middleware.

Terminology is in a different class to technology. If you want to use a foreign language, you must at least learn the basic vocabulary. You then have a chance of making yourself understood and of understanding some of what is said; the grammar can come later. This is also true of CTI. However, I can only apologize for the confusion of terminology that continues to proliferate. This book's glossary might help, but there are many glossaries of CTI and they certainly do not contain identical definitions for the same terms.

Some people do try very hard to establish some commonality of use, but this is an uphill battle in a fast growing industry—especially one in which various factions are still jostling for position and have a real interest in maintaining and sometimes spreading confusion. This situation has continued throughout the 1990s. Even the term CTI has been high-jacked. In the first edition of this book I advised readers as follows: Just remember that CTI is the general term covering the whole topic, that there are two types of CTI—first-party CTI and third-party CTI—and that CTI depends on a protocol for linking the two worlds of computers and telephony. All of that was once true—but the coverage of CTI expanded enormously. First- and third-party integration are becoming an almost minor part of the whole as voice processing, image processing, and video are taken under the CTI wing.

It is the world of voice processing that has caused the most confusion in CTI. Voice processing, precisely as its name suggests, involves some processing of the voice—conversion to a suitable form for storage at one extreme, detailed analysis for actually recognizing individual words at the other. Voice processing is not necessary in an integrated system, but it can be very useful in certain applications. Voice processing deals with the content of a call, as all media processing technologies do. Integration is concerned with call control.

It is IVR that caused the most confusion here—and this is the function that also provides the most powerful addition to CTI applications. The problem is that IVR can—with some difficulty—provide a number of CTI functions without the need for an intelligent protocol to link the switch and computer. Chapter 4 analyzes each CTI function in detail to demonstrate the overlap, the manner in which they can be implemented, and the differences. Always remember that the two are not in conflict—even if the suppliers might be. IVR is an essential element in many CTI implementations, but there are others in which it is either irrelevant or unsuitable. Integration is mostly concerned with improving the efficiency of people whose business it is to handle customers over the telephone. IVR is concerned with automating the handling of customers—that is, removing agents from the task wherever possible. IVR suppliers were predominantly hardware suppliers; CTI suppliers were predominantly software suppliers. However, nothing stands still. During the 1990s, the voice processing suppliers rapidly transformed themselves into software and service companies (with some exceptions). They did this by moving from proprietary hardware to the use of PC-based systems. They also began to see the logic of creating integrated servers that deal with call and media processing. Many of them purchased software from leading CTI enabler producers—and so they became bona fide integration suppliers.

9.6 Openness and the Production of CTI Applications

Although the first edition of this book contained a section on openness, it is one of the topics that has changed beyond all recognition. Openness became, during the latter half of the 1990s, the stick with which to beat the established telephony suppliers. Their systems were "closed, inflexible and costly." The open solution which sang the chorus of PC-based suppliers, is built on Windows NT operating within an industry strength PC chassis containing cards linked by the H.100 bus. But what about the software?

It is because the supply of CTI is mostly concerned with the supply of software or the integration of software that a whole chapter of this book was devoted to application creation. However, there is a general and increasing tendency to move away from programming and toward the use of existing programs, modules, and objects. In the CTI world, the most important interfaces to the application provider are not the protocols that run between the switch and computer, nor the interfaces to the PC cards that perform media processing. The most important interfaces are the APIs. These represent the entire world of call and media control as far as the application provider is concerned, and the functions provided by the interfaces decide what can and cannot be achieved by a given business application.

It is because of the importance of the API to the application provider that this becomes the focal point of the openness issue in software. The openness issue is not new. At its most basic, it is all about the customers' ability to choose who supplies the application software. Once again, at its most basic, the issue is all about the environment that the applications run in—the operating environment. The earliest applications were programmed in machine code—application and machine were intimately linked. High-level languages, such as Fortran, Algol, and, nowadays, C, removed some of that linkage but not all. The programs were still machine-dependent and, in particular, operating system-dependent. The operating system can be regarded as an API through which the resources of the computer can be accessed and through which application programs communicate with each other and the outside world. The dependency that exists between the computer hardware and its operating system is broken to some extent by the use of a standard operating system. At the end of the 1990s, that operating system was likely to be UNIX or Windows NT.

Just as openness began to progress nicely, CTI threw a spanner in the works. CTI introduced other pieces of hardware into the picture—the switch and the media cards! Moreover, with this comes a new set of verbs—things like Make Call, Transfer Call, Record, Play, etc. Not the sort of commands that you would expect to find in UNIX or Windows! Thus, the command set is extended by establishing an API.

Unfortunately, APIs are generally proprietary. CTI supplied in this way is not only switch-dependent, it is also language- and operating system-dependent. The first dependency can be minimized by using a small subset of verbs that most switches support; the other dependencies are a little more difficult. Microsoft's third attempt at TAPI does achieve language independence, by utilizing an object-oriented approach such that the interface is entirely message-based. On the other hand, JTAPI achieves operating system independence, at least in the client environment, by emulation. The only approach that achieves language and operating system independence is SCTP, but at the time of writing it is little used.

Currently, there is no clear route to a standard API that covers call control and media control. The ECTF has done sterling work in producing the S.100 family of media control APIs. It is also working on JTAPI as the "portable API." Ideally, it will bring these together. However, even if it does, conflict between a truly open standard and the market bias toward TAPI from Microsoft will not be resolved.

On the application generation front, development was rapid. I personally examined 25 different graphical application generators in 1995. These were targeted at the voice processing market, but many of them have been expanded to cover call control. Other suppliers have produced generators with both call and media control designed in from the outset. The Vicorp QuickScript Tool is one such example. It contains specific icons for transferring a call from an IVR to an agent and vice versa. Meanwhile, solutions such as the CallPath and DirectTalk Java Beans from IBM removes dependence on any particular application generator. Any visual builder capable of assembling beans can be used. The beans provide a graphical representation of objects that can be linked to provide a particular application. IBM supplies the telephony and voice automation beans, but any available beans can be used to link to the database and other applications.

9.7 The Market and the Benefits

There are lots of brake pedals associated with the CTI market and lots of accelerators. Of course, if you are going downhill you do not need an accelerator to go faster—you simply need to remove the brake! This is analogous to the history of CTI. It also raises that often-repeated question: Why didn't CTI accelerate from the standing start of 1969? Clearly, CTI was not poised on the brow of the hill in the late 1960s. Parked at the foot of the hill with the brakes on and the accelerator (gas pedal) broken is a more apt description for those days. The product announcements of the 1980s, together with the many factors pinpointed in Chapter 7, gave the CTI car its accelerator and propelled it to the top of the hill. Completion of the first issue of the standards, together with the need for business to become much more efficient during a world recession, was enough to remove the final brake as far as the call center was concerned. The entry of Microsoft and Novell certainly had some effect in removing that brake—as did the rear guard action by the producers of PC telephony cards, Dialogic in the van. The latter led to the establishment of the ECTF, which quickly became an effective standardization body and a focus for the media-processing sector of the expanded CTI world. All of this produced noise and clamor. Exaggerated claims were made, the industry expanded, and the promoters made a fortune. Finally, the word did get around to end users—the people who bought the stuff and the billion dollar annual revenue accolade was achieved. CTI was a part of life.

The most difficult area of the market to broach has been that of desktop CTI. It is accepted that the gains in productivity and therefore business efficiency have been sufficient to fuel the introduction of CTI into the call center market. It is also accepted that the call center market is itself limited in extent. However, the projections for the CTI market provided in this book do not anticipate a limitation of market size. This is because the so called productivity market or desktop market will grow to take up the slack. This market is mainly concerned with improving personal productivity and not directly associated with the agent efficiency benefits described for CTI. The productivity market was slow to start, and it is generally accepted that this was due to the fact that no standard applications platform existed upon which to develop desktop productivity tools and applications. It was thought that Microsoft's TAPI or Novell's TSAPI would cause a breakthrough here, but the effect of the entry of these two players was limited. Nevertheless, TAPI has encouraged software developers to produce individual and group productivity applications for this market. There is every indication that CTI will become a commodity item through desktop integration.

9.7.1 The User's View of CTI

People often say to me, even after a seminar on CTI, "But what does the user really see of this stuff?" By user, they usually mean the customer who calls into a call center. My answer is, not a lot. If you call a help desk that employs CTI, you might be impressed by the fact that they already seem to know a lot about you and the nature of your inquiry. You might be impressed that, when you are transferred to someone else, that person seems to know just as much about you as the first person. You may be impressed that they do call you back when they say they will, and you might come away with the general feeling that you have been treated as a person rather than as nuisance to be disposed of as quickly as possible. You might even feel that you have been disposed of as quickly as possible—and pleased that you have. You might have your call answered by a machine of course—and be annoyed that the company cannot be bothered to provide a real person with whom to speak. However, if you are calling at two o'clock in the morning, you might be very happy to do your business via IVR. In 1998, two of the major phone based banks in the United Kingdom observed that 50% of their callers chose to contact the IVR system rather than wait for an agent.

If you are an agent using a system with CTI, you might feel that you are more on top of the job, that you have all the right information at your finger tips, and that the people you deal with are more pleasant because they are not kept waiting. You might feel more confident about transferring people to the experts in another group, because you do not have to tell them everything that you discussed with the caller so far. You might even be earning more money because you are taking more orders!

If you are the boss, you might feel you know more about what's going on. Instead of that great pile of call statistics you used to get and used to ignore, you might now look forward to a monthly report that tells you where the orders are coming from, how long the average reservation call takes, how many calls are simply information requests, and so on.

If you are a knowledge worker, you might think it is a great step forward to be able to make calls from your computer screen simply by pointing the mouse at a person's name. You might find it particularly helpful that you can book a call in advance and the computer will set it up for you. You might find that your personal productivity has increased now that you can give the computer a list of people you want to call and it will keep trying them until it finds someone who answers. You might find it helpful that your PC keeps a record of the cost of calls you make against each project that you are engaged in. You might also find that it is much more convenient to see all of your messages listed in the in-tray on your computer screen—be they voice, text, or fax. Furthermore, you might be very happy to find that it is as easy to reply to the sender of a message as it is to call them back.

Hence the users, whoever they may be, do benefit from CTI—even if they do not notice it.

9.8 Delivering CTI

All the benefits come to nothing if CTI cannot be delivered on time, at the estimated cost, with the specified performance capabilities, and so on. This is CTI system engineering. It depends on the availability of good implementation tools, such as a well-functioned APIs and the aids to create and maintain CTI applications. Performance and reliability are very real issues. Often, in this book and whenever I give presentations, I emphasize the very real difference between reliability and performance expectations in the world of telephony and computers. When you enter your place of work on a Monday morning, you might switch on your computer terminal or PC. You might also lift the phone, get a dial tone, make a call, replace the receiver—then look at your screen to see whether the computer is ready to begin! You expect the computer to take its time loading software or whatever it does when you switch on, but you expect the telephone system to work the moment you lift the handset. Probably, you do not even wait to hear whether it is ready or not before you begin dialing. Similarly, outages and failures are more or less the norm for the average computer system (nonstop computers excepted), but a telephone system failure is a disaster.

All of this has to be considered when the two worlds are linked. Telephone systems are renowned for their reliability and inflexibility. Computer systems are renowned for their flexibility and unreliability. CTI should produce systems that are flexible and reliable but may produce systems that are inflexible and unreliable—the worst of all worlds.

If the telephone system becomes dependent on the computer system for the control of calls (in application-controlled routing, for example), overall performance must be as good as the switches in processing those calls. This means that the computer's performance must be predictable so that the delays experienced by the telephone user are the same as those normally experienced and therefore expected.

In CTI system engineering, the telephone and computer systems do really become one system, and they have to be analyzed, tested, and managed as one system. In implementing the combined system, all the good practice that marks good project management must be brought to bear, possibly with even more attention to the details of compatibility, overall system performance, and so forth.

Returning to the history of CTI, the Delphi telephone answering system was described in Chapter 1. It was first installed in 1978 and provided an early example of media-processing CTI use in a telephone answering bureau. The designers did not use a conventional computer for the database application, even though there were plenty of minicomputers around in the mid 1970s. Why was it designed in the way it was? The reason that a commercial computer was rejected for the database role was performance. The computer could not be relied upon to bring up the greeting data on the agent's screen within the required time. Developments since then have certainly delivered computer systems that can provide the required performance to deliver screen information that is in synchronism with the delivery of the call. However, you may also recall that the innovative Delphi system had a single call manager dealing with the computing and telephony tasks and that this simplified the entire business of synchronizing call and data events. The management systems and software tools that will provide this level of synchronization are still lacking today, so the argument for a single integrated system to deal with computing and telephony is still compelling. This leads nicely into the topic of merging—merging at the desktop, at the server, and at network level will be covered in Section 9.10.

9.9 The Patent Situation

Does anyone own CTI? Patents may be a gold mine for people who manage to register an idea before it becomes commercialized. However, they can have a very distorting effect on the growth of the market and its subsequent deployment. Witness the effect that early patents on voice mail have had in that market. Although in the final analysis the market has grown in that area, there have been a lot of time-consuming and expensive litigatory proceedings and a distortion of the market, particularly with regard to commercial liaison, distribution routes, and so on.

Although no one has laid claim to fundamental patents covering CTI as yet, this is an area always to be watched. To the layperson's eye, there is nothing unique or patentable in CTI, be it first-party or third-party integration, automation or messaging. The concepts that have been explored in this book are by no means new, and they represent an integration of existing worlds rather than invention of a new one. Even so, the layperson's view of intellectual property and patentability is not always correct, and this area should be regarded with some circumspection.

Though I have not done an exhaustive search of patents that may relate to CTI, I have noted a few patent grants that may have a direct bearing on the growth and structure of the future CTI market place. A relevant patent was filed by IBM in 1987. It is patent number 4,805,209 and relates to simultaneous transfer of calls and data. This is what the abstract says:

> In many business applications, data about a client is created and entered on an agent's data terminal during a teleconference with the client. Often, it is necessary to transfer the client to a specialist during the course of the conversation. This invention describes a method of transferring the call and the data terminal information associated with the call to any available phone extension with an associated data terminal. A computerized branch exchange is used to transfer the call and pass to the host program the phone source extension and the destination extension for the transfer. The host program looks up the source and destination extensions in a phone to terminal file, determines the network address of the data terminal involved, and transfers the appropriate host application terminal display to invoke a transfer of display information. The host application sends the data terminal information to the destination data terminal display in conjunction with the transferred phone call.

It is rather surprising that such a general patent should be granted—as far as I know, the grant is limited to the United States and there are no examples of IBM pursuing any abuse of the patent through the courts. Worded as above, it does seem to cover a lot of applications that are already in the field.

Another patent of some relevance was granted to a company called Unifi Communications Corporation; it is also a U.S. patent. (Unifi later changed its name to Teloquent.) Patent 5036535 was granted in 1992. It concerns a "switchless automatic call distribution system." This embodies an interesting concept—using the telephone network to distribute calls to many agents at any location via a call router. In other words, the call router, a host computer, and some form of data network are arranged to provide a fully distributed ACD function. Although this performs CTI-like functions, in that the call router is obviously a computer and is controlling the rerouting of calls to distributed agents, the relevance to CTI is not as clear as that demonstrated by the IBM patent. The Unifi patent, in the time-honored manner of many patents, makes a number of claims—86 in all. Some of these do suggest a simultaneous transfer of telephone calls and data across the networks.

In 1993, InterVoice obtained a patent for "call back." This covers the situation in which an incoming call arrives at a call center and the queue length for inbound agents is excessive. The patent covers the ability to examine queue length and to use a VRU to ask the incoming caller to key in their telephone number, or simply to use the CLI, if known. The incoming caller is then offered the option of being called back when the agent is free.

A selective search of the patent file in early 1998 revealed the list provided in Table 9.2.

Patent 5526411 describes an integrated hand-held portable telephone and personal computing device. It is summarized in the following paragraph.

> An integrated portable phone and personal computing device has a hand-held body structure that allows the device to be comfortably held as a telephone receiver, yet is configured in such a way as not to decrease the computing capabilities of the device. The body structure provides for a display screen and input unit on a front surface of the device with a speaker and a microphone located on a side surface of the device. The arrangement for the body structure increases the usable surface area for both the display screen and input unit by providing for a combined display screen and input unit on the front surface. An activation key on the side of the unit inhibits data entry on the input unit when the device is being held as a telephone receiver without the need for a cover over the input unit. The arrangement for the body structure also decreases the overall width of the device by providing for a structure that expands the width of a portion of the side surface around the speaker to a width that is larger and more comfortable as an ear piece for a telephone receiver.

This mainly concerns physical integration; the patent was granted in 1996. More relevant is patent 5655015, Computer-telephone integration system,

Table 9.2
CTI Patents

Patent No.	Patent Title
5655015	Computer-telephone integration system
5642410	Call processor for a computer telephone integration system
4686699	Call progress monitor for a computer telephone interface
5161180	Call interceptor for emergency systems
5588049	Method for the automatic insertion of removal of a calling number identification blocking prefix from within a telephone number in a personal computer based telephone management system
5526411	Integrated hand-held portable telephone and personal computing device
5548581	System and method for making connection acceptance/rejection decisions in a communication system
5535204	Ringdown and ringback signaling for a computer-based multifunction personal communications system
5327490	System and method for controlling call placement rate for telephone communication systems
5189632	Portable personal computer and mobile telephone device
4751728	Telephone call monitoring
4604499	Computer telephone access security processor

granted to Maryann Walsh and Paul Gasparro in 1997. The patent abstract is as follows.

> A computer telephone integration system includes a computer integrated with a telephone system. The computer has a plurality of independent application programs that are operable thereon and a plurality of files containing data that are retrievable by any of the plurality of independent application programs. The computer also includes a plurality of first commands that are disposed on the computer to retrieve and run any of the plurality of independent application programs and a plurality of second commands to access any of the files containing the data with any of the plurality of independent application programs. The computer further includes a call processor disposed on the computer that allows a user of the system to associate the first commands and at least one of the second commands with call information received from the telephone system at the reception of an incoming call or with call information provided by a user of

> the system at the initiation of an outgoing call. The call processor automatically executes the first commands and the second command upon receipt of the call information to retrieve and to run the application program and to access the file containing the data with the application program at the reception of an incoming call or at the initiation of an outgoing call.

Difficult to read as this is, it is clearly broad and directly describes screen popping.

Some patent areas are very detailed. In the media processing area, the compression codecs are often based on patented ideas. A fundamental patent covering the basic ideas of both CELP and MPLPC is now owned by AT&T [4]. The claims of this patent are so broad that arguably all CELP and MPLPC codecs are covered. Although the old AT&T Bell Labs often applied for patents with the intention of cross-licensing its rights with other companies, in recent years Lucent has been more aggressively pursuing patent licensing revenues. In fact, Lucent has indicated its intention to charge unit-based royalties for certain uses of the basic CELP patent.

No doubt there are other relevant patents around. Although it is difficult to believe that the concept of CTI itself is a patentable invention, there is a large body of intellectual property building up. As yet, I have no record of general CTI patents being pressed or of companies paying royalties on a specific patent. There does not seem to be a CTI equivalent to the voice messaging patent obtained by VMX in the 1980s. Nevertheless, this is a space to be watched very closely—certainly the holders of some of the patents mentioned will be doing so.

In fact, there is one area of CTI in which intellectual property is hotly pursued and that is in the predictive dialing area. Skip Cave, who worked for a company that was later taken over by Davox, established a patent in 1979 concerning call progress monitoring. His patent describes the detection of the click that occurs when a called party hangs up or answers. This is an essential function in the provision of predictive dialing, particularly the detection of call answer. Other inventors have tackled the same problem by detecting the removal of ring tone—but this is slower than the click method.

9.10 Merging Desktops: The PC as a Telephone

Historically, the only electronic item on the desk was the telephone. Later this was joined, in some instances, by a dumb terminal. By the end of the 1980s, the most common desktop device besides the telephone was the PC. In a CTI environment, that computer can be linked to the telephone system in either a first-party manner or a third-party manner. There is scope for great confusion here. From a first-party point of view, the computer can actually connect to the telephone, or line—in which case the telephone connects to it. If the latter is the case, the telephone provides a data channel and the computer deals entirely with data functions and is not capable of providing voice functions. If the former is the case, the computer may well—through a suitable single-channel voice card—provide voice functionality, such as intelligent telephone answering. But this is not the source of confusion.

The source of confusion is this: Most PCs operate in the client server mode, that is, they are not free-standing computers with terminals connected to them; they are distributed computing elements connected via a network (normally a LAN) to like elements and servers. The most common server function is that of a file server that provides a central database for the business. The client server architecture allows all sorts of other functions to be provided from servers, including printing, access to the external world, and access to other LANs and external databases. In the CTI world, the telephony function can be supplied as a server within a LAN. Thus, all the PCs connected to the LAN could access telephonic functions via the server. This is client server CTI. The confusion that is very likely to occur is between the client server method of providing CTI and the straightforward first-party approach. The client server approach is clearly a third-party solution. It requires a CTI link between the server and the switch and then distributes that functionality across the LAN. The first party approach, meanwhile, requires that each computer emulate a telephone in its dealings with the switch.

So, what are the application-related differences between the first-party approach to CTI and the third-party client/server approach? Much of this was considered in Chapter 3, and the conclusion was that there is very little difference except for the following.

1. There may be restrictions in functionality imposed by either approach with regard to the telephony functionality available at the PC. These restrictions will reflect lack of functionality in the first-party signaling protocol or in the third-party protocol.
2. The first-party implementation does not generally allow the computer to observe the status of other lines; inevitably, as a single-line device, it cannot observe the overall activity of the switch or of the agents of interest within the switch. However, key system and multiline feature phone implementations can provide this capability within a first-party implementation.

3. First-party implementations cannot easily provide the CTI routing functions or voice and data call association.
4. The whole area of managing first-party implementations versus third-party implementations is somewhat analogous to the business of managing distributed computing systems versus centralized computing systems. The problems that can occur in a distributed environment are amplified in third-party CTI, in which a fixed association between telephones and terminals must be maintained, at least for the period that the user is logged in. Similarly, the area of management reporting becomes both more difficult and less reliable when dealing with distributed first-party CTI systems.

For all of this, the first-party implementation does have advantages, particularly where the number of people within a business area that computer telephony integration serves is small. Integration at the desktop has given birth to a new term—the Desktop Area Network (DAN). The DAN is essentially the bringing together of the telephone and the PC at the desk. The DAN can be accomplished by first-party integration via a PC card or a data module within the telephone. With the introduction of the USB, the DAN has become a reality.

The next stage to consider is, of course, the physical merging of the PC and phone—or is it? Physical integration is motivated by cost and convenience rather than functionality. A merged PC telephone is unlikely to have any functions that are unavailable in a good first party discrete implementation. It also has some downsides: lack of reliability and lack of choice. For example, merged solutions usually depend upon the correct functioning of the PC and they do not allow you to choose the color of your telephone.

9.10.1 PC Phone Options

The PC phone can be achieved in many different ways; however, the basic input and output mechanisms—the microphone and speaker/receiver—have to be provided in some way. Some of the possibilities for the PC phone, together with pros and cons, are described as follows:

- Use the PC's sound card. This makes sense for IP telephony, though even here there are cards available that allow the use of an ordinary telephone. The problem encountered when using the sound card is that it lacks privacy. Many people are not happy to hear other people's calls or to have their own overheard.

- Plug a handset/headset into a PC card. This overcomes the privacy problem and can work with any telephone network—provided that the card has the correct external interface. As usual, the card may provide other functions besides telephony (e.g., telephone answering).
- Special keyboards with handset attachment. Such things do exist. They are neat, but not popular. Keyboards are generally very cheap—special ones are not.
- Plug into an intercept. This is not exactly a PC phone since the intercept is external, but this approach is popular in the call center. Headsets are plugged into a simple terminal adapter that is still capable of receiving calls if and when the PC fails.

There are many other ways of creating a PC telephone. Some include a place to hang the handset; this is important for normal telephony since the hookswitch function has to be provided. Another possibility here is that the phone takes on the function of the PC. There have been many "Smartphone" products over the years. They have a reasonably sized screen and a small keyboard. However, since they lack the flexibility of the PC, they have not proven very successful.

9.11 Merging Servers: The PC-Based CTI System

Here, the switch and server merge and usually drag in the voice and fax processing functions with them. Figure 9.1 shows how one producer envisaged the fit of its CT-server into a business. The server is shown connected through an existing PBX—it could of course have been connected directly to the telephone network.

There are three options here. Either the switch swallows the server, the server swallows the switch, or some new device swallows both. Some switches did swallow a PC. They provided a plug-in PC capability—that is, a PC card can be embedded within the PBX or ACD. It is a solution that was first used, to my knowledge, in the CXC Rose, an innovative PBX design of the early 1980s. It has been regularly rediscovered since then. It has not proven popular. The PC has to be built to the equipment practice used within the switch and is therefore a special item. The major benefit is that the plugged-in PC is closely coupled with the switch; no CTI link is required.

Much more popular is the server swallows switch option. There are many ways in which this can be implemented. The main approaches are described in Sections 9.11.1 and 9.11.2 and are listed as follows:

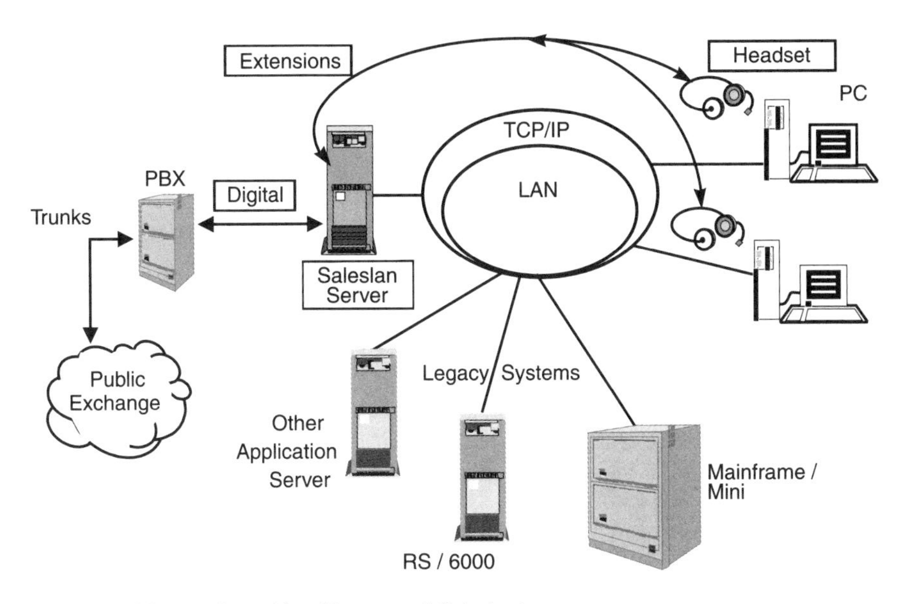

Figure 9.1 CT-server in position. (*Courtesy of:* Saleslan.)

- Self-contained card;
- Specialized cards;
- Off-the-shelf cards.

Designing a new box for computing and telephony has been tried many times. This solution is not open. Both the hardware and the software used are proprietary. However, products designed in this way are often the cheapest, and they do connect into the open world of the LAN and the PC. Some solutions provide more integration than the PC-based solutions—incorporating bridges and routers, for example.

9.11.1 Server-Independent Architectures

The self-contained card approach is characterized by the NetPhone card, first introduced in 1996. This card plugged into a PC and provided support for six trunks and 18 extensions per card, and up to six boards could be plugged into a system. It provides all the basic functions of a PBX, and many of the advanced ones. Everything, including the call processing software, is on the board. Hence a failure in the PC's processor or the software that is running it should not affect the PBX facility. The card is, of course, dependent on the PC's power supply—but that is about all. It provides a range of functions including the following:

- The use of ordinary analog phones;
- Nonblocking, digital switching architecture;
- Onboard line surge protection and ring voltage generator;
- SCbus interface that allows connection to other NETphone boards and other resource cards, if required;
- Calling line identity detection;
- Customizable auto-attendant;
- Screen-based administration on-site or remote.

The telephony functions were comprehensive, including the following special facilities:

- Call hold;
- Call transfer;
- Call conferencing;
- Call waiting with beep;
- Call forwarding;
- Call pick-up;
- Call-group covering;
- Call-group hunting;
- Call queuing.

No need for a CTI link here of course—but the way in which the server processor interfaces with the card had to be defined. The solution here was to provide drivers for both TAPI and TSAPI. Applications can then communicate with the card's telephony functions via these APIs through drivers supplied by the manufacturer. This enables a range of desktop and server-resident telephony applications including NetPhone's screenphone and voice mail applications. All of this is available with no extra hardware, of course. Many suppliers followed NetPhone's example, showing that packing all of this technology onto a PC card was not that difficult.

What sort of integration is this? Clearly, it is third-party client server—but what category? The nearest match is third-party primary. There is no specific

CTI link; this solution uses the interface that exists by virtue of the card's connection with the server, the PC bus.

9.11.2 Server-Dependent Architecture

For the two remaining categories of PC-based solutions, the call processing software does live on the server's processor. The solutions are much the same. One set of cards is plugged into the PC to support extensions, and one or more is provided for trunk interfacing. Resource cards are used to provide special functions—fax distribution, for example. The major difference is that, in some cases, the suppliers have designed and manufactured their own line interface cards (usually for extension and/or trunk interfacing). In others, the suppliers use off-the-shelf line cards from independent suppliers (e.g., Dialogic, NMS, and Excel).

Initially the traditional switch suppliers did not seem to react to the young upstarts who first entered this market (Altigen, NexPath, Saleslan, etc.). However, they had their eyes to the future and were actively considering the use of PC-based technology. The first to produce a solution was Mitel with its MediaPath Server, announced in 1996. Others followed more slowly. Rockwell announced that it was doing something with Dialogic, but it was not until 1998 that the "something" was released. Called Transcend [5], it did not seem at all novel at first sight. It seemed to be just another CT-server solution. Much was made of its openness and adherence to standards. This meant that it ran on Windows NT, used Dialogic-based cards and software, and used ECTF standards. However, under detailed examination, it is a rather clever mixture of the closed and open.

In the call center world the ACD is clearly closed. The hardware and the software belong to the supplier—Rockwell in the case of the Spectrum. The user buys the system but is highly restricted in the areas where change can be made. The user can configure the tables that control the distribution of calls and can add extra trunks and agent interfaces—and that's about it. It is usually enough. On the other side of the divide, the supplier develops the hardware and software and generally tests it to destruction. It also provides intensive service support—at a price. The ACD is often called proprietary, which it is within the box. However, it is usually designed to standard external interfaces, and this allows it to network to the telephone system and connect to other ACDs.

Transforming this into an open CT-server usually implies that the ACD software must be rewritten for this new processing environment. To rewrite this software is crazy. It has consumed hundreds of man-years to produce and at least as much again to test. There are whole teams of engineers trained in that software and many tools around to help produce and configure it. It is constantly improved for the existing ACD installed base. To move to an open system seems to be throwing out the baby with the bath water.

Rather than do this, Rockwell moved its Spectrum software onto a PC card-running UNIX—without any significant changes. The card was plugged into a server together with Dialogic telephony cards. The driver part of the Spectrum software was changed so that it controlled the Dialogic PC cards rather than the Spectrum switch. The server itself worked through the ECTF's S.100 application programming interface and into Dialogic's CT Media implementation of it. This controlled things like speech recording, announcements, intelligent queuing, and voice response. Calls can be passed to the Spectrum card and handed back.

This product moved Rockwell into the low end of the call center market where system administration must be simplicity itself. Transcend achieves this through a Windows-based approach. Software wizards allow the administrator to set up new groups, add new agents, and so on. The wizards have many standard options so that minimal administrator training is necessary.

9.11.3 Boxed Solutions

It is a little unfair to call these things proprietary. Shallow examination would leave the impression that they are similar, if not identical to, the PC-based solutions. They are administered from PCs and connect to PCs. For all we know, there is a PC in the box. The fact that we do not know makes these systems closed. Take the Small Office eXchange that appeared in the late 1990s from Australia. It was a closed box providing telephony and LAN access for up to 64 users. It acted as PBX, voice messaging system, LAN hub, router, and shared modem. It provides access to the PSTN and ISDN. For the unsophisticated, small end user, it seemed the perfect solution. All that was needed to make the whole thing work were the telephones, PCs, and normal twisted pair wiring.

Another product available at the same time was TouchWave from WebSwitch, a U.S. company. It too was a closed box supporting up to 32 extensions and 16 trunks. It was based on an undisclosed real-time operating system. WebSwitches could be linked locally or remotely through an IP network. Users could employ an ordinary phone, a phone plus PC, or a PC with headset. The latter uses VoIP, the others use conventional telephony. The system provides Windows-based telephony and voice mail. It also provides one number contact via hot desking.

Such devices are strong competitors to the PC-based solutions, even though they may be relatively less scaleable and flexible. They are also more difficult to move to other countries where the telephone interfaces are different and connection approval may be needed.

9.12 Merging Networks: IP Telephony and All That

Merging networks was not originally a CTI issue. It goes on at a much lower level. There are three approaches:

- Data can be carried over the circuit switch networks that were designed for voice.
- Voice and video can be carried over the packet networks that were designed for computer data.
- A new network technology can be introduced that is designed to carry both.

The first is very common; a modem or ISDN connection is the obvious example. It works well but is neither efficient nor flexible. Data is limited to the channel bandwidth/bit rate and the circuit is held whether data is transmitted or not. If, as is likely, the call is being charged for on a time basis this can be an expensive solution.

The second, sending voice over packet, has been mooted for some time but only became feasible as good speech compression technology became available and high-speed packet networks were deployed. Here quality is traded for efficiency. Many packet networks are either owned by the user or charged for at a fixed access rate. In both cases, there is no usage charge. Hence, it is possible to bypass the telephone network, which charges by the minute.

The third solution has no drawbacks—except that the technology, ATM, is new and therefore not widely used and expensive. It is also considered to be more complex to implement. However, it really does combine the characteristics of circuit-switched networks with the efficiency of packet networks. For all of that, it became the target of brutal criticism and the stage had clearly been stolen by Internet telephony as the 1990s drew to a close.

9.12.1 Voice Over Internet Protocol

By 1998, packetized voice had built up an enormous head of steam. First implementations were over frame relay networks where "spare" bandwidth could be used for voice. However, it was Internet telephony that really grabbed people's attention—primarily because it was "free." This led to a wider and more intelligent interest in *Voice over Internet protocol* (VoIP)-based networks [5].

Remember that IP networks are simply packet networks based on the IP part of the TCP/IP protocol suite that forms the basis of the Internet itself. IP networks consist of multiple routers linked together. The routers store the packets and then forward them to the most appropriate output links. The links themselves range from 64 Kbps links over twisted pair to 155 Mbps links or greater over fiber. IP is a datagram-based approach that provides best efforts to get the packets through and, therefore, offers no guarantee of quality. Network delay is determined by the number of links and routers traversed by each packet. Since packets are self-steering, dependent on their IP address, they can take different routes; therefore, delay is variable. On top of this, the amount of queuing at each router will depend upon how busy the network is. All of this means that delay is significant and variable. This is not good for telephony. A normal telephone uses a circuit that is fixed for the duration of a call. This ensures low and fixed delay.

Implementing packet-based voice is a relatively simple process. It is shown in Figure 9.2. First, a compressor is used to reduce the bit rate to at least one eighth of the normal 64 Kbps. Then a sufficient amount of speech is collected to build into a packet. Only speech is packetized; silence is suppressed, and this reduces the average bit rate still further. Each packet has a header attached to it. This is equivalent to an envelope in the mail system. It allows the packets to travel across the network to their destination. An example from Micom uses compression based on the ITU G.729 standard, which reduces the digitized voice stream from 64 Kbps to 8 Kbps. Here the IP header overhead increases the 8 Kbps to 15 Kbps. However, silence suppression reduces the 15 Kbps down to an average 6 Kbps.

At the other end, the packets are passed through a jitter buffer—to remove the variable delay in packet arrival. They are then unwrapped and decompressed. Note that all of these steps imply delay and that compression itself reduces quality. The overall delay is generally large so echo cancellation is

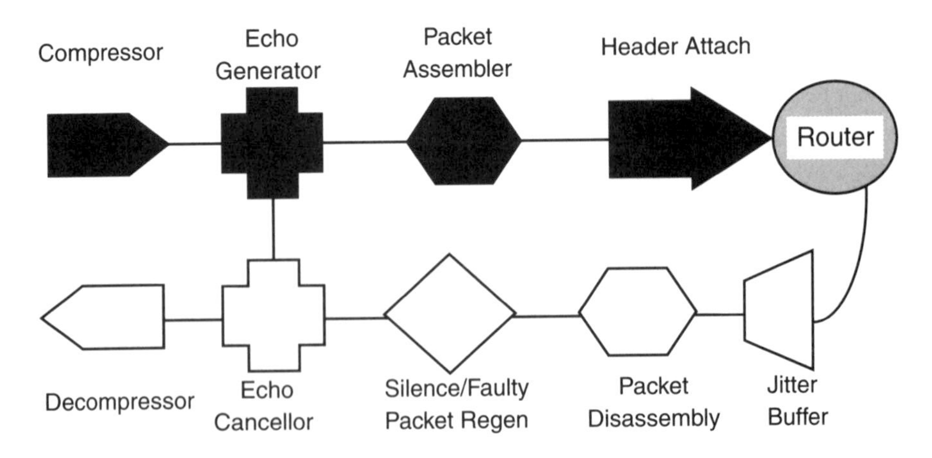

Figure 9.2 Speech packetization and recovery process.

required—just as it is in a satellite-based telephone call. Even with echo cancellation, excessive delay does make conversation difficult. (Some perception of the delays encountered is shown in Figure 9.3.) Furthermore, the removal of silence sounds like a good idea but in practice does not sound too good. This is because background noise is present when a person is talking, but when silence is suppressed, the background noise is not transmitted. Here the techniques used in digital cellular are employed. Comfort noise is played to the receiver when silence is received. This is sometimes called background noise regeneration. Inevitably, some packets will be lost in transit. If the following packet is in the jitter buffer, then interpolation can be used to regenerate the missing packet. If not then, once again the techniques used in digital cellular can be adopted. Here, the sound that was last received is played—at a slightly lower volume. This continues until the silence level is reached or a packet is received.

Following the introduction of the first Internet phones, an immense amount of imaginative work was brought to bear in improving the quality of VoIP calls. Clever though it may be, you can't beat the physics. VoIP is of significantly poorer quality than normal, circuit-based telephony. There are only two ways to significantly improve it, vastly increasing the capacity of the network and removing the delay through the routers or reserving bandwidth for speech.

Returning to Figure 9.3, the figures for delay are clear enough, but what do they mean? Does delay matter? Of course it does. If the delay exceeds 30 ms or so, echo becomes a significant problem. This can generally be dealt with by echo cancellation. This leaves pure delay.

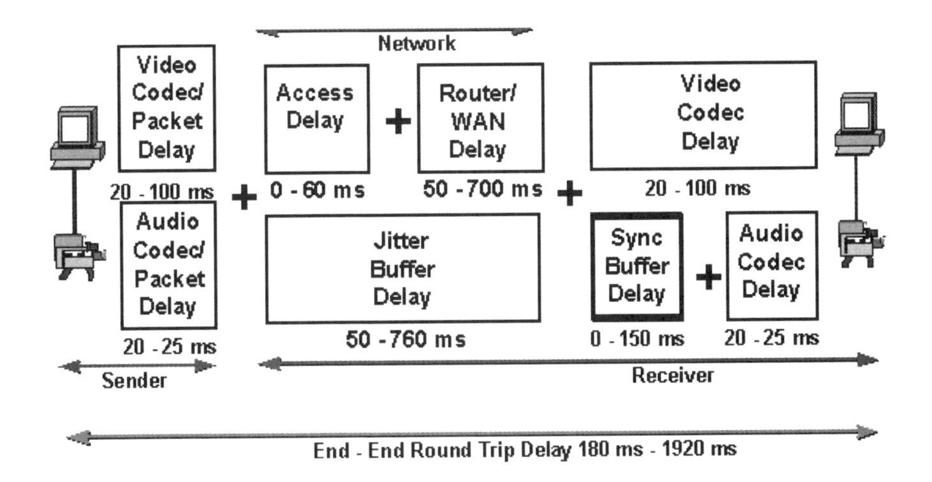

Figure 9.3 Delay contributions to VoIP. (*Source:* Cisco.)

Early work on the subjective effect of delay was stimulated by satellite telephony, which introduces a delay of around one-quarter of a second in each direction. Once again, the first problem to be tackled was echo. The effect of delay without any echo was measured in the mid 1960s [6]. The experiments indicated that "mean number of confused situations during the first 30 seconds of conversations" rises almost linearly with delay. It reaches unity at about 900 ms of one-way delay. That means that there is almost bound to be some confusion if delay is at the top end of that predicted by Figure 9.3. However, it should be noted that the confusion declines as the call progresses. The mean number of confusions at the satellite delay figure (250 ms) is around 0.3. Of course, these figures do not take into account the effect of packet loss and silence regeneration. Neither do they take into account the effect of compression. Conventional pulse code modulation at 64 Kbps yields a mean opinion score of around 4.3, whereas 8 Kbps-compressed speech drops this to 3.8. In other words, the majority of people would regard uncompressed speech as somewhat above good, whilst regarding compressed speech as somewhere between fair and good.

In summary then, the quality of packetized speech is likely to be significantly poorer than that experienced over a circuit, but is likely to be a lot cheaper because it bypasses the telephone network. Telephony bypass is not the only reason for adopting VoIP, although it is the short-term driver. In the long term, a packet-based termination has a major attraction. It provides single-point access to a network that can support any media simultaneously with voice. It allows collaboration during the call, and it provides flexible bandwidth.

9.12.2 IP Phones and Gateways

The basic Internet "Nerdphone" is a PC with soundcard and some downloaded software. This implementation has its uses, but most people want to use the ordinary phone. This is possible. One of the first devices to tackle this was PhoneJack. This consisted of a PC card that takes over the packetization process from the sound card. It had a socket for an ordinary phone. The phone's keypad could be used to set up calls, and the phone rings when an Internet phone call comes in. This helps, but users were interested in calling other people who have ordinary phones and in using their normal phone line to make calls via the Internet. Enter the gateway. It bridges the ordinary telephone network to the Internet and therefore allows callers to bypass the telephone network by breaking into the Internet.

VocalTec introduced one of the first gateways in 1996 [7]. It was called the VocalTec Internet Phone Telephony Gateway Server, and the company produced Internet Phone software for Microsoft and Netscape Web browsers at the same time. In normal operation, the gateway connected to the telephone network on one side and to the Internet on the other. VocalTec's development changed the perception of Internet telephony from the nerdphone to some thing serious. It allowed an ordinary phone user to break in to the Internet in order to make cheap calls.

To do this, users first dialed the phone number of an Internet gateway server. They were then prompted by an automated greeting inviting them to key in the number required. The system then announced "Internet Phone call, please hold," while the call was being connected. The caller then completed the call to the other party, hung up if the other party was not available, or left a voice mail message on the other party's voice mail if equipped. Delays were claimed to be minimal and audio quality to be excellent.

Security was provided by enabling access to the gateway server only to authorized incoming or outgoing callers, to authorized extensions on the company's phone system, or to telephone users with a valid authorization or PIN number. This means that to make a phone call, many users had to key the number of the gateway, key their PIN, and then key the number that they wanted. This was hardly an improved interface, but some of the dialing could be automated.

Gateway owners could specify which countries or area codes could be called and the hours of service for each gateway. For billing, troubleshooting, or maintenance purposes, full call status, progress, and other information was provided. This included details such as incoming or outgoing, the date, time and duration, the source and destination of the call, the call originator's name, and the reason for the end of the call. From this it can be seen that managing a gateway is similar to managing a telephone exchange. Later solutions had a centralized manager. The Clarent Account Manager, for example, could handle up to 200 gateways through the IP network. It handled billing, gateway selection, and call detail recording.

The original Vocaltec gateway included a 486 or faster PC running the gateway software, a voice board from Dialogic, and Windows NT workstation software. A choice of voice compression schemes was available. The VocalTec software was designed to handle up to four lines per system. The cost was $3,995 for a single line.

A few years later, gateways had expanded. Natural MicroSystems produced Fusion, a three-card gateway that did most of the packetization process at card level. It supported 24 lines.

The early days of doing-it-yourself were replaced by the dawn of standards. Phonet produced a H.323-compliant PacketPhone card and an

EtherGate gateway for corporate IP networks rather that Internet-based telephony. Micom's V/IP Phone/Fax gateway also targeted corporate solutions but used the G.729 standard for speech compression. It supplied overflow and back-up capabilities via the normal telephone network.

Some of the gateways began to offer PBX-like facilities. The T-2000 gateway from e-Net was based on the use of ordinary phones, and most PBX facilities were available via the phone or a Windows screen. It used standard Q.931 signaling and worked with any network—LAN, frame relay, and ATM.

9.12.3 Standards and Gatekeepers

Although there are still many solutions to VoIP—both proprietary and standards-based—the trend is clearly toward implementations based on the ITU's H.323 standard entitled, *Visual Telephone Systems and Equipment for Local Area Networks that Provide a Non-Guaranteed Quality of Service.* Although the standard is targeted at LANs, it clearly addresses "best efforts" networks such as those based on IP, whether WAN or LAN. The standard covers many things including signaling across the channel and formatting data for the channel. It specifies a range of standard approaches to compression, including G.723. This standard is most used in VoIP; and is capable of compressing down to 5.3 Kbps.

The standard also addresses management through the specification of a gatekeeper. The basic roles of a gatekeeper are defined as follows [8]:

- Call control and call routing;
- Basic telephony services such as directory services and PBX functions (e.g., call transfer, call forwarding);
- Controlling H.323 bandwidth usage to provide quality of service and protect other critical network applications from H.323 traffic;
- Total network usage control;
- System administration and security policies.

Table 9.3 indicates those functions that a gatekeeper has to do and those that are optional. The first function is mostly concerned with translating phone numbers into IP addresses. The gatekeeper has control over a zone that defines the things that it manages. The things can be gateways, terminals, or conference units.

Table 9.3
Gatekeeper Functions

Required Gatekeeper Functions	Optional Gatekeeper Functions
Address translation	Call control signaling
Admissions control	Call authorization
Bandwidth control	Bandwidth management
	Call management

9.13 The Future of CTI

Sitting in a bar in the early 1990s, raking over the coals of a recent CTI conference, someone said to me, "When we have pulled all these systems together with CTI and there is nothing else to do, what are we then going to have conferences about?" I was speechless for a while. I wasn't sure whether he was joking or serious, but he was serious, so I explained a little of what I foresaw as a long and exciting evolution.

CTI, I said, is but a start in a long process of integration that began some time ago and will probably never end. The uses to which a good platform can be put are limited by the imagination, not necessarily by the technology. Most platforms are launched with just a few cost-justifiable applications in mind. If the platform becomes popular, more and more people start to use it, and those people bring along their own ideas. If the platform is very useful, they will think up applications that were never dreamed of originally, and so the uses for the platform expand. The introduction of the PC in the early 1980s is a good case in point. No one at that time could possibly have imagined all the applications that would be implemented on that platform. No one could have imagined the sheer ubiquity of the PC—and the PC story still has a long way to run.

When I look back at that little speech, some eight years later, I feel good. CTI has expanded in ways that I could not have predicted. However, the point that I made has come true. The CTI platform has developed beyond the scope that our imaginations could then grasp. Moreover, the bandwagon has grown and grown. I am regularly surprised by the number of people who have CTI in their job title. Similarly, the market for CTI-related conferences has grown exponentially.

Someone once told me that early market research at the beginning of this century indicated that the market for the motor car would plateau at around

100,000 vehicles worldwide! That prediction was based on the fact that vehicles at that time were rather difficult to drive, maintain, and operate. It was, therefore, assumed that all motor cars would be driven by chauffeurs and that there were a limited number of people who could and would be trained to operate this very new form of transport. I think there may be similar stories buried in the history of computing, especially predictions that did not recognize the ascendancy of the PC.

It was reasoning of this sort that convinced me that platforms that supported both telephony and computing were going to become widespread and that we could not conceive of all the possible uses to which they would be put in the future—but where do we go from here? Most of the things that I wrote about as futures in the first edition are now innovations of the past. The merged solutions are quite tangible, though the market for them remains small at the end of the 1990s. Is there a CTI beyond the scope of integration—physical and functional—that has been described in this book? I think not. The fully integrated call center shown in Figure 9.4 from the first edition is a fact of life at the time of writing. I can think of many call centers that are just like this.

In that first edition I asked the obvious question, why have all these separate subsystems? Why not bring them all into one? The response then was that this was too big a question to address initially—but closer integration may be sensible for the voice elements of the integrated call center. All of that has happened, and, as described in the merging servers section, things have moved well beyond that. So what next?

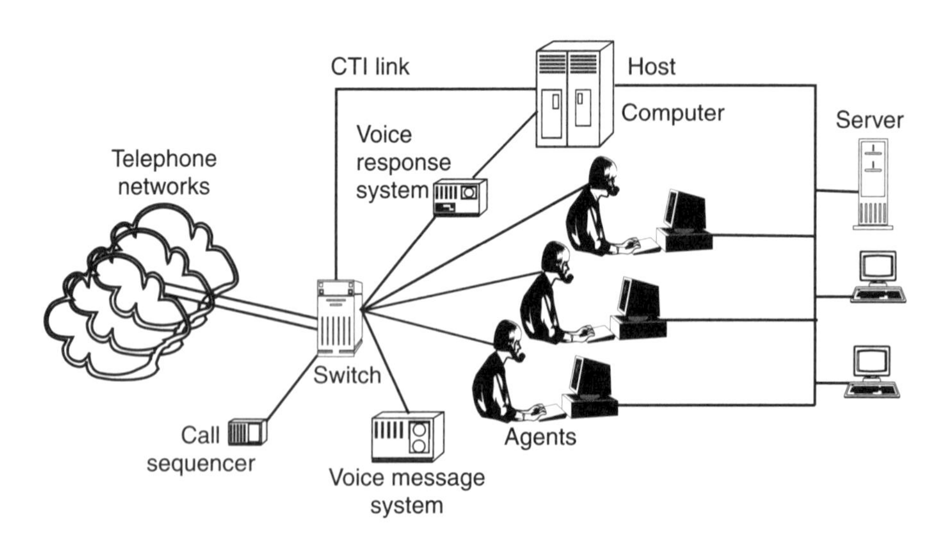

Figure 9.4 The fully integrated call center.

When all is integrated and CTI really has become complete and total integration, is there a further future? Of course there is—but it is not a CTI future. By then, CTI will have done its job. It will not be an issue. There will be outstanding issues: issues that concern the viability of speech recognition and speech understanding when applied to difficult tasks, issues concerning the ease with which we can flip between a machine-mediated session and a person, issues that compare the ability of a human helper with a machine backed by artificial intelligence, and issues that focus upon the limitations of a wired-versus-wireless terminal. In all of this, CTI will not be an issue; it will be a given. Just as people cease to marvel at the availability of electricity or the telephone or radio, integration will be a fact of life.

To consider the call center of the future in this perspective, the writing is on the wall. The large call centers will progressively become network-based, the small ones will use PC-based integrated systems. The combination of people, process, and technology will tilt toward technology—as the technology becomes good enough to take over the role of the agent. Just as the industrial revolution was followed by automation, the call center revolution is followed by automation. If there is process then there is progress toward automation.

Marshall McClewan is generally reckoned to be the man who foresaw the global village. CTI is a piece in the puzzle that enables the global village. One last thought: in the global village, who challenges whom to cricket?

9.13.1 The Last Word

In the first book that I wrote—on voice systems—I ended by stating that "when all is said and done, a lot more is said than done." This was particularly relevant in a book about voice processing. The most appropriate and relevant thing about CTI is integration, so in the first edition of this book, I misquoted the Bible by writing, "Go forth and integrate." I even went further in my misquoting spree by advising readers that, "You have to integrate to accumulate." On the occasions that I was asked to sign the first edition I usually included the phrase, "Everything comes together with integration." And so it does, and has, and continues to do so.

References

[1] "Computer Telephone Integration: Markets, Products, and Suppliers," Tern Systems, 1992.

[2] Computer Telephone Integration Business-Solutions Catalog, ACTAS, 1993.

[3] *CTI Informer,* Issue 5, December 1997.

[4] "VoIP Codecs and the Issue of Intellectual Property," Voxware, Inc., vox@voxware.com., 1998.

[5] *CTI Informer,* Issue 7, March 1998.

[6] Richards, D. L., *Telecommunication By Speech,* Butterworths, 1973.

[7] *VINE Newsletter,* ISSN 0966-9922, Issue 55, September 1996.

[8] *CTI Magazine,* February 1998.

CTI Glossary

automatic call distributor (ACD) A switch that queues and distributes incoming calls fairly so that the longest waiting call goes to the next available operator, and the longest waiting operator gets the next available call. ACDs can also be used for making outgoing calls; they produce lots of call statistics and support both agents and supervisors.

ACD agent *See* agent.

application-controlled routing for incoming calls (ACRI) Computer-controlled distribution of calls.

application-controlled routing for outgoing calls (ACRO) Computer-controlled distribution of outgoing calls; used by power dialers, etc.

applied computerized telephony (ACT) application programming environment for CTI once supplied by Hewlett Packard.

Alliance of Computer Telephone Application Suppliers (ACTAS) Industry forum specializing in CTI in the United States. Subgroup of North American Telecommunications Association (NATA); now part of MMTA.

Association of Computer Telephone Integration Users and Suppliers (ACTIUS) Industry forum specializing in CTI in Europe.

adaptive differential PCM (ADPCM) Speech waveform compression algorithm (normally 32 Kbps).

analog display services interface (ADSI) American standard developed by Bellcore; a pragmatic approach to providing signaling between a local public exchange and the telephone. It builds on the automatic number identification (ANI) feature available from many switches in the United States.

asymmetric digital subscriber line (ADSL) High-speed digital transmission across the twisted pair—for video-on-demand.

agent A person whose job is to handle outgoing or incoming telephone calls (e.g., an agent in an ACD group).

agent group Group of ACD operators who normally deal with one type of call.

advanced mobile phone system (AMPS) Analog cellular system invented by Bell Laboratories. Used in the United States and many other countries.

analog telephone network Network based on the transport of voice band signals.

automatic number identification (ANI) A number passed across the public telephone network that identifies the calling party's directory number. ANI is usually transmitted in analog form. *See also* CLI.

American National Standards Institute (ANSI) Standards body within which the T1S1 committee has standardized SCAI.

application programming interface. (API) A boundary between an application program and a computer control program. A CTI API, for example, allows an application program to invoke program calls to request that functions be performed by an attached telephony switch.

advanced peer-peer communications (APPC) High-level protocol that allows remote applications to communicate.

application A business-oriented task—e.g., order processing.

application generators Programs that allow applications to be developed without the use of code.

application sharing Parties to a call can manipulate an active application in one of the participant's computers and see the results.

application software The program or suite of programs that implements a task.

application-to-application level interface Protocol that allows applications to communicate directly—e.g., APPC.

asynchronous data Data that occurs in bursts and is not often particularly sensitive to varying delay.

AT commands A protocol between a modem and terminal to enable auto-dialing, etc. from a modem.

asynchronous transfer mode (ATM) Technology based on the transmission of fixed sized cells or packets that use a priority mechanism to allow the effective carriage of synchronous and asynchronous data.

attribute routing Using attributes of a call and agents to route the call to the most appropriate agent.

audio conferencing Conference carried out by telephone.

audio graphics The combination of telephony and collaborative working.

audiotext Access of voice-based data across the telephone network.

auto-attendant A system that automatically routes calls through a PBX.

autodialing A process in which an application program makes a telephone call or calls. Though an agent may initiate the call, the digits are transmitted from the computer.

bandwidth The amount of frequency spectrum required by a signal.

base station Fixed cellular transmitter/receiver.

basic rate ISDN The lowest bit rate ISDN connection. Supports two traffic channels and one signaling channel.

bit map image General term for a scanned image; formally describes where each pixel is represented by one bit but also used to describe greyscale and color images.

bridges, LAN A mechanism for connecting LANs of a similar type, selectively transferring frames of data.

broadband ISDN (B-ISDN) High-speed ISDN based on ATM technology.

bulletin board Area of a communicating computer that contains shared data; generally can be read by all comers.

call center A central point at which all incoming calls are handled by a group of agents on a controlled sequential basis. Most call centers are built around an ACD and computer system.

call detail recording (also known as call detail records or call information logging) Record of telephone call activity, such as the time the call began, its duration, or the number called.

call distribution Routing a call to the correct agent group—and ensuring that the longest waiting operator receives the next call.

call model A definition of the transitions a telephone call can and cannot make. A formal description of call set up and clear.

call processing The setting up, routing, connection, and clearing of telephone calls; often applied to the software in a telephone switch that performs this function.

call progress monitoring A facility to monitor and interpret the progress of an outbound telephone call (e.g., dialing, ringing, engaged, unobtainable, or answered).

call waiting A feature that allows a busy caller to be made aware of, and pick up, a new call arrival.

CallPath *See* CSA.

call applications manager (CAM) The application programming environment for CTI once supplied by Tandem.

computer assisted telephony (CAT) A general term used by, for example, GPT (Siemens) to describe CTI; hence iCAT for the iSDX PBX.

call-based data selection (CBDS) Use of telephone call data (e.g., CLI) to select the data to be displayed on the agent screen.

Comite Consultatif International Telegraphique et Telephonique (CCITT) The international standards body for telecommunications up until 1992; now ITU-T.

coordinated call monitoring (CCM) Performance monitoring and reporting based on call and application data.

code division multiple access (CDMA) A multiplexing method based on spread spectrum and the use of digital codes for channel separation.

cell relay Packet-based system using small, fixed-sized packets.

cellular networks Mobile telephony networks based on radio and frequency reuse.

code-excited linear predictive coding (CELP) A speech compression mechanism based on vocoding; standardized at 16 and 8 Kbps.

central office U.S. term for the local public telephone exchange or switch.

Centrex Centrex requires different software in the local exchange to create PBX-like facilities for groups of lines.

chalkboard In CMC, the chalkboard is an area of screen that is shared and can be written to and pasted to by any participants in a call—just like a physical chalkboard, blackboard, or whiteboard.

chalkboard tools Pens, erasers, cut and paste mechanisms, etc. for manipulating a shared chalkboard.

call information logging (CIL) Recording of telephone call activity, such as the time the call began, its duration, or the number called.

circuit-switched network A network that establishes and maintains a fixed path or circuit for the duration of a call.

computer-integrated telephony (CIT) The application programming environment for CTI once supplied by DEC (Compaq).

calling line identity (CLI) Identity of a calling party—the telephone number of the caller.

client operating system Operating system in a client computer, e.g., Windows 3.1 in a PC.

client server An architecture for the distribution of computing functions over a LAN.

computer-mediated communication (CMC) The use of an application computer to control multimedia interactive and message-based communication to provide more effective ways of doing things.

CMC upgrade kit Components and software required to transform a normal PC into a CMC workstation.

CMC workstation A multimedia workstation capable of communicating with remote workstations, usually offering collaborative working, integrated messaging, and video.

central office (CO) Switch to which ordinary telephones and PBXs connect (United States).

co-channel interference In cellular networks, frequency channels are reused. Channels using the same frequencies therefore interfere with each other.

coder and decoder (codec) Device that transforms analog signals into digital and vice versa.

collaborative applications Applications that can be shared during a communications session; applications that support group working.

common channel signaling In telephony, where signals mediating call set up, etc. are carried over a channel that is distinct from that over which the call is carried. The common channel is usually shared by a number of traffic channels.

communicating kiosk Information kiosk that allows user to contact central facility—for help or to purchase goods, for example.

communicating whiteboard *See* chalkboard.

companding Compression and expansion in, for example, PCM.

compound documents Documents containing media other than simply text: e.g., text and speech.

computer conferences Text-based conversations in non-real time involving many parties. Each party can read existing and add new messages to a list of messages on a particular topic stored in a database. (Lotus Notes or Internet Usenet are example platforms.)

computer conversations Text-based conversations in near real time involving two parties—chat.

computer network A network of communicating computers.

computer telephony *See* CTI. Sometimes used to cover the automation area of CTI only.

concatenated speech A phrase produced by playing a series of separately recorded words.

conference bridge Equipment that combines the communications of three or more networked parties into a conference. Arrangements for who gets the floor and what participants hear or see vary according to product.

connection-based transfer Packet transfer over virtual circuits.

connectionless transfer Packet transfer using datagrams.

contention on LANs When two or more users access the LAN at the same time.

continuous speech Speech as we normally produce it.

common object request broker architecture (CORBA) A standard for defining objects that can be linked and embedded.

computer PBX interface (CPI) An early CTI protocol from Northern Telecom (now Nortel).

CallPath services architecture (CSA) The application programming environment for CTI supplied by IBM.

computer-supported cooperative working (CSCW) Working together at a distance via shared workspace, usually in the text domain.

carrier sense multiple access/collision detection (CSMA/CD) The multiple access method used in Ethernet LANs.

computer-supported telephony. (CST) *See* CTI.

computer-supported telecommunications applications (CSTA) Term originally used by ECMA to describe its CTI standards activities.

computer telephony integration (CTI) The functional integration of computing and telephony.

D channel The signaling channel in an ISDN link (can be used for user packets in some cases).

desktop area network (DAN) The interconnections necessary for desktop integration.

data compression Reducing the bit rate or file size by removing redundancy.

data-conferencing *See* computer conferencing and application sharing.

datagram A packet that finds its own destination through a packet network.

distributed computer telephony group (DCT) An industry grouping that attempted to standardize some aspects of CTI.

discrete cosine transform (DCT) Transformation used to aid the compression of digital video signals.

direct dialing in (DDI) A facility to enable callers to dial a public network number (containing the internal number required) and be connected to the number without the intervention of the PBX operator (DID in North America).

desktop conferencing A term used to describe videotelephony and PC-based videoconferencing.

desktop integration Usually used to describe the use of CTI, etc. for personal productivity.

deterministic traffic Synchronous data as, for example, produced by a codec.

dialogues Interactions in speech between a voice processing system and a user.

digital cellular Cellular systems that use digital speech for transmission over the radio path.

digital signal processing Processing signal in the digital domain, i.e., as numbers.

digital telephone network Circuit-switched network based on 64-Kbps clear channels.

digitization Conversion of analog signals into digital signals.

dynamic-link library (DLL) Microsoft term for service procedures that can be linked at run time.

direct memory access (DMA) Within a computer, a method of transferring data directly to and from RAM without the involvement of the processor.

directory number or dialed number (DN) The address used to identify a party to a telephone call.

dialed number identification service (DNIS) Information passed through the public telephone network identifying the *called* telephone number.

document conferencing *See* computer conferencing and application sharing.

document image processing Scanning in and subsequent archival and handling of document image files.

digital signal processor (DSP) Special processor for high-speed signal processing.

data transport (DT) Use of CTI link to carry data—computer data input from telephone keyboard, for example.

dual tone multifrequency (DTMF) DTMF signaling is being adopted increasingly to replace pulse dialing. Each numeral is transmitted as a pair of concurrent audio frequencies. In addition to being used to establish connections, DTMF signaling may be used to exercise control during a call (e.g., in conjunction with an IVR device).

dumb switch *See* matrix switch.

dumb terminal A terminal that is fully dependent on a host computer.

E1 multiplex General term to describe the European 32-channel digital multiplex (*see also* T1), which has a bit rate of 2.048 Mbps.

echo cancellation Removal of unwanted echo signals by characterizing echo path and generating inverse signal.

ECMA Standards body responsible for the development of CSTA.

Enterprise Computer Telephony Forum (ECTF) One more supplier-initiated group formed to encourage CTI interworking.

e-mail The electronic equivalent of the postal service. (*See also* X.400.)

embedding *See* object linking and embedding.

emulation of terminals Where a PC or an interactive voice response system appears to a host to be a normal dumb terminal.

encryption Encoding data in such a way that anyone without the correct key cannot eavesdrop.

Ethernet A LAN architecture that provides multiple access by first monitoring the LAN for other users, then backing off if collision occurs.

European Telecommunications Standards Institute (ETSI) European standards body; replaced CEPT.

expansion bus The bus to which PC expansion cards connect.

expansion slots The position into which PC expansion cards are plugged.

extended industry standard architecture (Extended ISA bus) 32-bit version of the PC AT bus.

facsimile recognition Optically examining fields within received faxes.

fading Signal fluctuation in radio transmission.

fax modem A modem used to send faxes directly from a PC; has fax set up protocol in addition to the normal modem functions.

fax overflow Diverting a fax to, for example, a voice messaging machine when the fax machine is busy.

fax selection and response Using a voice response system to select a fax of interest and have it delivered.

fax store and forward Taking a fax signal into a system for later delivery or broadcast to many recipients or for collection.

Federal Communications Commission (FCC) U.S. regulatory body.

fiber-distributed data interface (FDDI) Fiber LAN standard (100 Mbps).

frequency division multiple access (FDMA) Sharing bandwidth between channels by dynamically allocating part of the spectrum to each.

file conversion Changing the format of a file.

file transfer Moving a file from one computer to another—usually via a network.

fractal compression A technique for reducing the size of, say, images that uses the fractal approach of identifying a root function from which the image can be generated.

frame relay Packet switching standard mainly used for LAN interconnect; a simplified, faster, larger form of X.25.

frame, video One complete scan of a picture.

frequency division multiplexing Providing separate channels by permanently allocating frequency bands.

file transfer protocol (**FTP**) Rules for exchanging files of information.

functional integration An integration that changes the way that things are done.

gateway In communications, usually a server dedicated to providing access to a particular network.

Geoport Apple's line intercept.

geostationary satellite Satellite that is placed in such an orbit that it seems stationary when viewed from the Earth.

GO-MVIP Initiative to achieve international standardization for MVIP.

graphic formats Different methods of representing graphic objects and storing them.

graphic processing Manipulating graphics on a screen or transforming graphic files.

graphics Using a vector representation (simply defined shapes such as lines and circles) of a picture.

graphics interchange format A common format used to represent graphic images.

group hunting In telephony, an incoming call hunts over a group of telephones to find a free one, for example.

groupware Software to improve group productivity.

global system for mobile communications (**GSM**) A digital standard for the operation of interworking cellular systems.

graphical user interface (**GUI**) Interface that uses screen interaction via the mouse and pictures rather than straight text.

GUI flowcharting Graphical approach to application generation.

H.320 The ITU-T umbrella standard for narrowband videotelephony over ISDN. H.323 is a similar umbrella for LAN/Internet.

handoff In cellular, being moved from one cell to another.

handover *See* handoff.

high-speed digital subscriber line (HDSL) High-speed transmission over the twisted pair without repeaters. Used to deliver E1, T1, etc.

high definition TV (HDTV) A long awaited spectacle.

home location register (HLR) A database in a mobile network containing location details of mobiles against their telephone numbers.

homeworking Working from home via the use of CMC.

host application Shared business application, usually resident on a mainframe.

hypertext mark-up language (HTML) Open document format used on the Web.

hypertext transfer protocol (HTTP) Used to transfer data within the Web.

hub In LANs, the thing from which twisted pairs to the terminals radiate.

hypertext access Mechanism where highlighted text can be activated to obtain more information on the highlighted topic.

Institution of Electrical and Electronic Engineers (IEEE) U.S. institution that, amongst other services, produces LAN standards.

Interactive Multimedia Association (IMA) Established in 1988, the IMA aims to promote multimedia and to reduce the barriers to its use (Tel: +1 410 626 1380).

image A scanned or bit map representation of a picture.

image compression Reducing the size of a scanned representation.

image formats The various representations available for storing and retrieving images.

image processing Manipulating image data or transforming it into other media.

impairments Things that lead to the distortion/contamination of a signal during transmission.

International Multimedia Teleconferencing Consortium (IMTC) Industry grouping to promote multimedia communication standards.

intelligent network (IN) Telephone network with central embedded computer (SCP) that is programmed to provide value-added services.

information highway A network giving access to information sources.

information kiosks Large things with screens and keyboards—often located at airports or railway stations; actually a computer with database and access software. *See* communicating kiosk.

infrared transmission Using the infrared spectrum for communication.

integrated messaging Presentation of multimedia message within a PC based in-tray. For example e-mail, voice mail, and fax mail.

intelligent agent *See* intelligent software agents.

intelligent link Third-party link between computer and switch that carries command and status information.

intelligent peripheral In an intelligent network, an interface system, usually an interactive voice response system, that allows a user to interact with an SCP to program, for example, call destinations.

intelligent signaling Signaling, usually packet-based, that carries information beyond the basic necessary for call set up.

intelligent software agents Software that acts for its owner, negotiating with other intelligent agents according to a set of rules programmed by the owner.

intelligent transfer A feature of a CTI system whereby the host controls the simultaneous transfer of a telephone call and associated data between agents or between a VRU and an agent.

interactive conversation Conversation over a duplex path with minimal transmission delay.

interactive voice response Using a voice processing system to transform a telephone into a computer terminal.

Internet A public information highway with information servers, etc.

Iridium Global cellular system based on multiple LEOs, to be available before the millennium.

industry standard architecture bus (ISA bus) Independent specification of the PC AT expansion bus.

integrated services digital network (ISDN) A digital telephone network with well-defined local interfaces that provide digital transmission and intelligent signaling.

isochronous services Those requiring fixed and low delay—e.g., speech and some video.

isoENET A LAN product that supports synchronous and asynchronous services.

International Telecommunications Union-Telecommunications Sector (ITU-T) The international standards body for telecommunications, based in Geneva.

interactive voice response (IVR) An IVR system uses prerecorded voice responses to provide information in response to DTMF or voice input from a telephone caller. The sequence of messages may be determined dynamically by this additional input.

Joint Photographic Experts Group (JPEG) ISO/ITU group evolving compression standards for still photographic images.

Java telephony API (JTAPI) The portable API from Sun and its partners

key system An automatic equivalent of the "key and lamp unit." Provides PBX-like functions without an operator for less than 50 extensions.

local area network (LAN) Network for connecting computers within a building or campus; usually single-user access and usually only suitable for data transfer.

LAN bridging *See* bridge LAN.

local exchange (LE) Switch to which ordinary telephones and PBXs connect (Europe).

low-Earth orbiting satellite systems (LEOs) Communications satellites that orbit near the Earth relative to geostationary satellites. LEOs move relative the Earth's surface.

link protocol, CTI The set of rules and verbs used to communicate across a CTI link—CSTA is an example.

link, CTI Physical connection between telephone switch and computer/server to enable CTI.

linking and embedding *See* object linking and embedding.

mail box Storage allocated to a mail user within a host system to store messages and configuration details.

messaging API (MAPI) Microsoft specification for application access to messaging functions.

massively parallel computers Powerful computers with many processors—essential for dispensing video-on-demand.

matrix switch A fully programmable telephony switch in which every feature can be placed under host computer control. Also known as *dumb* or *specialty* switches.

micro channel architecture (MCA bus) IBM alternative to the ISA bus; now obsolescent.

media conversion Transforming one media into another—e.g., text into speech.

media processing Performing some action on digitized media—e.g. video compression.

media servers Media stores connected to an access network—e.g., video server on a LAN.

Mediabus Specification to allow the interconnection of PC cards to provide video transfer, etc.

metropolitan area networks High-speed connection between LANs over medium distances.

musical instrument digital interface (MIDI) Interface specification to allow the exchange and control of digital audio signals between electronic instruments and, for example, a PC.

multipurpose Internet messaging extensions (MIME) Method of embedding nontext attachments in Internet mail.

mixed media documents See compound document.

Multimedia Telecommunications Association (MMTA) U.S.-based trade group with an interest in CTI. *See* ACTAS.

mobile computing The ability to use a computer "anywhere." Usually based on a radio modem and packet radio system.

motherboard The main board within a PC. Usually contains at least the processor, its support chips, and RAM.

multimedia PC specification (MPC specification) Industry standards supported by a group of suppliers headed by Microsoft.

Motion Picture Experts Group (MPEG) ISO/ITU group evolving compression standards for moving video images.

mobile switching center (MSC) The telephone exchanges to which base stations are connected. MSCs route calls from the fixed to the mobile network and vice versa.

multicast Broadcast of data, particularly across the Internet, wherein the sender sends just one stream to a downstream node, and that node then sends one stream to each interested subnode and so on; overcomes the inefficiency of point-to-point sessions for video or voice.

multidimensional routing *See* attribute routing.

multimedia At least two of the following media: graphics, voice, image, text, and video.

multimedia call center Call center that supports, for example, image processing in addition to voice.

multimedia documents *See* compound document.

multimedia network A network capable of carrying multimedia.

multimedia workstation Workstation that provides personal computing and supports at least one of the following: voice, video.

multimode fiber Optical fiber of limited bandwidth with a relatively large inner core.

multiparty In telephony, a call that involves more than two parties—the features that are used to control this.

multiplexer Device that can send multiple information streams down one channel and derive multiple information streams from one channel.

multiplexing Sending multiple information streams down one channel and deriving multiple information streams from that one channel.

multipoint conferencing unit *See* conference bridge.

multipoint control units *See* conference bridge.

multitasking The capability to support more than one *active* task in a computer.

multi-user The capability for a computer to support more than one user, in addition to multitasking.

multivendor interface protocol (MVIP) A specification that allows the connection of PC cards via a synchronous bus capable of carrying digital speech.

narrowband ISDN Basic or primary rate ISDN.

network computer (NC) A diskless workstation that obtains its applications from the network.

network-based voice messaging A service provided by telcos by embedding a voice messaging system within their networks. At its simplest, this service replaces the answering machine.

neural networks Processing elements that are similar in structure to the brain.

Nordic mobile telephone system (NMT) The Scandinavian analog cellular system, used in many nonScandinavian countries.

non-real time In communications, networks with delays that are sufficient to prevent immediate responses are suitable only for non-real time services. In this definition, e-mail is non-real time, whilst voice telephony is real time.

object linking and embedding In, for example, a word processing document, you may wish to include a graphic object, a picture. Linking the object will simply provide the link to the picture, which has a separate existence and can be separately updated. Embedding it makes it part of the document but stores with it the instructions necessary to invoke the graphics application.

object orientation Application creation through the use of well-defined code objects that communicate only via messages.

optical character recognition (OCR) Use of software to convert a file of scanned text to text. Image to text conversion.

open document architecture (ODA) Standard method of constructing compound documents.

OLE technology *See* object linking and embedding.

open technology Technology based upon freely available standards or specifications.

Open Software Foundation (OSF) Organization that provides open standards for UNIX and higher level application support functions.

open system interconnection model (OSI model) Architectural model to encourage standard methods of communication software layering; formulated by ISO.

packet-switched network Network in which information holds a circuit only whilst there is information to be transmitted. Data is bursty and is sent as packages that contain destination information and management data.

PBX computer teaming (PACT) General term for CTI used by Siemens.

packet assembler disassembler (PAD) Device that connects a character-based terminal to a packet network.

party An addressable participant in a telephone or multimedia call.

passive bus In basic rate ISDN, the means of resolving contention between up to eight terminals attempting to access the two available channels.

private automatic branch exchange (PBX) A telephone exchange owned and operated by a private company; same as PBX.

PC AT bus PC expansion bus supporting 16-bit data transfer and into which PC cards, for example a modem card, can be plugged.

PC expansion bus Generally, a bus within the PC into which extra cards can be plugged. These cards can then interchange data with the PC processor and memory.

peripheral component interconnect (PCI) An Intel-specified PC expansion bus that supports 32- and 64-bit working at significantly higher transfer rates than the AT bus.

pulse code modulation (PCM) Standard method of converting analog signals to digital for use within telephone networks.

personal computer memory card interface association (PCMCIA) A miniature PC expansion bus used for credit card-sized expanders for laptops.

personal conferencing specification (PCS) An Intel specification for CMC workstations.

personal computer television (PCTV) Combined TV and communicating workstation.

personal digital assistant (PDA) Palmtop computer with personal efficiency applications—e.g., Apple Newton.

PCM Expansion Bus (PEB) A Dialogic specification that allows the connection of PC cards via a synchronous bus capable of carrying digital speech.

permanent virtual circuits Predefined routes in a packet network that cannot be altered directly by the user.

PBX host interfacing (PHI) General term for CTI once used by Probe Research.

phonemes Small items of sound from which we build words.

pixel (picture element) Smallest discrete element in a video frame or scanned image. Represented by one bit for simple encoding through to 24 bits for full color.

power dialing The power dialing controller has a list of numbers to be called, a number of outgoing telephone lines, and a group of agents. It launches as many telephone calls as it possibly can and, as soon as answer is detected on a particular line, attempts to connect the call to an agent. If no agent is free it drops the answered call and launches another.

predictive dialing Predictive dialing is very similar to power dialing but is far more subtle. The difference is that, rather than launch a mass of telephone calls regardless of agent availability, the controller uses a pacing algorithm such that the rate of launch is based upon the probability of answer and the probability of an agent being free.

premium rate charging In telephony, the telco charges the caller at a high rate and splits the revenue with the called party.

preview dialing A list of customers to be called is maintained by the computer system, and the call list or data pertaining to the next person to be called is presented to the agent. The agent then initiates the call using screen or manual dialing.

primary rate ISDN ISDN delivery of 24 or 30 channels.

private networks Networks operated by the user.

private wire or circuit A connection between two sites rented by the owner and for the owner's sole use.

progressive dialing As power dialing but at least one agent must be free before a call is launched.

protocol Defined set of rules for communicating across a link.

public switched telephone network (PSTN) The telephone network to which private telephones, PBXs, ACDs, and key systems are connected.

post, telegraph, and telephone administration (PTT) Government-owned telco.

pulse dialing Also known as *loop disconnect dialing.* The process by which numbers dialed, or the keys pressed on certain types of push-button telephones, are converted into a series of pulses to be counted and recognized by the telephone system as the number or access code required.

pulse recognition Detection of dial pulses (or what remains of them) across the telephone network.

pulse signaling *See* pulse dialing.

quarter common intermediate format video (Q-CIF) Reduced quality video frame represented by half the usual number of pixels in the vertical and horizontal direction.

QSIG A European standard for telephone signaling across digital private networks.

quantization distortion Distortion produced when converting a signal from analog to digital. A result of using a fixed number of bits per sample.

radio spectrum Available radio frequency bands and their allocation.

regional bell operating companies (RBOCs) The U.S. telcos.

real time Response with negligible delay from a computer or across a communication network.

regeneration of digital signals Detection of a digital stream and its reproduction with noise removed and levels restored.

repeater In a LAN, a device that relays *all frames* from one segment to another.

replication In groupware, the updating of distributed computer conferencing databases.

routers Devices that route packets of information over and between a variety of different networks.

record, send, play, forward, reply (RSPFR functions) Basic voice mail functions.

S.100 Media-access API defined by the ECTF; originally based on Dialogic's SCSA. There is a whole family of related standards.

sampling rate Rate at which an analog system is sampled when converting it into digital.

switch computer application interface (SCAI) Term used by ANSI to describe its CTI standards activities.

screen-based telephony (SCB) Using the computer screen and keyboard to make and observe the progress of telephone calls; sometimes called computer assisted autodialing or screen dialing.

signal computing bus (SCbus) A Dialogic specification that allows the connection of PC cards via a synchronous bus capable of carrying digital speech.

service control point (SCP) Computer embedded in the telephone network to create an intelligent network (*see* IN).

screen dialing Dialing from an object of the screen (e.g. name, number) and receiving an indication of call progress on the screen. *See* SCB.

screen pop Presenting information that relates to a call onto the screen before the call is answered.

screen sharing See application sharing and chalkboard.

signal computing system architecture (SCSA) An overall architecture formulated by Dialogic and its followers that covers most of CMC.

simple computer telephony protocol (SCTP) An Internet-based protocol for linking switches to computers.

synchronous digital hierarchy (SDH) A specification for multilevel multiplexing over optical fiber; more flexible than plesiochronous multiplexing.

separate signaling channel *See* common channel signaling.

server A computer that provides services to other computers—e.g., shared access to a corporate database.

set-top box Control box on TV to select video-on-demand and related services.

standard general mark-up language (SGML) A standard representation for formatted text documents.

shared whiteboard *See* chalkboard.

short messaging In mobile, the ability to send limited-length messages to the mobile phone and vice versa.

signal processor A device that operates on digital signals.

signal-to-noise ratio A measure of signal quality.

signaling system In telephony, the protocol used to convey call routing information, etc.

subscriber's interface module (SIM) A smart card that plugs into a mobile phone to make it yours.

single-mode fiber Optical fiber of very high bandwidth, having a very small inner core.

system network architecture (SNA) IBM definition for computer networking.

Soft ACD An ACD that is made up of a dumb switch and separate computer. The call processing and business applications run in the same computer.

sound card In a PC, the card that interfaces the PC to the analog world of sound through microphone, speakers, etc.

sound processing The processing of audio signals for a sound card.

speaker recognition Recognizing the person who is speaking rather than what he or she is saying.

speaker verification *See* speaker recognition.

speech synthesis Producing speech from a machine—possibly from text.

service provider interface (SPI) Microsoft interface specification for devices that wish to support TAPI.

spoofing In ISDN, maintaining a computer session whilst dropping the link when there is no activity.

star services In telephony, special features available via the PSTN—forwarding or call waiting, for example.

store and forward In messaging, placing a message in a computer store, examining the destination address, and then reading it out onto the correct route.

super highway The thing we are all waiting for and that everyone else seems to have.

supervisor In a call center, a supervisor manages a number of agents. The supervisor has special rights and can for example monitor the agents' transactions and change agent status.

switch A generic word for telephone exchanges—a PBX, for example.

switched virtual circuits Predefined routes in a packet network that can be altered directly by the user for each communication session.

switching system A system that steers information channels from one port to another.

synchronous services Services that require continuously allocated channel capacity and fixed, short delay; more precisely isochronous.

T1 Basic U.S. digital multiplex; name derives from the fact that it is supposed to be carried on twisted pair (T) and it is the first level of multiplex (1). Supports 24 channels but steals bits for signaling.

telephony API (TAPI) The Microsoft approach to CTI. TAPI provides Windows interface for accessing telephony and multimedia functions.

telecommunications applications for switches and computers (TASC) ITU CTI link protocol standard, now discontinued.

transmission control protocol/ Internet protocol (TCP/IP) Internet and general standard for information transfer across a packet network.

time division multiple access (TDMA) Dynamically sharing a transmission channel between a number of users by allocating fixed segments of time to each user.

telephone company (telco) The operator of a PSTN.

telebusiness *See* teleservicing.

telecommuting Using telecommunications to avoid travel.

telemarketing Selling via the telephone (*see also* teleservicing).

telemedicine Medical examination and treatment via CMC.

telephone exchange A telephone switching system—often called a switch.

telephony An odd term used to describe things to do with the telephone and telephone system.

teleservicing Term used to describe the use of telephony in conjunction with access to computer-held information to conduct business applications such as telemarketing, customer service, debt collection, sales order processing, etc.

teleshopping Shopping via CMC.

teleworking Generally working away from a central facility by using communication networks.

terminal adapter In ISDN, a mechanism for connecting a non-ISDN terminal to an ISDN interface.

text-to-speech Converting a text string to a synthetic voice announcement.

thin client Where a PC or NC simply provides presentation functions and the applications run on a server.

time division multiplexing Sharing a transmission channel between a number of users by allocating fixed segments of time to each user.

time slot A regularly occurring segment of time that is allocated to a user within a time division multiplex.

Tmap A program that can convert between TAPI and TSAPI

token passing In LANs, a method of multiple access that allows only the current holder of a token to use the network.

tone signaling *See* DTMF.

transducer A thing that converts a signal from one form to another—e.g., a microphone converts sound pressure to electrical signals.

transmission The movement of information across a network.

trunk In telephony, a channel between the PBX and LE, or between LEs.

telephone services API (TSAPI) The Novell approach to CTI. TSAPI provides a client server approach to accessing telephony functions via NetWare.

unicode standard A 16-bit character representation that extends ASCII into many other languages.

voice and data call association (VDCA) Simultaneous transfer of telephone calls and data (*see also* intelligent transfer).

vector graphics *See* graphics.

Versit Industry group formed to improve interworking in the CTI. Produced the CTI Encyclopedia.

video bridges *See* conference bridges.

video codec Device for converting the analog video signal into digital and vice versa.

video compression algorithms Procedures for reducing the bit rate of the digital video signal.

video mail Sending and receiving video messages.

video processing Manipulating the video signal in the digital domain—e.g., compressing and expanding it.

video server *See* media server.

video signal Generally the signal produced from a TV camera.

videoconferencing Multiparty videotelephony.

video-on-demand Selection and playing of a video recording on the instant. The recording will normally be delivered over a communication network.

videophone A telephone that provides voice and video capability.

videotelephony Adding video to a telephone call.

virtual circuits Predefined routes in a packet network.

virtual reality Using technology to create an environment that appears to be real but is not. This technology is virtually here.

visitor location register Databases in a mobile network containing details of mobiles that have wandered from their home location.

visual programming Using a GUI interface for application generation.

vmail *See* voice mail.

voice and data modem A modem that combines speech compression and analog to digital conversion in order to send speech and data simultaneously over a single analog channel.

voice editing Using a media-processing program to trim and improve voice announcements.

voice mail The voice equivalent of e-mail.

voice navigation Using voice commands to control a computer.

voice print Voice pattern used in speaker recognition.

voice processing General term embracing voice messaging, IVR, audiotex, and the general storage, compression, synthesis, and recognition of speech.

voice server Voice processing system connected to a LAN.

voice over Internet protocol (VoIP) Packetized voice transmitted over the IP protocol.

Voice profile for Internet mail (VPIM) Standard method for embedding voice message into Internet mail.

virtual private networks (VPNs) Closed telephone user groups operating over the PSTN. Provided by the telco as an alternative to the private network.

voice processing unit (VPU) A system that allows information to be obtained, or telephone calls to be completed, by interaction with a machine.

voice response unit (VRU) A self-contained voice system.

wide area network (WAN) In computer networks, a packet network used to link geographically separated offices—usually X.25-based. More generally any communication network capable of covering large distances.

waveform encoding Speech compression based on approaches that ignore the mechanism of speech production—e.g., ADPCM.

waveguide A pipe for guiding radio waves.

Web *See* WWW

whiteboard, shared *See* chalkboard.

wireless LAN LAN based on radio or infrared communications.

wireless modem Modem that provides an interface to a packet radio system.

wireless PBX PBX based on radio connections to the telephone extensions.

wireline network In the cellular world this is the name used for the PSTN.

workflow rules Programmable definitions of the manner in which forms, etc. move around the business to complete a transaction.

Windows open service architecture (WOSA) Microsoft's method for getting the application writer and the hardware device supplier to do the work—by providing defined interfaces for each.

World Wide Web (WWW) Internet-based information service.

WYSIWIS what you see is what I see

About the Author

Rob Walters runs Satin Information Services, a company that provides consulting, training, and publishing services in the voice and computer telephony integration world. He is one of the world's leading exponents of the theory and practice of computer telephony integration (CTI), as well as an independent consultant, author, and speaker who advises corporations and institutions on an international basis.

Rob Walters studied Electrical and Electronic Engineering, his academic career culminated in the award of a Master's Degree in Telecommunications from Essex University. His commercial and technical backgrounds enable him to identify the opportunities presented by convergent technology, and to make recommendations on strategy and tactics unbiased by connections with hardware or software suppliers.

As a seminar leader and speaker on the subject, Rob Walters has given numerous lectures on the techniques and disciplines which CTI embraces: voice processing, telecommunications, call center technology, communicating multimedia and Internet telephony.

Founder of the Association of CTI Users and Suppliers—ACTIUS—and editor of the CTI Informer newsletter, he is the author of a number of books and reports. In 1993 he wrote Computer Telephone Integration, the first book to be published on the topic. Rob also produced the Financial Times management report on CTI and the recent book from Wiley, CTI in Action. He is a long term member of the Institute of Electrical Engineers and the Institute of Acoustics. Contact Rob at rob@satin.demon.co.uk or visit his Web site at www.demon.co.uk/satin.

Index

ACD. *See* Automatic call distributor
ACRI. *See* Application-controlled routing for incoming calls
ACRO. *See* Application-controlled routing for outgoing calls
ACTAS. *See* Alliance of Computer Telephone Application Suppliers
ACTIUS. *See* Association of Computer Telephone Integration Users and Suppliers
Activex, 95
Adaptive differential pulse code modulation, 161
Addressing, 97, 144–146
ADPCM. *See* Adaptive differential pulse code modulation
ADSI. *See* Analog display signaling interface
Aiken, Howard, 12
Aircall, 43
Alcatel, 198, 212, 225
Alliance of Computer Telephone Application Suppliers, 47, 384
American National Standards Institute, 47, 352–54, 365–73, 388
Analog telephone, 53–59, 85, 209
Analog-to-digital conversion, 62–65
ANI. *See* Automatic number identification
ANSI. *See* American National Standards Institute
Answering point, 81
API. *See* Application program interface
Application availability, 298–99
Application-controlled routing for incoming calls, 23, 155–57, 175–77, 240, 243, 308, 385
Application-controlled routing for outgoing calls, 23, 155–57, 175–77, 240, 243, 385
Application-enabling layer, 226–28
Application program interface, 126–27, 186–90, 390, 393
Applications
  basics, 21–24, 92–95
  call handling, 242–43, 246–47
  categories, 240–41
  creation, 86, 181–87, 223–26
  desktop communications, 242
  integration potential, 239–40, 299
  need for, 238–39
  power dialers, 244–46
  software, 22, 110
Application Software Integration Support Tools, 219
AS/400 Telephony Applications Services, 44–45

ASAI protocol, 353
ASIST. *See* Application Software Integration Support Tools
Association and networking, 137–38
Association of Computer Telephone Integration Users and Suppliers, 47, 288–89, 314, 384
Asterisk key, 57–58
Asynchronous circuit switching, 41
Asynchronous transfer mode, 103–4
AT&T, 42, 45, 47, 48, 198, 353, 365, 366, 401
ATM. *See* Asynchronous transfer mode
ATM Forum, 103, 351
AT set. *See* Attention set
Attachment interface, 111–13
Attendant action, 357–58
Attention set, 111, 117–18, 210
Aurora FastCall, 253–256
Autoattendant, 150
Autocalling (autodialing), 5–6
Automated banking, 261
Automated call handling, 243–44
Automatic call distributor, 43, 75–76, 127, 246–47, 335–36, 374, 387, 399, 407
Automatic number identification, 274, 347
Automation, 5

Babbage analytical engine, 11–12
Balance circuit, 58
Bandwidth, 101
Banking application, 261, 284–86
Basic telephone, 27
B channel, 102
Bell Telephone Laboratories, 8, 12, 35
Benefit analysis, 304–15
Best Data ACF 5000, 116
Betulander switch, 8
Bidirectional signaling, 64, 241
Binary synchronous communication, 32
Bit rate, 96
Bits per second, 101
Blended calling, 241, 246
Blocking, 74
Bookseller application, 269–70
Bothway calling, 241
Boxed solution, 408
Bridge, 22–23
British Airways, 277–78
British Telecom, 40–43, 45–46
Broadband integrated services digital network, 104
Browser, 51, 97
BSC. *See* Binary synchronous communication
BT Customer Communications Unit, 275
Burton, Jim, 47
Bus capacity, 339–40
Business computer application, 89
Business diversification, 296
Business process automation, 282
Butlins, 270–71

C programming language, 94, 187, 219, 224, 235
Call answering, 82
Call applications manager, 191, 274
Call-based data selection, 23, 43, 156, 174–75, 228, 240, 243, 296, 308, 385
Callbox 2, 118–19
Callbox XM, 115–16
Call center, 23–24, 380, 407, 416–17
  applications, 240–41
  bureau operation, 296
  costs, 294–95, 304
Call center bypass, 328
Call completion, 82
Call control, 111–13, 206–7
Call forwarding, 81–82
Call from Teleint, 248–49
Calling line identity, 121, 347, 399
Call manipulation service, 216
Call modeling, 83–85
Call monitoring, 157
Call parking, 59
CallPath Services Architecture, 191–98, 203, 222
Call processing, 83–85
Call routing, 243, 246–47, 336
Call state, 133, 210–11, 359–62, 367
Call transfer, 82
Call waiting, 82

CallWare Phonetastic, 251–53
CAM. *See* Call applications manager
Campus emergency application, 264, 271–72
Car company application, 265–66
Car theft prevention application, 283–84
Cave, Skip, 37, 401
CBDS. *See* Call-based data selection
CCITT. *See* Comite Consultatif Internationale de Telephonie et de Telegraphie
CCM. *See* Coordinated call monitoring
CCS. *See* Century-call-second
CELP. *See* Code excited linear predictive
Central office, 76, 127
Central processing unit, 88
Centrex, 77, 127, 130, 264, 365, 373
Century-call-second, 77–79
Charity services application, 266–67
Circuit-switched network, 95–96, 104
CIT. *See* Computer-integrated telephony
Citel, 116–17
Classic Hawaii, 274
CLI. *See* Calling line identity
Client/server computing, 128
CO. *See* Central office
Coaxial cable, 67–69, 130
Cobol, 94, 187
Code compiler, 93–94
Code excited linear predictive, 401
Collaborative working, 377, 389
Collections agency application, 281–83
Colossus, 12
Comdial, 198
Comite Consultatif Internationale de Telephonie et de Telegraphie, 112
Command link, 134–35
Common channel signaling, 72
Communications stack, 130
Companding, 64
Compatibility, application, 299
Compeer to primary protocol conversion, 141
Competitive edge, 295–96, 304, 309
Component specifications, 329–30
Component testing, 330–31
Compound document, 94
Compression. *See* Voice compression
Computers
  basics, 87–90
  first-party CTI, 110
  software, 92–95
  technology evolution, 11–18
  terminals, 91–92
  third-party CTI, 128
  types, 89–91
  voice system to, 165
  See also Networking, computer
Computer-integrated telephony, 47, 191
Computer-mediated communications, 5
Computer-supported telecommunications application, 199, 201, 227, 335, 352–55, 362–64, 370–73
Computer-supported telephony, 47
Computer telephony, 5, 49
Computer telephony integration, 2
  defined, 5, 17, 47, 49, 288–90, 391
  features, 4–7
  future, 415–17
  history, 380–85
  taxonomy, 28–29
  terminology, 7, 390–92
Computer Telephony Ltd. ST32, 116
Computing protocol model, 133–34
Concentrator, 79
Conferencing, 82
Connection, 133
Convergent industries, 44–47, 387–90
Conversion, protocol, 139, 141
Coordinated call monitoring, 23, 157, 178, 240, 243, 307, 345–46, 385
CORBA, 95
Cost, as market inhibitor, 303–4, 309–10
Cost justification, 326–29
Cost variables, 318–20
CPI. *See* Computer PBX interface
CPU. *See* Central processing unit
Crossbar switch, 8, 34
Crosspoint, 73
CSA. *See* CallPath Services Architecture
CST. *See* Computer-supported telephony
CSTA. *See* Computer-supported telecommunications application
CT. *See* Computer telephony

CT-Connect, 191, 206, 227, 390
CTI. *See* Computer telephony integration
Culture, as market inhibitor, 303
Customer service, importance of, 292–93, 304, 312
Customer service application, 278–79
CXC system, 42

DAN. *See* Desktop-area network
DART. *See* Data Analysis and Reporting Tool
Data Analysis and Reporting Tool, 230
Database synchronism, 344–45
Data collection by telephone, 33–34
Data phone, 117–18
Datapoint, 199
Dataset, 214
Data transfer, 23, 157, 232, 240, 243, 385
Data transport, 177–78
D channel, 102–3, 117, 121
DDE. *See* Dynamic data exchange
DDI. *See* Direct dialing in
Dealer board, 76–77
Debt collection, 312
DEC, 45, 47, 92, 227, 353
Delay, 332, 336–40, 410–12
Delivery, 396–97
Delphi Delta systems, 35–36, 43, 397
Denmark, 347
Desktop-area network, 24, 403
Desktop communications, 239–40, 242
Desktop integration, 24–25
Desktop standardization, 296–97
Device driver, 94
Dialed number, 215
Dialed number information service, 175
Dialer mode definitions, 244
Dialing technology, early, 36–38
Dialogic, 49, 227, 384, 408
Dial tone, 113
Digital multiplexed interface, 42, 352–53
Digital network architecture, 98
Digital private network signaling system, 353
Digital signal processor, 101
Digital telephony, 8, 10, 13
  characteristics, 62–65
  switching/signaling, 44, 70–74
  transmission, 65–68
Digital-to-analog conversion, 95
Digit monitoring, 209
Dimensioning, 16–17
DirecTalk, 394
Direct dialing in, 74, 82
Directory number. *See* Dialed number
DirectX interface, 222
Discrete system element, 28
Distributed client/server computing, 98
DLL. *See* Dynamic-link library
DMI. *See* Digital multiplexed interface
DN. *See* Dialed number
DNA. *See* Digital network architecture
DNIS. *See* Dialed number information service
DPNSS. *See* Digital private network signaling system
Driver, computer, 188
DSP. *See* Digital signal processor
DT. *See* Data transfer; Dial tone
DTMF. *See* Dual-tone multifrequency
Dual-tone multifrequency, 57, 161
Dumb switch, 76, 135, 246
Dumb terminal, 92, 95
Dynamic data exchange, 256
Dynamic identifier, 359
Dynamic-link library, 206, 209–10

E1 transmission, 69, 72
ECMA. *See* European Computer Manufacturers Association
ECTF. *See* Enterprise Computer Telephony Forum
EEC. *See* European Economic Community
Electromechanical switch, 8, 34
Embedded packet switching, 41–42
Enabler layer, 226–28
End-to-end testing, 330–31
Enhanced profile, 312
Enhanced telephone, 27
ENIAC, 12
Enterprise Computer Telephony Forum, 203, 352, 384, 393
Erlang's formula, 78
Ethernet, 97, 104, 130

ETSI. *See* European Telecommunications Standards Institute
Euro integrated services digital network, 121
Europe, 69, 96, 161
  regulation, 347–49, 351, 353–65
European Computer Manufacturers Association, 42, 47, 48, 133, 199, 335, 351, 353–65
European Economic Community, 347
European Telecommunications Standards Institute, 121
Executel, 39
Expectations, as market driver, 291–92

FastCall, 228, 253–56, 384
Faxing, 162–63, 165–66
FDM. *See* Frequency division multiplexing
FE. *See* Functional element
Feature phone, 59–62, 111, 116, 118, 215
File transfer protocol, 97
Financial Times survey, 307–8
Firewall, 97
First-party basic computer telephony integration, 27
First-party computer telephony integration, 25–28, 55, 60, 109, 154, 206, 212, 343, 352
  applications, 110, 402–403
  architecture, 109–13, 330
  attachment interface, 111–113
  line signaling, 120–122
  networking, 122–126
  signaling intercept, 113–120
First-party enhanced computer telephony integration, 27, 28, 111, 117, 121
Fixed digital link, 96
Fortran, 94, 187
Forwarding. *See* Call forwarding
4GL. *See* Fourth-generation language
Fourth-generation language, 94, 187, 224
Frequency division multiplexing, 69
Fujitsu Business Communications Systems, 198, 211
Functional element, 370
Functional integration, 5
  evolution, 35–36
  explained, 14–15
  linkage required, 18–21
  overlap conditions,
Functional interface, network computer, 92
Functional message, 135–38
Future, computer telephony integration, 415–17

G.704 multiplex, 69
Gatekeeper, 414–15
Gateway, 41, 412–14
Genesys, 212, 229–30, 233
Germany, 30–32
Global system for mobile communication, 105
GPT Corporation, 198
Graphical application, 225–26
Group PhoneWare, 256–58
"Grunt mode" detection, 37
GSM. *See* Global system for mobile communication

Harvard Mask 1, 12
Hayes command set, 111, 118, 210
HCI. *See* Host Command Interface
HDLC, 130
Help desk, 268–69
History, computer telephony integration, 380–85
Holiday bookings services, 270–71
Host Command Interface, 46, 214
Host terminal, 95
Host-to-host communications, 98
Housekeeping message, 137
HTML. *See* Hypertext mark-up language
HTTP. *See* Hypertext transfer protocol
Hunt group, 81
Hypertext mark-up language, 96
Hypertext transfer protocol, 374

IBM
  as industry standard, 296–97, 398
  CallPath, 191–98, 203, 222, 374, 394
  early technology, 29–31, 34, 44–45, 48, 59, 92
ICD. *See* Intelligent Call Delivery
Identifier attribute, 359, 367
Ignorance, as market inhibitor, 301–3
Image processing, 163–64

Implementation, 324–25
formal methodology, 325–31
risk and containment, 331–33
user concerns, 333–36
IN. *See* Intelligent network
Inbound call handling, 35–36, 80–81, 83, 241–43
Information manager, 119–20
Information technology, 186
Intecom, 47
Integrated circuit, 12
Integrated message center, 264, 271–72, 356–58
Integrated messaging, 166–168, 247
Integrated services digital network, 85–86, 95, 102–4, 117–18, 121, 130, 163, 170, 352, 360, 374, 409
Integrated system support, 334
Integrated voice data terminal, 38–40
Integration
desktop, 24–25
physical, 5, 14–15, 95, 104, 129–30
technology, 108, 164–66
See also Functional integration
Integration expertise, as market inhibitor, 303
Intel, 48, 389
Intelligent Call Delivery, 230
Intelligent network, 29, 86, 126, 373
Interactive voice response, 152–60, 178–79, 199, 228, 240, 243, 284–85, 295, 385, 390, 392
Interconnect, 198
Interdata 7/16, 34
International Standards Organization, 98, 169
International Telecommunications Union, 65, 69, 96, 102, 121, 169, 170, 351, 360, 373, 376, 389, 410, 414
Internet
as market driver, 297–98
growth, 13, 49
protocols, 96–97
Internet Engineering Task Force, 300, 351
Internet protocol, 130, 409–14
Internet protocol phone, 412–14
Interposition, 233–35
InterVoice, 399
Interworking, 98–100, 164–68, 332–33
Intranet, 97
Investment management applications, 280–81
IP. *See* Internet protocol
ISDN. *See* Integrated systems digital network
ISDN/DMI Users' Group, 353
ISO. *See* International Standards Organization
IT. *See* Information technology
ITU. *See* International Telecommunications Union
IVDT. *See* Integrated voice data terminal
IVR. *See* Interactive voice response

Japan, 69, 373
Java, 94, 225, 235
JavaTel, 48, 219–22
Java Telephony Application Program Interface, 48, 220–22, 352, 374, 384, 393
Job satisfaction, 313–14
JTAPI. *See* Java Telephony Application Program Interface

Key, feature phone, 61
Keystroke macro, 228
Key system, 76, 111, 116–17

LAN. *See* Local area network
Layered architecture, 98–100, 130, 138
LD-CELP. *See* Low delay-code excited linear predictive
LE. *See* Local exchange
Line interface variants, 74
Line signaling, 120–22
Linkage standardization, 352–75
Link management, 344
Lloyds TSB PhoneBank, 285–86
Local area network, 13, 97–98, 122–26, 166, 171, 402, 414
Local area network emulation, 104
Local exchange, 76–77
Logical unit, 98
Loss measurement, 67
Loudness rating, 67

Low delay-code excited linear predictive, 161
LU. *See* Logical unit

Macintosh, 43, 203, 297
Mailbox, 150
Mainframe, 12, 29, 31, 90
Management, implementation, 343–46
Managing customer expectations, 323
MAPI. *See* Messaging Application Program Interface
Market, computer telephony integration
  definition, 287–90
  drivers, 290–300
  expansion, 394–96
  inhibitors, 300–4
  quantization, 315–21
ME. *See* Messaging exchange
Media monitoring, 209
Media processing, 222
Merged system, 28, 334
Merging desktops, 401–4
Merging networks, 409–15
Merging servers, 404–8
Meridian Business Telephone, 118
Meridian Link, 344
Message-based signaling, 121–22
Message storage computer, 87
Messaging, 93, 135–38, 357
  protocol types, 135–38
  standards, 376–77
  unified, 166–68
Messaging Application Program Interface, 222, 250
Messaging exchange, 179, 232, 240, 243, 385
Mezza system, 40–41, 119–20
Microprocessor, 13–14, 19, 39, 101
Microsoft, 48, 91, 95, 116, 222–23, 304, 389
Messaging Application Program Interface, 222, 250
  Phone, 249–51
  Telephony Application Programming Interface, 48, 206–14
Middleware, 228–35, 238–39, 252, 254, 390
MIME. *See* Multipurpose Internet messaging extensions
Minicomputer, 12, 90
Mitel, 42, 44–47, 84, 118, 198
  MITAI, 214–19
Mobile network, 104–5
Modem, 95, 210, 409
Modem autodialing, 111
Modularization, 391
Morita, Akio, 238, 240
Moving Pictures Experts Group, 169–70
MPEG. *See* Moving Pictures Experts Group
MPLPC, 401
Multiline feature phone, 61–62
Multimedia, 105, 168, 247
Multiplexing, 68–69
Multiply/accumulate, 101
Multipurpose Internet messaging extensions, 376–77
MVIP specification, 339

Natural Microsystems, 384
Natural MicroSystems Fusion, 413
NC. *See* Network computer
NetMeeting, 389
NetPhone, 306
Netscape Navigator, 92
NetWare Loadable Module, 200
NetWare Telephony Services, 48, 198, 200–1
Network computer, 92
Network, public, 10–11, 13, 95–96
Networking, computer
  architecture, 98–100
  Internet/intranets, 96–97
  local area, 13, 97–98, 104, 122–26, 166, 171, 402, 414
  merging networks, 409–15
  third-party, 335–36
  wide area, 95–96, 98, 414
Network numbering, 144
Nexos, 36
Nissan USA, 265–66
NLM. *See* NetWare Loadable Module
Nonblocking switch, 74
Noncall-oriented service, 216–17
"Nonstop" computer, 90–91

Nortel, 84, 118, 202–3, 344, 365
North American regulation, 69, 347, 366
Northern Telecom, 41, 42, 45, 353
Novell, 48, 198–206, 304, 354, 395
Novell Open Telephony Association, 199

OAI. *See* Open application interface
Object Management Group, 95
Object orientation, 94, 220
Objects in system design, 84–85
Octel/Lucent, 258–60
Octel Unified Messenger, 167
Olivetti, 47
Open application interface, 272
Open environment, 185, 198–99, 214, 246, 333, 391, 392–94, 407
Open systems interconnection, 98–100, 351
Operating system, 5, 93
Operational checking, 134–35
Operational efficiency, 294–95
Operator assistance, 272–74
Optical fiber, 67–69, 71
OSI. *See* Open systems interconnection
Outbound call handling, 36–37, 83, 241, 243

PABX. *See* Private automatic branch exchange
Packet switching, 41–42, 96, 104, 170, 409–12
Packet Transport Equipment, 41
Paging, 43
Partnership, 299
Patents, 397–401
PBX. *See* Private branch exchange
PC. *See* Personal computer
PCM. *See* Pulse-code modulation
Performance, 13, 332
  bus capacity, 339–40
  delay, 336–39
  reliability implications, 342–43
  throughput, 340–41
Personal computer
  as telephone, 401–4
  as voice processing platform, 48
  cost-effectiveness, 97
  evolution, 13, 89–90
  merging servers, 404–8
Personal computer card, 115–17, 299–300, 403, 407–8
Personal computer card suppliers, 49
Phone from Microsoft, 249–51
PhoneJack, 412
Phonetastic, 251–53, 384
Phonet Internet phone, 413–14
Physical integration, 5, 14–15, 95, 104, 129–30
Physical unit, 98
Physical voice/data integration, 13
Pitney Bowes Credit Corporation, 281–83
Plain old telephony service, 111, 116, 121, 374
Plessey, 34, 42, 353
Plug-in card, 115
Porsche Cars, 268–69
Portability, 332–33
POTS. *See* Plain old telephony service
Pound key, 57–58
Power dialing, 36–37, 244–46, 310–12
Predictive dialer, 37, 38, 244, 239, 311
Preview dialing, 244
Private automatic branch exchange, 76
Private branch exchange
  evolution, 17, 29–34, 42, 45, 59–60, 353
  feature variants, 74–75
  integrated systems, 41–47
  physical integration, 95
Private network, 13, 127
Process control computer, 87
Processor power, 100–101
Processor-to-processor link, 27–28
Process priority, 93
Productivity improvement, 327
Progressive dialing, 244
Proprietary protocol, 138–41
Protocol, third-party, 130
  converters, 139, 141
  defined, 131–32
  models, 132–34
  proprietary, 138–41
  requirements, 134–38
Protocol converter, 46
Protocol data unit, 98

PSTN. *See* Public switched telephone network
PU. *See* Physical unit
Public network, 10–11, 13, 95–96
Public switched telephone network, 27, 70, 74, 85, 130, 272–74
Pulse-code modulation, 64, 161
Pulse signaling, 57–58, 82

Q.931 standard, 121–22
Q.Sys, 256–58
Quantization, 63–64
QuikScript, 225

Radio link, 68
RAM. *See* Random-access memory
Random-access memory, 88
Real-time system, 16
Redwood telephone system, 44
Regent system, 46
Regulation, as market inhibitor, 301
Regulatory legislation, 346–49
Relational database, 94
Reliability, system, 331–32, 342–43
Remote access, 95
Request message, 135–36
Request/response-based protocol, 135
Revenue increase, 312–13
Rhetorex, 384
Ring back, 82
Risk and containment, 331–33
Rockwell, 407–8
Rolmphone, 44
Rose PBX, 42
Rostrvm middleware, 230–32, 235
Royalblue Rostrvm, 230–32, 235
RS-232 standard, 111, 117–18, 130
RSPFR functions, 150

Sales call cost, 295
SAPI. *See* Speech Application Program Interface
Satellite link, 68
SCAI. *See* Switch computer application interface
Scanned document, 163–64
SCB. *See* Screen-based telephony
Scheduler, 93
Schema study, 305–6
SCL. *See* Service creation language
Screen-based telephony, 23–24, 155–56, 174, 228, 240, 242–43, 251, 385
Screen popping, 174–75
SCTP. *See* Simple computer telephony protocol
SDX Corporation, 118, 187
Security Pacific National Bank, 261–64
Serial interface, 111, 129
Server, 97–98, 404–8
Server-dependent architecture, 407–8
Server-independent architecture, 405–7
Service control point, 86
Service creation language, 224–25
Service invocation, 356–57
Shannon, Claude, 12
Signaling
  bidirectional, 64
  common channel, 72
  digital, 70–74
  evolution, 26–27, 55–59
  feature phone, 61–62
  line, 120–122
  types, 72–73
Signaling intercept, 113–20
Signaling System 7, 86, 336
Signaling System No. 1, 121
Simple computer telephony protocol, 374–76
Simple Mail Transfer Protocol, 97, 376–77
Single-line feature phone, 60–62
SKF data collection system, 33
Skynet 2000, 283–84
SL1C system, 41
Small Office eXchange, 408
Smart routing, 243
SMDS. *See* Switched multimegabit data service
SMTP. *See* Simple Mail Transfer Protocol
SNA. *See* Systems network architecture
Soft automatic call distributor, 246–47
SoftTalk, 228
Software dialer, 246
Software solutions, 48
  computer application, 92–95
  CTI applications, 186–87

Solution availability, 298–99
Space switching, 73–74
Speaker-independent application, 161–62
Special plain ordinary telephone, 118
Special Telephone Systems, 43
Spectrum, 407–8
Speech Application Program
  Interface, 222, 251
Speech quality. *See* Transmission quality
Speech recognition. *See* Voice recognition
Speech Recognition Application Program
  Interface, 222
SPOT. *See* Special plain ordinary telephone
SQL. *See* Standard query language
SRAPI. *See* Speech Recognition Application
  Program Interface
SS7. *See* Signaling System 7
SSCP. *See* System services control point
Stand-alone unit, 118–19, 168
Standardization, 69, 112, 182, 199, 335, 377
  as market inhibitor, 300
  gatekeeper role, 414–15
  importance of, 349–52
  interworking, 100, 332–33
  linkage, 47, 352–75
  messaging standards, 376–77
  problems with, 388–89
  regulatory legislation, 346–49
  voice compression, 169–70
  voice over packet, 375–76
Standard query language, 94
Stanza system, 45–47
Star switch topology, 70
Startalker, 115
State attribute, 359
Static identifier, 359
Status link, 134
Status message, 27, 136–37
Step-by-step exchange, 8
Stibitz, George, 12
Strowger telephone exchange, 7–9
STS. *See* Special Telephone Systems
Summa Four, 219
Sun Microsystems, 48
Superset, 215
Switch computer application
  interface, 352–54, 365–73, 388
Switch controller, 17
Switched multimegabit data service, 96
Switch farm, 330
Switching, 7–9
  digital, 70–74
  dimensioning, 16–17
  exchange types, 74–77
  first-party CTI, 110
  space, 73–74
  protocol model, 132–33
  software, 83
  third-party CTI, 127–28, 144–46
  time, 73–74
  voice system to, 164–65
Switch normalization, 199, 333
SwitchServer, 191
SX2000 switch, 44, 46–47
Sydis integrated system, 40
Syntax checking, 134–35
Syntellect, 390
Syracuse University, 264–65
System 2150, 34
System/360 computer, 33
System engineering, 323–24
System services control point, 98
Systems network architecture, 98

T1 transmission, 69
Tandem switching, 77
TAPI. *See* Telephony Application
  Programming Interface
TASC. *See* Telecommunications applications
  for switches and computers
Taxonomy, computer telephony
  integration, 28–29
TBS. *See* Telephone Broadcasting Systems
TCP/IP. *See* Transmission control
  protocol/Internet protocol
TDM. *See* Time division multiplexing
TDS. *See* Telephone delivery system
TE. *See* Trunk exchange
Telecommunications applications for
  switches and computers, 373
Telecom Tan, 43, 46
Teleint Call, 248–49
Telemarketing, 42–43, 46, 295
Telephone answering market, 149

Telephone answering bureau, 35–36
Telephone Broadcasting Systems, 37
Telephone delivery system, 262
Telephones/telephony
analog, 53–59, 62–64, 85, 209
as signaling intercept, 117–18
basic and enhanced, 27
compared to computing, 13–18
features, 80–84
history, 7–11
feature telephones, 59–62, 111, 116, 118, 215
traffic measurement, 77–80
See also Digital telephony
Telephony Application Programming Interface, 48, 116, 206–14, 227, 250, 317, 374, 384, 393, 395
Telephony service provider interface, 206, 210
Telephony Services Application Program Interface, 48, 198–206, 227, 251, 282, 354, 384
Telephony Services Early Implementers Program, 198–99
Teleprocessing line handling, 31–33
TeleTech Marketing System, 269
Telnet, 97
Terminal, 111, 128
Third-party compeer, 28, 45
Third-party computer telephony integration, 25–28, 55, 126–30, 154, 206, 212–13, 304
applications, 402–3
component specifications, 329–30
networking, 144–46, 335–36
solution types, 141–44
standardization, 352
Third-party dependent computer telephony integration, 28, 34, 135, 246
Third-party primary computer telephony integration, 28, 44
Throughput, 332, 340–41
Time division multiplexing, 69
TimeLife Libraries, 269–70
Time switching, 73–74
Tmap, 203
Token Ring network, 97
Tone monitoring, 209
Tone signaling, 56–58, 161
TouchWave, 408
Tour operator services, 274–76
TPLH. *See* Teleprocessing line handling
Traffic. *See* Telephone traffic
Transaction processing computer, 87
Transcend, 407
Transistor, 8, 12
Transmission evolution, 10
Transmission control protocol/Internet protocol, 96–97, 100, 130, 199, 351, 374
Transmission quality
analog, 56–59
compression techniques, 10, 64, 67, 160–61, 169–70
digital, 63–69
multiplexing, 68–69
objective measurement, 67
subjective assessment, 66–67
impairments to, 10, 59
Travel agent applications, 277–78
Trends, technology, 100–2
Trunk exchange, 76–77, 80, 85
TSAPI. *See* Telephony Services Application Program Interface
TSPI. *See* Telephony service provider interface
TTMS. *See* TeleTech Marketing System
Turing, Alan, 12
Twisted-pair cable, 67
Two-motion selector, 8
Two-wire system, 58

Unified messaging, 166–68, 389
Unified Messenger, 167, 258–60
Unimodem, 210, 250
Unisys, 45
United Kingdom, 32, 34, 36, 347
Universal messaging, 247
UNIX, 40, 214, 393, 408
UnixWare, 203
User concerns, 333–36, 395–96
User feature, 80–81
User requirement definition, 326–29

V.24 standard, 130
V.25 bis standard, 112, 117
Vanguard Communications
Corporation, 326
VDCA. *See* Voice and data call association
Vicorp QuickScript Tool, 394
Videoconferencing, 170–71
Video mail, 171
Video processing, 168–71
Videotelephony, 170
Virtual termination, 329
VistaLink, 390
Visual Age, 225
VMX voice systems, 38, 401
VocalTec Internet Phone software, 412–13
Voice and data call association, 23, 43, 155, 177, 213, 240, 243, 307–8, 385
Voice compression, 10, 64, 67, 160–61, 169–70
Voice mail, 150–51
Voice messaging, 38, 148–51
Voice over Internet protocol, 409–12
Voice over packet, 375
Voice processing, 391–92
  defined, 147–48
  integration interface, 164–66
  technology evolution, 48–49, 160–62
Voice profile for Internet mail, 377
Voice quality. *See* Transmission quality
Voice recognition, 160–61, 222
Voice response, 38
Voice response unit, 151–52, 156, 399
Voice synthesis, 160, 162
Voice verification, 160–62, 223
VoIP. *See* Voice over Internet protocol
VoxWare, 375
VPIM. *See* Voice profile for Internet mail
VRU. *See* Voice response unit

WAN. *See* Wide area network
Wang Corporation, 47
Washing machine, 14
Welsh Water, 278–79
Wide area network, 95–96, 98, 414
WIL. *See* Workstation Integration Link
Windows, 48, 91, 203, 206, 210, 212, 392, 393
Workgroup, 24
Workstation Integration Link, 138
World Vision Telecommunications Center, 266–67
World Wide Web, 96, 328

X.21 standard, 96, 117, 118
X.25 standard, 96, 130
X.400 standard, 98–99

Ztel PNX, 42

The Artech House Telecommunications Library

Vinton G. Cerf, Series Editor

Access Networks: Technology and V5 Interfacing, Alex Gillespie

Advanced High-Frequency Radio Communications, Eric E. Johnson, Robert I. Desourdis, Jr., et al.

Advanced Technology for Road Transport: IVHS and ATT, Ian Catling, editor

Advances in Computer Systems Security, Vol. 3, Rein Turn, editor

Advances in Telecommunications Networks, William S. Lee and Derrick C. Brown

Advances in Transport Network Technologies: Photonics Networks, ATM, and SDH, Ken-ichi Sato

An Introduction to International Telecommunications Law, Charles H. Kennedy and M. Veronica Pastor

Asynchronous Transfer Mode Networks: Performance Issues, Second Edition, Raif O. Onvural

ATM Switches, Edwin R. Coover

ATM Switching Systems, Thomas M. Chen and Stephen S. Liu

Broadband: Business Services, Technologies, and Strategic Impact, David Wright

Broadband Network Analysis and Design, Daniel Minoli

Broadband Telecommunications Technology, Byeong Lee, Minho Kang and Jonghee Lee

Cellular Mobile Systems Engineering, Saleh Faruque

Cellular Radio: Analog and Digital Systems, Asha Mehrotra

Cellular Radio: Performance Engineering, Asha Mehrotra

Cellular Radio Systems, D. M. Balston and R. C. V. Macario, editors

CDMA for Wireless Personal Communications, Ramjee Prasad

Client/Server Computing: Architecture, Applications, and Distributed Systems Management, Bruce Elbert and Bobby Martyna

Communication and Computing for Distributed Multimedia Systems, Guojun Lu

Communications Technology Guide for Business, Richard Downey, et al.

Community Networks: Lessons from Blacksburg, Virginia, Andrew Cohill and Andrea Kavanaugh, editors

Computer Networks: Architecture, Protocols, and Software, John Y. Hsu

Computer Mediated Communications: Multimedia Applications, Rob Walters

Computer Telephony Integration, Second Edition, Rob Walters

Convolutional Coding: Fundamentals and Applications, Charles Lee

Corporate Networks: The Strategic Use of Telecommunications, Thomas Valovic

The Definitive Guide to Business Resumption Planning, Leo A. Wrobel

Desktop Encyclopedia of the Internet, Nathan J. Muller

Digital Beamforming in Wireless Communications, John Litva and Titus Kwok-Yeung Lo

Digital Cellular Radio, George Calhoun

Digital Hardware Testing: Transistor-Level Fault Modeling and Testing, Rochit Rajsuman, editor

Digital Switching Control Architectures, Giuseppe Fantauzzi

Digital Video Communications, Martyn J. Riley and Iain E. G. Richardson

Distributed Multimedia Through Broadband Communications Services, Daniel Minoli and Robert Keinath

Distance Learning Technology and Applications, Daniel Minoli

EDI Security, Control, and Audit, Albert J. Marcella and Sally Chen

Electronic Mail, Jacob Palme

Enterprise Networking: Fractional T1 to SONET, Frame Relay to BISDN, Daniel Minoli

Expert Systems Applications in Integrated Network Management, E. C. Ericson, L. T. Ericson, and D. Minoli, editors

FAX: Digital Facsimile Technology and Applications, Second Edition, Dennis Bodson, Kenneth McConnell, and Richard Schaphorst

FDDI and FDDI-II: Architecture, Protocols, and Performance, Bernhard Albert and Anura P. Jayasumana

Fiber Network Service Survivability, Tsong-Ho Wu

Future Codes: Essays in Advanced Computer Technology and the Law, Curtis E. A. Karnow

Guide to Telecommunications Transmission Systems, Anton A. Huurdeman

A Guide to the TCP/IP Protocol Suite, Floyd Wilder

Implementing EDI, Mike Hendry

Implementing X.400 and X.500: The PP and QUIPU Systems, Steve Kille

Inbound Call Centers: Design, Implementation, and Management, Robert A. Gable

Information Superhighways Revisited: The Economics of Multimedia, Bruce Egan

Integrated Broadband Networks, Amit Bhargava

International Telecommunications Management, Bruce R. Elbert

International Telecommunication Standards Organizations, Andrew Macpherson

Internetworking LANs: Operation, Design, and Management, Robert Davidson and Nathan Muller

Introduction to Document Image Processing Techniques, Ronald G. Matteson

Introduction to Error-Correcting Codes, Michael Purser

An Introduction to GSM, Siegmund Redl, Matthias K. Weber and Malcom W. Oliphant

Introduction to Radio Propagation for Fixed and Mobile Communications, John Doble

Introduction to Satellite Communication, Second Edition, Bruce R. Elbert

Introduction to T1/T3 Networking, Regis J. (Bud) Bates

Introduction to Telecommunications Network Engineering, Tarmo Anttalainen

Introduction to Telephones and Telephone Systems, Second Edition, A. Michael Noll

Introduction to X.400, Cemil Betanov

LAN, ATM, and LAN Emulation Technologies, Daniel Minoli and Anthony Alles

Land-Mobile Radio System Engineering, Garry C. Hess

LAN/WAN Optimization Techniques, Harrell Van Norman

LANs to WANs: Network Management in the 1990s, Nathan J. Muller and Robert P. Davidson

The Law and Regulation of Telecommunications Carriers, Henk Brands and Evan T. Leo

Minimum Risk Strategy for Acquiring Communications Equipment and Services, Nathan J. Muller

Mobile Antenna Systems Handbook, Kyohei Fujimoto and J. R. James, editors

Mobile Communications in the U.S. and Europe: Regulation, Technology, and Markets, Michael Paetsch

Mobile Data Communications Systems, Peter Wong and David Britland

Mobile Information Systems, John Walker

Networking Strategies for Information Technology, Bruce Elbert

Packet Switching Evolution from Narrowband to Broadband ISDN, M. Smouts

Packet Video: Modeling and Signal Processing, Naohisa Ohta

Performance Evaluation of Communication Networks, Gary N. Higginbottom

Personal Communication Networks: Practical Implementation, Alan Hadden

Personal Communication Systems and Technologies, John Gardiner and Barry West, editors

Practical Computer Network Security, Mike Hendry

Principles of Secure Communication Systems, Second Edition, Don J. Torrieri

Principles of Signaling for Cell Relay and Frame Relay, Daniel Minoli and George Dobrowski

Principles of Signals and Systems: Deterministic Signals, B. Picinbono

Private Telecommunication Networks, Bruce Elbert

Pulse Code Modulation Systems Design, William N. Waggener

Radio-Relay Systems, Anton A. Huurdeman

RF and Microwave Circuit Design for Wireless Communications, Lawrence E. Larson

The Satellite Communication Applications Handbook, Bruce R. Elbert

Secure Data Networking, Michael Purser

Service Management in Computing and Telecommunications, Richard Hallows

Signaling in ATM Networks, Raif O. Onvural, Rao Cherukuri

Smart Cards, José Manuel Otón and José Luis Zoreda

Smart Card Security and Applications, Mike Hendry

Smart Highways, Smart Cars, Richard Whelan

SNMP-Based ATM Network Management, Heng Pan

Successful Business Strategies Using Telecommunications Services, Martin F. Bartholomew

Super-High-Definition Images: Beyond HDTV, Naohisa Ohta, et al.

Telecommunications Deregulation, James Shaw

Television Technology: Fundamentals and Future Prospects, A. Michael Noll

Telecommunications Technology Handbook, Daniel Minoli

Telecommuting, Osman Eldib and Daniel Minoli

Telemetry Systems Design, Frank Carden

Teletraffic Technologies in ATM Networks, Hiroshi Saito

Toll-Free Services: A Complete Guide to Design, Implementation, and Management, Robert A. Gable

Transmission Networking: SONET and the SDH, Mike Sexton and Andy Reid

Troposcatter Radio Links, G. Roda

Understanding Emerging Network Services, Pricing, and Regulation, Leo A. Wrobel and Eddie M. Pope

Understanding GPS: Principles and Applications, Elliot D. Kaplan, editor

Understanding Networking Technology: Concepts, Terms and Trends, Mark Norris

UNIX Internetworking, Second Edition, Uday O. Pabrai

Videoconferencing and Videotelephony: Technology and Standards, Richard Schaphorst

Voice Recognition, Richard L. Klevans and Robert D. Rodman

Wireless Access and the Local Telephone Network, George Calhoun

Wireless Communications in Developing Countries: Cellular and Satellite Systems, Rachael E. Schwartz

Wireless Communications for Intelligent Transportation Systems, Scott D. Elliot and Daniel J. Dailey

Wireless Data Networking, Nathan J. Muller

Wireless LAN Systems, A. Santamaría and F. J. López-Hernández

Wireless: The Revolution in Personal Telecommunications, Ira Brodsky

World-Class Telecommunications Service Development, Ellen P. Ward

Writing Disaster Recovery Plans for Telecommunications Networks and LANs, Leo A. Wrobel

X Window System User's Guide, Uday O. Pabrai

For further information on these and other Artech House titles, including previously considered out-of-print books now available through our In-Print-Forever™ (IPF™) program, contact:

Artech House
685 Canton Street
Norwood, MA 02062
781-769-9750
Fax: 781-769-6334
Telex: 951-659
email: artech@artech-house.com

Artech House
Portland House, Stag Place
London SW1E 5XA England
+44 (0) 171-973-8077
Fax: +44 (0) 171-630-0166
Telex: 951-659
email: artech-uk@artech-house.com

Find us on the World Wide Web at:
www.artech-house.com